Dynamic Inequalities On Time Scales

Ravi Agarwal • Donal O'Regan • Samir Saker

Dynamic Inequalities On Time Scales

Ravi Agarwal
Department of Mathematics
Texas A&M University–Kingsville
Kingsville, TX, USA

Donal O'Regan
School of Mathematics, Statistics,
and Applied Mathematics
National University of Ireland
Galway, Ireland

Samir Saker
Department of Mathematics
Mansoura University
Mansoura, Egypt

ISBN 978-3-319-36404-9   ISBN 978-3-319-11002-8 (eBook)
DOI 10.1007/978-3-319-11002-8
Springer Cham Heidelberg New York Dordrecht London

Mathematics Subject Classification: 34K11, 34C10, 34K20

© Springer International Publishing Switzerland 2014
Softcover reprint of the hardcover 1st edition 2014

Printed on acid-free paper

Springer is part of Springer Science+Business Media (www.springer.com)

Ravi P Agarwal: To Sadhna, Sheba, and Danah

Donal O'Regan: To Alice, Aoife, Lorna, Daniel, and Niamh

Samir H Saker: To Mona, Meran, Maryam, Menah, and Ahmed

Preface

All analysts spend half their time hunting through the literature for inequalities which they want to use and cannot prove.

G.H. Hardy.

The study of dynamic inequalities on time scales has received a lot of attention in the literature and has become a major field in pure and applied mathematics. This book is devoted to some fundamental dynamic inequalities on time scales such as Young's inequality, Jensen's inequality, Hölder's inequality, Minkowski's inequality, Steffensen's inequality, Čebyšev's inequality, Opial's inequality, Lyapunov's inequality, Halanay's inequality, and Wirtinger's inequality.

The book on the subject of time scale, i.e., measure chain, by Bohner and Peterson [51] summarizes and organizes much of time scale calculus. The three most popular examples of calculus on time scales are differential calculus, difference calculus, and quantum calculus (see Kac and Cheung [89]), i.e, when $\mathbb{T} = \mathbb{R}$, $\mathbb{T} = \mathbb{N}$, and $\mathbb{T} = q^{\mathbb{N}_0} = \{q^t : t \in \mathbb{N}_0\}$ where $q > 1$. There are applications of dynamic equations and inequalities on time scales to *quantum mechanics, electrical engineering, neural networks, heat transfer, combinatorics, and population dynamics.* A cover story article in New Scientist [141] discusses several possible applications. In population dynamics the dynamic equations can be used to model insect populations that are continuous while in season, die out in say winter, while their eggs are incubating or dormant, and then hatch in a new season, giving rise to a nonoverlapping population.

This book presents a variety of integral inequalities. We assume the reader has a good background in time scale calculus. The book consists of six chapters. In Chap. 1 we present preliminaries and basic concepts of time scale calculus, and in Chap. 2 we discuss and prove dynamic inequalities on time scales such as Young's inequality, Jensen's inequality, Holder's inequality, Minkowski's inequality, Steffensen's inequality, Hermite–Hadamard inequality, and Čebyšv's inequality. Opial type inequalities on time scales and their extensions with weighted functions will be discussed in Chap. 3. In Chap. 4 we present some inequalities of Lyapunov type for some dynamic equations, and in Chap. 5 we employ the shift operators $\delta_\pm$ to construct delay dynamic inequalities on time scales and use them to derive Halanay type inequalities for dynamic equations on time scales. Using Halanay's inequalities and the properties of exponential function on time scales, we establish new conditions that lead to stability for nonlinear dynamic equations. Finally in Chap. 6 we discuss Wirtinger-type inequalities on time scales and their extensions.

We wish to express our thanks to our families and friends.

Kingsville, TX, USA   Ravi Agarwal
Galway, Ireland   Donal O'Regan
Mansoura, Egypt   Samir H. Saker

Contents

Chapter 1

Preliminaries

The essence of mathematics lies in its freedom.
Georg Cantor (1845–1915).
As for everything else, so for a mathematical theory: beauty can be perceived but not explained.
Arthur Cayley (1821–1895).

From a modeling point of view it is realistic to model a phenomenon by a dynamic system which incorporates both continuous and discrete times, namely, time as an arbitrary closed set of reals. It is natural to ask whether it is possible to provide a framework which allows us to handle both dynamic systems simultaneously so that we can get some insight and a better understanding of the subtle differences of these two systems. The recently developed theory of "dynamic systems on time scales" or dynamic systems on measure chains (by a measure chain we mean the union of disjoint closed intervals of $\mathbb{R}$) offers a desired unified approach.

This chapter contains some preliminaries, definitions, and concepts concerning time scale calculus. The results in this chapter will cover delta, nabla, and diamond-α derivatives and integrals.

© Springer International Publishing Switzerland 2014   1
R. Agarwal et al., *Dynamic Inequalities On Time Scales*,
DOI 10.1007/978-3-319-11002-8_1

1.1 Delta Calculus

For the notions used below we refer the reader to the books [51, 52] which summarize and organize much of time scale calculus. A time scale is an arbitrary nonempty closed subset of the real numbers. Throughout the book, we denote the time scale by the symbol $\mathbb{T}$. For example, the real numbers $\mathbb{R}$, the integers $\mathbb{Z}$, and the natural numbers $\mathbb{N}$ are time scales. For $t \in \mathbb{T}$, we define the forward jump operator $\sigma : \mathbb{T} \to \mathbb{T}$ by $\sigma(t) := \inf\{s \in \mathbb{T} : s > t\}$. A time-scale $\mathbb{T}$ equipped with the order topology is metrizable and is a K_σ-space; i.e., it is a union of at most countably many compact sets. The metric on $\mathbb{T}$ which generates the order topology is given by $d(r; s) := |\mu(r; s)|$, where $\mu(.) = \mu(.; \tau)$ for a fixed $\tau \in \mathbb{T}$ is defined as follows. The mapping $\mu : \mathbb{T} \to \mathbb{R}^+ = [0, \infty)$ such that $\mu(t) := \sigma(t) - t$ is called the graininess.

When $\mathbb{T} = \mathbb{R}$, we see that $\sigma(t) = t$ and $\mu(t) \equiv 0$ for all $t \in \mathbb{T}$ and when $\mathbb{T} = \mathbb{N}$, we have that $\sigma(t) = t + 1$ and $\mu(t) \equiv 1$ for all $t \in \mathbb{T}$. The backward jump operator $\rho : \mathbb{T} \to \mathbb{T}$ is defined by $\rho(t) := \sup\{s \in \mathbb{T} : s < t\}$. The mapping $\nu : \mathbb{T} \to \mathbb{R}_0^+$ such that $\nu(t) = t - \rho(t)$ is called the backward graininess. If $\sigma(t) > t$, we say that t is right-scattered, while if $\rho(t) < t$, we say that t is left-scattered. Also, if $t < \sup \mathbb{T}$ and $\sigma(t) = t$, then t is called right-dense, and if $t > \inf \mathbb{T}$ and $\rho(t) = t$, then t is called left-dense. If $\mathbb{T}$ has a left-scattered maximum m, then $\mathbb{T}^k = \mathbb{T} - \{m\}$. Otherwise $\mathbb{T}^k = \mathbb{T}$. In summary,

$$\mathbb{T}^k = \begin{cases} \mathbb{T} \backslash (\rho \sup \mathbb{T}, \sup \mathbb{T}], & \text{if} \quad \sup \mathbb{T} < \infty, \\ \mathbb{T}, & \sup \mathbb{T} = \infty. \end{cases}$$

Likewise $\mathbb{T}_k$ is defined as the set $\mathbb{T}_k = \mathbb{T} \backslash [\inf \mathbb{T}, \sigma(\inf \mathbb{T})]$ if $|\inf \mathbb{T}| < \infty$, and $\mathbb{T}_k = \mathbb{T}$ if $\inf \mathbb{T} = -\infty$.

For a function $f : \mathbb{T} \to \mathbb{R}$, we define the derivative f^Δ as follows. Let $t \in \mathbb{T}$. If there exists a number $\alpha \in \mathbb{R}$ such that for all $\varepsilon > 0$ there exists a neighborhood U of t with

$$|f(\sigma(t)) - f(s) - \alpha(\sigma(t) - s)| \leq \varepsilon |\sigma(t) - s|,$$

for all $s \in U$, then f is said to be differentiable at t, and we call α the delta derivative of f at t and denote it by $f^\Delta(t)$. For example, if $\mathbb{T} = \mathbb{R}$, then

$$f^\Delta(t) = f'(t) = \lim_{\Delta t \to 0} \frac{f(t + \Delta t) - f(t)}{\Delta t}, \text{ for all } t \in \mathbb{T}.$$

If $\mathbb{T} = \mathbb{N}$, then $f^\Delta(t) = f(t+1) - f(t)$ for all $t \in \mathbb{T}$. For a function $f : \mathbb{T} \to \mathbb{R}$ the (delta) derivative is defined by

$$f^\Delta(t) = \frac{f(\sigma(t)) - f(t)}{\sigma(t) - t},$$

if f is continuous at t and t is right-scattered. If t is not right-scattered then the derivative is defined by

$$f^\Delta(t) = \lim_{s \to t} \frac{f(\sigma(t)) - f(s)}{t - s} = \lim_{t \to \infty} \frac{f(t) - f(s)}{t - s},$$

provided this limit exists. A useful formula is

$$f^\sigma = f + \mu f^\Delta \quad \text{where} \quad f^\sigma := f \circ \sigma.$$

A function $f : [a, b] \to \mathbb{R}$ is said to be right-dense continuous (rd-continuous) if it is right continuous at each right-dense point and there exists a finite left limit at all left-dense points, and f is said to be differentiable if its derivative exists. The space of rd-continuous functions is denoted by $C_r(\mathbb{T}, \mathbb{R})$. A time scale $\mathbb{T}$ is said to be regular if the following two conditions are satisfied simultaneously:

(a). For all $t \in \mathbb{T}$, $\sigma(\rho(t)) = t$.

(b). For all $t \in \mathbb{T}$, $\rho(\sigma(t)) = t$.

Remark 1.1.1 *If* $\mathbb{T}$ *is a regular time scale, then both operators are invertible with* $\sigma^{-1} = \rho$ *and* $\rho^{-1} = \sigma$.

The following theorem gives the product and quotient rules for the derivative of the product fg and the quotient f/g (where $gg^\sigma \neq 0$) of two delta differentiable functions f and g.

Theorem 1.1.1 *Assume* $f; g : \mathbb{T} \to \mathbb{R}$ *are delta differentiable at* $t \in \mathbb{T}$. *Then*

$$
\begin{align}
(fg)^\Delta &= f^\Delta g + f^\sigma g^\Delta = fg^\Delta + f^\Delta g^\sigma, \tag{1.1.1} \\
\left(\frac{f}{g}\right)^\Delta &= \frac{f^\Delta g - fg^\Delta}{gg^\sigma}. \tag{1.1.2}
\end{align}
$$

By using the product rule, we see that the derivative of $f(t) = (t - \alpha)^m$ for $m \in \mathbb{N}$, and $\alpha \in \mathbb{T}$ can be calculated as

$$f^\Delta(t) = ((t - \alpha)^m)^\Delta = \sum_{\nu=0}^{m-1} (\sigma(t) - \alpha)^\nu (t - \alpha)^{m-\nu-1}. \tag{1.1.3}$$

As a special case when $\alpha = 0$, we see that the derivative of $f(t) = t^m$ for $m \in \mathbb{N}$ can be calculated as

$$(t^m)^\Delta = \sum_{\gamma=0}^{m-1} \sigma^\gamma(t) t^{m-\gamma-1}.$$

Note that when $\mathbb{T} = \mathbb{R}$, we have

$$\sigma(t) = t, \quad \mu(t) = 0, \quad f^\Delta(t) = f'(t).$$

When $\mathbb{T} = \mathbb{Z}$, we have

$$\sigma(t) = t + 1, \quad \mu(t) = 1, \quad f^\Delta(t) = \Delta f(t).$$

When $\mathbb{T} = h\mathbb{Z}$, $h > 0$, we have $\sigma(t) = t + h$, $\mu(t) = h$,

$$f^{\Delta}(t) = \Delta_h f(t) = \frac{(f(t+h) - f(t))}{h}.$$

When $\mathbb{T} = \{t : t = q^k, \ k \in \mathbb{N}_0, \ q > 1\}$, we have $\sigma(t) = qt$, $\mu(t) = (q - 1)t$,

$$f^{\Delta}(t) = \Delta_q f(t) = \frac{(f(qt) - f(t))}{(q - 1)t}.$$

When $\mathbb{T} = \mathbb{N}_0^2 = \{t^2 : t \in \mathbb{N}\}$, we have $\sigma(t) = (\sqrt{t} + 1)^2$ and

$$\mu(t) = 1 + 2\sqrt{t}, \ f^{\Delta}(t) = \Delta_0 f(t) = (f((\sqrt{t} + 1)^2) - f(t))/1 + 2\sqrt{t}.$$

When $\mathbb{T} = \mathbb{T}_n = \{t_n : n \in \mathbb{N}\}$ where (t_n) are the harmonic numbers that are defined by $t_0 = 0$ and $t_n = \sum_{k=1}^{n} \frac{1}{k}, n \in \mathbb{N}_0$, and we have

$$\sigma(t_n) = t_{n+1}, \ \mu(t_n) = \frac{1}{n+1}, \ f^{\Delta}(t) = \Delta_1 f(t_n) = (n+1)f(t_n).$$

When $\mathbb{T}_2 = \{\sqrt{n} : n \in \mathbb{N}\}$, we have $\sigma(t) = \sqrt{t^2 + 1}$,

$$\mu(t) = \sqrt{t^2 + 1} - t, \ f^{\Delta}(t) = \Delta_2 f(t) = \frac{(f(\sqrt{t^2 + 1}) - f(t))}{\sqrt{t^2 + 1} - t}.$$

When $\mathbb{T}_3 = \{\sqrt[3]{n} : n \in \mathbb{N}\}$, we have $\sigma(t) = \sqrt[3]{t^3 + 1}$ and

$$\mu(t) = \sqrt[3]{t^3 + 1} - t, \ f^{\Delta}(t) = \Delta_3 f(t) = \frac{(f(\sqrt[3]{t^3 + 1}) - f(t))}{\sqrt[3]{t^3 + 1} - t}.$$

For $a, b \in \mathbb{T}$, and a delta differentiable function f, the Cauchy integral of f^{Δ} is defined by

$$\int_a^b f^{\Delta}(t)\Delta t = f(b) - f(a).$$

Theorem 1.1.2 Let $f, g \in C_{rd}([a,b], \mathbb{R})$ be rd-continuous functions, $a, b, c \in \mathbb{T}$ and $\alpha, \beta \in \mathbb{R}$. Then, the following are true:

1. $\int_a^b [\alpha f(t) + \beta g(t)] \, \Delta t = \alpha \int_a^b f(t)\Delta t + \beta \int_a^b g(t)\Delta t$,

2. $\int_a^b f(t)\Delta t = -\int_b^a f(t)\Delta t$,

3. $\int_a^c f(t)\Delta t = \int_a^b f(t)\Delta t + \int_b^c f(t)\Delta t$,

4. $\left| \int_a^b f(t)\Delta t \right| \leq \int_a^b |f(t)| \, \Delta t$.

An integration by parts formula reads

$$\int_a^b f(t)g^\Delta(t)\Delta t = f(t)g(t)|_a^b - \int_a^b f^\Delta(t)g^\sigma(t)\Delta t, \qquad (1.1.4)$$

and infinite integrals are defined as

$$\int_a^\infty f(t)\Delta t = \lim_{b\to\infty} \int_a^b f(t)\Delta t.$$

Note that when $\mathbb{T} = \mathbb{R}$, we have

$$\int_a^b f(t)\Delta t = \int_a^b f(t)dt.$$

When $\mathbb{T} = \mathbb{Z}$, we have

$$\int_a^b f(t)\Delta t = \sum_{t=a}^{b-1} f(t).$$

When $\mathbb{T} = h\mathbb{Z}$, $h > 0$, we have

$$\int_a^b f(t)\Delta t = \sum_{k=0}^{\frac{b-a-h}{h}} f(a+kh)h.$$

When $\mathbb{T} = \{t : t = q^k, k \in \mathbb{N}_0, q > 1\}$, we have

$$\int_{t_0}^\infty f(t)\Delta t = \sum_{k=0}^\infty f(q^k)\mu(q^k).$$

Note that the integration formula on a discrete time scale is defined by

$$\int_a^b f(t)\Delta t = \sum_{t\in(a,b)} f(t)\mu(t).$$

It is well known that rd-continuous functions possess antiderivatives. If f is rd-continuous and $F^\Delta = f$, then

$$\int_t^{\sigma(t)} f(s)\Delta s = F(\sigma(t)) - F(t) = \mu(t)F^\Delta(t) = \mu(t)f(t).$$

Now, we will give the definition of the generalized exponential functions and its derivatives. We say that $p : \mathbb{T}^\kappa \mapsto \mathbb{R}$ is *regressive* provided $1 + \mu(t)p(t) \neq 0$ for all $t \in \mathbb{T}^\kappa$. We define the set $\mathcal{R}$ of all regressive and rd-continuous functions. We define the set $\mathcal{R}^+$ of all positively regressive elements of $\mathcal{R}$ by $\mathcal{R}^+ = \{p \in \mathcal{R} : 1 + \mu(t)p(t) > 0, \text{ for all } t \in \mathbb{T}\}$. The set of all regressive functions on a time scale $\mathbb{T}$ forms an Abelian group under the addition $\oplus$ defined by $p \oplus q := p + q + \mu pq$. If $p \in \mathcal{R}$, then we can define the exponential function by

$$e_p(t,s) = \exp\left(\int_s^t \xi_{\mu(\tau)}(p(\tau))\Delta\tau\right), \quad \text{for } t \in \mathbb{T}, s \in \mathbb{T}^k, \tag{1.1.5}$$

where $\xi_h(z)$ is the cylinder transformation, which is defined by

$$\xi_h(z) = \begin{cases} \frac{\log(1+hz)}{h}, & h \neq 0, \\ z, & h = 0. \end{cases}$$

If $p \in \mathcal{R}$, then $e_p(t,s)$ is real-valued and nonzero on $\mathbb{T}$. If $p \in \mathcal{R}^+$, then $e_p(t,t_0)$ is always positive. Note that if $\mathbb{T} = \mathbb{R}$, then

$$e_a(t,t_0) = \exp\left(\int_{t_0}^t a(s)ds\right),$$

if $\mathbb{T} = \mathbb{N}$, then

$$e_a(t,t_0) = \prod_{s=t_0}^{t-1}(1 + a(s)),$$

and if $\mathbb{T} = q^{\mathbb{N}_0}$, then

$$e_a(t,t_0) = \prod_{s=t_0}^{t-1}(1 + (q-1)sa(s)).$$

If $p : \mathbb{T}^\kappa \mapsto \mathbb{R}$ is rd-continuous and regressive, then the *exponential function* $e_p(t,t_0)$ is for each fixed $t_0 \in \mathbb{T}^\kappa$ the unique solution of the initial value problem $x^\Delta = p(t)x$, $x(t_0) = 1$, for all $t \in \mathbb{T}$. We will use the following definition to present the properties of the exponential function $e_p(t,s)$. If $p, q \in \mathcal{R}$, then we define $\ominus p(t) = -p(t)/(1 + \mu(t)p(t))$ and $(p \ominus q)(t) = p(t) + q(t) + \mu(t)p(t)q(t)$ for all $t \in \mathbb{T}^\kappa$. The following properties are proved in [51].

Theorem 1.1.3 *If $p, q \in \mathcal{R}$ and $t_0 \in \mathbb{T}$, then*

- $e_p(t,t) \equiv 1 \quad and \quad e_0(t,s) \equiv 1;$

- $e_p(\sigma(t),s) = (1 + \mu(t)p(t))e_p(t,s);$

- $\frac{1}{e_p(t,s)} = e_{\ominus p}(t,s) = e_p(s,t);$

- $\frac{e_p(t,s)}{e_q(t,s)} = e_{p\ominus q}(t,s);$

- $e_p(t,s)e_q(t,s) = e_{p\oplus q}(t,s);$

- *if* $p \in \mathcal{R}^+$, *then* $e_p(t, t_0) > 0$ *for all* $t \in \mathbb{T}$.

- $e_p^\Delta(t, t_0) = p(t)e_p(t, t_0)$.

- $\left(\frac{1}{e_p(\cdot, s)}\right)^\Delta = -\frac{p(t)}{e_p^\sigma(\cdot, s)}$.

Lemma 1.1.1 *For a nonnegative* φ *with* $-\varphi \in \mathcal{R}^+$, *we have the inequalities*

$$1 - \int_s^t \varphi(u)\Delta u \leq e_{-\varphi}(t, s) \leq \exp\left\{-\int_s^t \varphi(u)\Delta u\right\} \text{ for all } t \geq s.$$

If φ *is rd-continuous and nonnegative, then*

$$1 + \int_s^t \varphi(u)\Delta u \leq e_\varphi(t, s) \leq \exp\left\{\int_s^t \varphi(u)\Delta u\right\} \text{ for all } t \geq s.$$

Remark 1.1.2 *If* $\lambda \in \mathcal{R}^+$ *and* $\lambda(r) < 0$ *for all* $t \in [s, t)_\mathbb{T}$, *then*

$$0 < e_\lambda(t, s) \leq \exp\left(\int_s^t \lambda(r)\Delta r\right) < 1.$$

Theorem 1.1.4 *If* $p \in \mathcal{R}$ *and* $a, b, c \in \mathbb{T}$, *then*

$$\int_a^b p(t)e_p(c, \sigma(t))\Delta t = -\int_a^b (e_p^\Delta(c, .)(t)\Delta t = e_p(c, a) - e_p(c, b).$$

Theorem 1.1.5 *If* $a, b, c \in \mathbb{T}$ *and* $f \in C_{rd}(\mathbb{T}, \mathbb{R})$, $a, b \in \mathbb{T}$ *such that* $f(t) \geq 0$ *for all* $a \leq t < b$, *then*

$$\int_a^b f(t)\Delta t \geq 0.$$

Lemma 1.1.2 *Let* $v \in C_{rd}^1(\mathbb{T}, \mathbb{R})$ *be strictly increasing and* $\tilde{\mathbb{T}} = v(\mathbb{T})$ *be a time scale. If* $f \in C_{rd}(\mathbb{T}, \mathbb{R})$, *then for* $a, b \in \mathbb{T}$,

$$\int_a^b f(x)v^\Delta(x)\Delta x = \int_{v(a)}^{v(b)} f(v^{-1}(y))\tilde{\Delta}y.$$

Throughout the book, we will use the following facts:

$$\int_{t_0}^\infty \frac{\Delta s}{s^\nu} = \infty, \text{ if } 0 \leq \nu \leq 1, \text{ and } \int_{t_0}^\infty \frac{\Delta s}{s^\nu} < \infty, \text{ if } \nu > 1,$$

and without loss of generality, we assume that $\sup \mathbb{T} = \infty$, and define the time scale interval $[a, b]_\mathbb{T}$ by $[a, b]_\mathbb{T} := [a, b] \cap \mathbb{T}$. The following results are adapted from [52].

Lemma 1.1.3 *Let* $f : \mathbb{R} \to \mathbb{R}$ *be continuously differentiable and suppose* $g : \mathbb{T} \to \mathbb{R}$ *is delta differentiable. Then* $f \circ g : \mathbb{T} \to \mathbb{R}$ *is delta differentiable and*

$$f^{\Delta}(g(t)) = f'(g(\zeta))g^{\Delta}(t), \quad for \quad \zeta \in [t, \sigma(t)]. \tag{1.1.6}$$

Lemma 1.1.4 *Let* $f : \mathbb{R} \to \mathbb{R}$ *be continuously differentiable and suppose* $g : \mathbb{T} \to \mathbb{R}$ *is delta differentiable. Then* $f \circ g : \mathbb{T} \to \mathbb{R}$ *is delta differentiable and the formula*

$$(f \circ g)^{\Delta}(t) = \left\{ \int_0^1 f'(g(t) + h\mu(t)g^{\Delta}(t))dh \right\} g^{\Delta}(t), \tag{1.1.7}$$

holds.

Lemma 1.1.5 *Assume the continuous mapping* $f : [r, s]_{\mathbb{T}} \to \mathbb{R}$, r, $s \in \mathbb{T}$, *satisfies* $f(r) < 0 < f(s)$. *Then there is a* $\tau \in [r, s)_{\mathbb{T}}$ *with* $f(\tau)f(\sigma(\tau)) \le 0$.

Lemma 1.1.6 *Let the mapping* $f : \mathbb{T} \to \mathbb{R}$, $g : \mathbb{T} \to \mathbb{R}$ *be differentiable and assume that*

$$|f^{\Delta}(t)| \le g^{\Delta}(t).$$

Then for r, $s \in \mathbb{T}$, $r \le s$,

$$|f(s) - f(r)| \le g(s) - g(r).$$

Assume $g : \mathbb{T} \to \mathbb{R}$ *be differentiable and* $g^{\Delta}(t) \ge 0$, *then* $g(t)$ *is nondecreasing.*

Definition 1.1.1 *We say a function* $f : \mathbb{T} \to \mathbb{R}$ *is right-increasing (right-decreasing) at* $t_0 \in \mathbb{T}^k$ *provided that*

(i) *if* $\sigma(t_0) > t_0$, *then* $f(\sigma(t_0)) > f(t_0), (f(\sigma(t_0)) < f(t_0))$,

(ii) *if* $\sigma(t_0) = t_0$, *then there is a neighborhood* U *of* t_0 *such that* $f(t) > f(t_0), (f(t) < f(t_0))$, *for all* $t \in U$, $t > t_0$.

Definition 1.1.2 *We say a function* $f : \mathbb{T} \to \mathbb{R}$ *assumes its local right-maximum (local right-minimum) at* $t_0 \in T$ *provided that:*

(i) *if* $\sigma(t_0) > t_0$, *then* $f(\sigma(t_0)) \le f(t_0), (f(\sigma(t_0)) \ge f(t_0))$,

(ii) *if* $\sigma(t_0) = t_0$, *then there is a neighborhood* U *of* t_0 *such that* $f(t) \le f(t_0), (f(t) \ge f(t_0))$, *for all* $t \in U$, $t > t_0$.

Theorem 1.1.6 *If* $f : \mathbb{T} \to \mathbb{R}$ *is* Δ-*differentiable at* $t_0 \in \mathbb{T}^k$ *and* $f^{\Delta}(t_0) > 0$, $(f^{\Delta}(t_0) < 0)$, *then* f *is right-increasing, (right-decreasing), at* t_0.

Theorem 1.1.7 *If* $f : \mathbb{T} \to \mathbb{R}$ *is* Δ-*differentiable at* $t_0 \in \mathbb{T}^k$ *and if* $f^{\Delta}(t_0) > 0 \, (f^{\Delta}(t_0) < 0)$, *then* f *assumes a local right-minimum (local right-maximum), at* t_0.

Theorem 1.1.8 *Suppose* $f : \mathbb{T} \to \mathbb{R}$ *is* Δ-*differentiable at* $t_0 \in \mathbb{T}^k$ *and assumes its local right-minimum (local right-maximum) at* t_0. *Then* $f^{\Delta}(t_0) \geq 0 (f^{\Delta}(t_0) \leq 0)$.

Theorem 1.1.9 *Let* f *be a continuous function on* $[a, b]_{\mathbb{T}}$ *that is* Δ-*differentiable on* $[a, b)$ *(the differentiability at* a *is understood as right-sided), and satisfies* $f(a) = f(b)$. *Then there exist* $\zeta, \tau \in [a, b)_{\mathbb{T}}$ *such that*

$$f^{\Delta}(\tau) \leq 0 \leq f^{\Delta}(\zeta).$$

Corollary 1.1.1 *Let* f *be a continuous function on* $[a, b]_{\mathbb{T}}$ *that is* Δ-*differentiable on* $[a, b)$. *If* $f^{\Delta}(t) = 0$ *for all* $t \in [a, b)_{\mathbb{T}}$, *then* f *is a constant function on* $[a, b]_{\mathbb{T}}$.

Corollary 1.1.2 *Let* f *be a continuous function on* $[a, b]$ *that is* Δ-*differentiable on* $[a, b)$. *Then* f *is increasing, decreasing, nondecreasing, and nonincreasing on* $[a, b]_{\mathbb{T}}$ *if* $f^{\Delta}(t) > 0, f^{\Delta}(t) > 0, f^{\Delta}(t) \geq 0$, *and* $f^{\Delta}(t) \leq 0$ *for all* $t \in [a, b)_{\mathbb{T}}$, *respectively.*

Theorem 1.1.10 *Let* f *and* g *be continuous functions on* $[a, b]$ *that are* Δ-*differentiable on* $[a, b)_{\mathbb{T}}$. *Suppose* $g^{\Delta}(t) > 0$ *for all* $t \in [a, b)$. *Then there exist* $\zeta, \tau \in [a, b)_{\mathbb{T}}$ *such that*

$$\frac{f^{\Delta}(\tau)}{g^{\Delta}(\tau)} \leq \frac{f(b) - f(a)}{g(b) - g(a)} \leq \frac{f^{\Delta}(\zeta)}{g^{\Delta}(\zeta)}.$$

1.2 Nabla Calculus

The corresponding theory for nabla derivatives was also studied extensively. The results in this section are adapted from [27].

Let $\mathbb{T}$ be a time scale, the backward jump operator $\rho : \mathbb{T} \to \mathbb{T}$ is defined by $\rho(t) := \sup\{s \in \mathbb{T} : s < t\}$. The mapping $\nu : \mathbb{T} \to \mathbb{R}_0^+$ such that $\nu(t) = t - \rho(t)$ is called the backward graininess. The function $f : \mathbb{T} \to \mathbb{R}$ is called nabla differentiable at $t_0 \in \mathbb{T}$, if there exists an $a \in \mathbb{R}$ with the following property: For any $\epsilon > 0$, there exists a neighborhood U of t, such that

$$|f(\rho(t)) - f(s) - a[\rho(t) - s]| \leq \epsilon |\rho(t) - s|$$

for all $s \in U$; we write $a = f^{\nabla}(t)$. For $\mathbb{T} = \mathbb{R}$, we have $f^{\nabla}(t) = f'(t)$ and for $\mathbb{T} = \mathbb{Z}$, we have the backward difference operator $f^{\nabla}(t) = \nabla f(t) = f(t) - f(t - 1)$.

A function $f : \mathbb{T} \to \mathbb{R}$ is left-dense continuous or *ld*-continuous provided it is continuous at left-dense points in $\mathbb{T}$ and its right-sided limits exist (finite) at right-dense points in $\mathbb{T}$. If $\mathbb{T} = \mathbb{R}$, then f is *ld*-continuous if and only if f is continuous. If $\mathbb{T} = \mathbb{Z}$, then any function is *ld*-continuous. The following theorem gives several properties of the nabla derivative.

Theorem 1.2.1 *Assume* $f : \mathbb{T} \to \mathbb{R}$ *is a function and let* $t \in \mathbb{T}$. *Then we have the following:*

1. *If f is nabla differentiable at t, then f is continuous at t.*

2. *If f is continuous at t and t is left scattered, then f is nabla differentiable at t with*
$$f^{\nabla}(t) = \frac{f(t) - f(\rho(t))}{\nu(t)}.$$

3. *If f is left-dense, then f is nabla differentiable at t iff the limit* $\lim_{s \to t} \frac{f(t) - f(s)}{t - s}$ *exists as a definite number, and in this case*
$$f^{\nabla}(t) = \lim_{s \to t} \frac{f(t) - f(s)}{t - s}.$$

4. *If f is nabla differentiable at t, then* $f(\rho(t)) = f(t) - \nu(t)f^{\nabla}(t)$.

Theorem 1.2.2 *Assume* $f, g : \mathbb{T} \to \mathbb{R}$ *are nabla differentiable at* $t \in \mathbb{T}$. *Then:*

(i) *The product* $fg : \mathbb{T} \to \mathbb{R}$ *is nabla differentiable at t, and we get the product rule*
$$(fg)^{\nabla}(t) = f^{\nabla}(t)g(t) + f^{\rho}(t)g^{\nabla}(t) = f(t)g^{\nabla}(t) + f^{\nabla}(t)g^{\rho}(t).$$

(ii) *If* $g(t)g^{\rho}(t) \neq 0$, *then* f/g *is nabla differentiable at t, and we get the quotient rule*
$$\left(\frac{f}{g}\right)^{\nabla}(t) = \frac{f^{\nabla}(t)g(t) - f(t)g^{\nabla}(t)}{g(t)g^{\rho}(t)}.$$

A function $F : \mathbb{T} \to \mathbb{R}$ is called a nabla antiderivative of $f : \mathbb{T} \to \mathbb{R}$ provided $F^{\nabla}(t) = f(t)$ holds for all $t \in \mathbb{T}$. We then define the nabla integral of f by
$$\int_a^t f(s)\nabla s = F(t) - f(a), \text{ for all } t \in \mathbb{T}.$$

If f and f^{∇} are continuous, then
$$\left(\int_a^t f(t, s)\nabla s\right)^{\nabla} = f(\rho(t), t) + \int_a^t f^{\nabla}(t, s)\nabla s.$$

One can easily see that every ld-continuous function has a nabla antiderivative. As in the case of the delta derivative we see that if $f : \mathbb{T} \to \mathbb{R}$ is ld-continuous and $t \in \mathbb{T}$, then
$$\int_{\rho(t)}^t f(s)\nabla s = \nu(t)f(t).$$

Theorem 1.2.3 *If $a, b, c \in \mathbb{T}$ and $\alpha, \beta \in \mathbb{R}$, and $f, g : \mathbb{T} \to \mathbb{R}$ are ld-continuous, then*

1. $\int_a^b [\alpha f(t) + \beta g(t)] \, \Delta t = \alpha \int_a^b f(t) \nabla t + \beta \int_a^b g(t) \nabla t$,

2. $\int_a^b f(t) \nabla t = - \int_b^a f(t) \nabla t$,

3. $\int_a^c f(t) \nabla t = \int_a^b f(t) \nabla t + \int_b^c f(t) \nabla t$,

4. $\left| \int_a^b f(t) \nabla t \right| \leq \int_a^b |f(t)| \, \nabla t$,

5. $\int_a^b f(t) g^\nabla(t) \nabla t = f(t) g(t) \big|_a^b - \int_a^b f^\nabla(t) g^\rho(t) \nabla t$,

6. $\int_a^b f^\rho(t) g^\nabla(t) \nabla t = f(t) g(t) \big|_a^b - \int_a^b f^\nabla(t) g(t) \nabla t$.

The relations between delta and nabla derivatives can be summarized as follows. Assume that $f : \mathbb{T} \to \mathbb{R}$ is delta differentiable on $\mathbb{T}^k$. Then f is nabla differentiable at t and
$$f^\nabla(t) = f^\Delta(\rho(t)),$$
for $t \in \mathbb{T}^k$ such that $\sigma(\rho(t)) = t$. If, in addition, f^Δ is continuous on $\mathbb{T}^k$, then f is nabla differentiable at t and $f^\nabla(t) = f^\Delta(\rho(t))$ holds for any $t \in \mathbb{T}_k$. Assume that $f : \mathbb{T} \to \mathbb{R}$ is nabla differentiable on $\mathbb{T}_k$. Then f is delta differentiable at t and
$$f^\Delta(t) = f^\nabla(\sigma(t)),$$
for $t \in \mathbb{T}_k$ such that $\rho(\sigma(t)) = t$. If, in addition, f^∇ is continuous on $\mathbb{T}_k$, then f is delta differentiable at t and $f^\Delta(t) = f^\nabla(\sigma(t))$ holds for any $t \in \mathbb{T}^k$.

We now give the definition of the generalized nabla exponential function. Assume that $p : \mathbb{T} \to \mathbb{R}$ is ld-continuous and $1 - p(t)\nu(t) \neq 0$ for $t \in \mathbb{T}_k$. We define the set $\mathcal{R}_\nu$ of all regressive and ld-continuous functions. We define the set $\mathcal{R}_\nu^+$ of all positively regressive elements of $\mathcal{R}_\nu$ by $\mathcal{R}_\nu^+ = \{p \in \mathcal{R} : 1 - \nu(t)p(t) > 0, \text{ for all } t \in \mathbb{T}\}$. The set of all ν-regressive functions on a time scale $\mathbb{T}$ forms an Abelian group under the addition $\oplus$ defined by $p \oplus_\nu q := p + q - \nu pq$. The explicit nabla exponential function is given by

$$\check{e}_p(t, s) = \exp\left(\int_s^t \overline{\xi}_{\nu(\tau)}(p(\tau)) \nabla \tau \right), \quad \text{for } t \in \mathbb{T}, \ s \in \mathbb{T}^k, \tag{1.2.1}$$

where $\overline{\xi}_h(z)$ is the cylinder transformation, which is defined by

$$\overline{\xi}_h(z) = \begin{cases} -\dfrac{\log(1 - hz)}{h}, & h \neq 0, \\ z, & h = 0. \end{cases}$$

For $t \in \mathbb{T}$, $s \in \mathbb{T}_k$, the exponential function $\check{e}_p(t, s)$ is the solution of the initial value problem

$$x^\nabla(t) = p(t)x(t), \quad t \in \mathbb{T}_k \text{ with } x(s) = 1.$$

The following theorem gives the properties of the exponential function $\check{e}_p(t, s)$. The theorem is adapted from Bohner and Peterson [52].

Theorem 1.2.4 *If* $p, q \in \mathcal{R}_\nu$ *and* $t_0 \in \mathbb{T}$, *then*

- $\breve{e}_p(t, t) \equiv 1,$ *and* $\breve{e}_0(t, s) \equiv 1;$

- $\breve{e}_p(\rho(t), s) = (1 - \nu(t)p(t))\breve{e}_p(t, s);$

- $\frac{1}{\breve{e}_p(t,s)} = \breve{e}_{\ominus_\nu p}(t, s);$

- $\frac{\breve{e}_p(t,s)}{\breve{e}_q(t,s)} = \breve{e}_{p \ominus_\nu q}(t, s);$

- $\breve{e}_p(t, s)\breve{e}_q(t, s) = \breve{e}_{p \oplus_\nu q}(t, s);$

- *if* $1 - p(t)\nu(t) \neq 0$, *then* $\breve{e}_p(t, t_0) > 0$ *for all* $t \in \mathbb{T}$.

- $\breve{e}_p^\nabla(t, t_0) = p(t)\breve{e}_p(t, t_0).$

- $\left(\frac{1}{\breve{e}_p(\cdot, s)}\right)^\nabla = -\frac{p(t)}{\breve{e}_p^\sigma(\cdot, s)}.$

1.3   Diamond-α Calculus

Now we introduce the diamond-α dynamic derivative and diamond-α dynamic integration. The comprehensive development of the calculus of the diamond-α derivative and diamond-α integration is given in [140]. Let $\mathbb{T}$ be a time scale and $f(t)$ be differentiable on $\mathbb{T}$ in the Δ and ∇ sense. For $t \in \mathbb{T}$, we define the diamond-α derivative $f^{\Diamond_\alpha}(t)$ by

$$f^{\Diamond_\alpha}(t) = \alpha f^\Delta(t) + (1 - \alpha)f^\nabla(t), \quad 0 \leq \alpha \leq 1.$$

Thus f is diamond-α differentiable if and only if f is Δ and ∇ differentiable. The diamond-α derivative reduces to the standard Δ-derivative for $\alpha = 1$, or the standard ∇ derivative for $\alpha = 0$. It represents a weighted dynamic derivative for $\alpha \in (0, 1)$.

Theorem 1.3.1 *Let* $f, g : \mathbb{T} \to \mathbb{R}$ *be diamond-α differentiable at* $t \in \mathbb{T}$. *Then*

(i). $f + g : \mathbb{T} \to \mathbb{R}$ *is diamond-α differentiable at* $t \in \mathbb{T}$, *with*

$$(f + g)^{\Diamond_\alpha}(t) = f^{\Diamond_\alpha}(t) + g^{\Diamond_\alpha}(t).$$

(ii). $f.g : \mathbb{T} \to \mathbb{R}$ *is diamond-α differentiable at* $t \in \mathbb{T}$, *with*

$$(f.g)^{\Diamond_\alpha}(t) = f^{\Diamond_\alpha}(t)g(t) + \alpha f^\sigma(t)g^\Delta(t) + (1 - \alpha)f^\rho(t)g^\nabla(t).$$

(iii). For $g(t)g^\sigma(t)g^\rho(t) \neq 0$, $f/g : \mathbb{T} \to \mathbb{R}$ *is diamond-α differentiable at* $t \in \mathbb{T}$, *with*

$$\left(\frac{f}{g}\right)^{\Diamond_\alpha}(t) = \frac{f^{\Diamond_\alpha}(t)g^\sigma(t)g^\rho(t) - \alpha f^\sigma(t)g^\rho(t)g^\Delta(t) - (1 - \alpha)f^\rho(t)g^\sigma(t)g^\nabla(t)}{g(t)g^\sigma(t)g^\rho(t)}.$$

Theorem 1.3.2 *Let* f, $g : \mathbb{T} \to \mathbb{R}$ *be diamond-α differentiable at* $t \in \mathbb{T}$. *Then the following hold:*

(i). $(f)^{\Diamond_\alpha \Delta}(t) = \alpha f^{\Delta\Delta}(t) + (1 - \alpha)f^{\nabla\Delta}(t)$,

(ii). $(f)^{\Diamond_\alpha \nabla}(t) = \alpha f^{\Delta\nabla}(t) + (1 - \alpha)f^{\nabla\nabla}(t)$,

(iii). $(f)^{\Delta\Diamond_\alpha}(t) = \alpha f^{\Delta\Delta}(t) + (1 - \alpha)f^{\Delta\nabla}(t) \neq (f)^{\Diamond_\alpha \Delta}(t)$,

(iv). $(f)^{\nabla\Diamond_\alpha}(t) = \alpha f^{\nabla\Delta}(t) + (1 - \alpha)f^{\nabla\nabla}(t) \neq (f)^{\Diamond_\alpha \nabla}(t)$,

(v). $(f)^{\Diamond_\alpha \Diamond_\alpha}(t) = \alpha^2 f^{\Delta\Delta}(t) + \alpha(1 - \alpha)[f^{\Delta\nabla}(t) + f^{\nabla\Delta}(t)]$
$+ (1 - \alpha)^2 f^{\nabla\nabla}(t) \neq \alpha^2 f^{\Delta\Delta}(t) + (1 - \alpha)^2 f^{\nabla\nabla}(t)$.

Theorem 1.3.3 *(Mean Value Theorem). Suppose that* f *is a continuous function on* $[a, b]_\mathbb{T}$ *and has a diamond-α derivative at each point of* $[a, b)_\mathbb{T}$. *Then there exist points* η, η' *such that*

$$f^{\Diamond_\alpha}(\eta')(b - a) \leq f(b) - f(a) \leq f^{\Diamond_\alpha}(\eta)(b - a).$$

When $f(a) = f(b)$, *then we have that*

$$f^{\Diamond_\alpha}(\eta') \leq 0 \leq f^{\Diamond_\alpha}(\eta).$$

Corollary 1.3.1 *Let* f *be a continuous function on* $[a, b]_\mathbb{T}$ *and has a diamond-α derivative at each point of* $[a, b)_\mathbb{T}$. *Then* f *is increasing if* $f^{\Diamond_\alpha}(t) > 0$, *decreasing if* $f^{\Diamond_\alpha}(t) < 0$, *nonincreasing if* $f^{\Diamond_\alpha}(t) \leq 0$ *and nondecreasing* $f^{\Diamond_\alpha}(t) \geq 0$ *on* $[a, b]_\mathbb{T}$.

Theorem 1.3.4 *Let* a, $t \in \mathbb{T}$, *and* $h : \mathbb{T} \to \mathbb{R}$. *Then, the diamond-α integral from* a *to* t *of* h *is defined by*

$$\int_a^t h(s)\Diamond_\alpha s = \alpha \int_a^t h(s)\Delta s + (1 - \alpha) \int_a^t h(s)\nabla s, \ \ 0 \leq \alpha \leq 1,$$

provided that there exists delta and nabla integrals of h *on* $\mathbb{T}$.

In general, we do not have

$$\left(\int_a^t h(s)\Diamond_\alpha s \right)^{\Diamond_\alpha} = h(t), \text{ for } t \in \mathbb{T}.$$

Example 1.3.1 ([31]) *Let* $\mathbb{T} = 0, 1, 2$, $a = 0$ *and* $h(t) = t^2$ *for* $t \in \mathbb{T}$. *This gives us that*

$$\left(\int_a^t h(s)\Diamond_\alpha s \right)^{\Diamond_\alpha} \bigg|_{t=1} = 1 + 2\alpha(1 - \alpha),$$

so that the equality above holds only when $\Diamond_\alpha = \Delta$ *or* $\Diamond_\alpha = \nabla$.

Theorem 1.3.5 *Let $a, b, c \in \mathbb{T}$, $\alpha, \beta \in \mathbb{R}$, and f and g be continuous functions on $[a, b] \cup \mathbb{T}$. Then the following properties hold:*

(1). $\int_a^b [\alpha f(t) + \beta g(t)] \, \Diamond_\alpha t = \alpha \int_a^b f(t) \Diamond_\alpha t + \beta \int_a^b g(t) \Diamond_\alpha t,$

(2). $\int_a^b f(t) \Diamond_\alpha t = - \int_b^a f(t) \Diamond_\alpha t,$

(3). $\int_a^c f(t) \Diamond_\alpha t = \int_a^b f(t) \Diamond_\alpha t + \int_b^c f(t) \Diamond_\alpha t.$

Example 1.3.2 *If we let $\mathbb{T} = \mathbb{R}$, then we obtain*

$$\int_a^b f(t) \Diamond_\alpha t = \int_a^b f(t) dt, \quad where \ a, \ b \in \mathbb{R},$$

and if we let $\mathbb{T} = \mathbb{Z}$, and $m < n$, then we obtain

$$\int_m^n f(t) \Diamond_\alpha t = \sum_{i=m}^{n-1} [\alpha f(i) + (1-\alpha)f(i+1)], \quad for \ m, \ n \in \mathbb{N}_0. \qquad (1.3.1)$$

Example 1.3.3 *If we let $\mathbb{T} = q^{\mathbb{N}}$, for $q > 1$ and $m < n$, then we obtain*

$$\int_{q^m}^{q^n} f(t) \Diamond_\alpha t = (q-1) \sum_{i=m}^{n-1} q^i \left[\alpha f(q^i) + (1-\alpha)f(q^{i+1}) \right], \quad for \ m, \ n \in \mathbb{N}_0, \qquad (1.3.2)$$

and if we let $\mathbb{T} = \{t_i : i \in \mathbb{N}_0\}$ such that $t_i < t_{i+1}$ and $m < n$, then we obtain the general case (which includes (1.3.1) and (1.3.2))

$$\int_{t_m}^{t_n} f(t) \Diamond_\alpha t = \sum_{i=m}^{n-1} (t_{i+1} - t_i) \left[\alpha f(t_i) + (1 - \alpha)f(t_{i+1}) \right], \ for \ m, \ n \in \mathbb{N}_0, \qquad (1.3.3)$$

Remark 1.3.1 *Note that if $f(t) \geq 0$ for all $t \in [a, b]_{\mathbb{T}}$, then $\int_a^b cf(t) \Diamond_\alpha t \geq 0$. If $f(t) \geq g(t)$ for all $t \in [a, b]_{\mathbb{T}}$, then $\int_a^b f(t) \Diamond_\alpha t \geq \int_a^b g(t) \Diamond_\alpha t \geq 0$, and $f(t) = 0$ if and only if $\int_a^b f(t) \Diamond_\alpha t = 0$.*

Corollary 1.3.2 *Let $t \in \mathbb{T}_k^k$ and $f : \mathbb{T} \to \mathbb{R}$. Then*

$$\int_t^{\sigma(t)} f(s) \Diamond_\alpha s = \mu(t)[\alpha f(t) + (1 - \alpha)f^\sigma(t)],$$

and

$$\int_t^{\rho(t)} f(s) \Diamond_\alpha s = \nu(t)[\alpha f^\rho(t) + (1 - \alpha)f(t)].$$

Recall a function $p : \mathbb{T} \to \mathbb{R}$ is called regressive provided $1 + \mu(t)p(t) \neq 0$ for all $t \in \mathbb{T}^k$. Note $\mathcal{R}$ denotes the set of all regressive and rd-continuous functions on $\mathbb{T}$. Similarly, a function $q : \mathbb{T} \to \mathbb{R}$ is called ν-regressive provided $1 - \nu(t)q(t) \neq 0$ for all $t \in \mathbb{T}_k$. Note $\mathcal{R}_\nu$ denotes the set of all ν-regressive and ld-continuous functions on $\mathbb{T}$. We consider two functions: $E_{p,\alpha}$ and $e_{p,\alpha}$ where $p \in \mathcal{R} \cap \mathcal{R}_\nu$ and $\alpha \in [0,1]$. For $p \in \mathcal{R} \cap \mathcal{R}_\nu$ and $\alpha \in [0,1]$, we define

$$E_{p,\alpha}(.,t_0) = \alpha e_p(.,t_0) + (1-\alpha)\check{e}_p(.,t_0); \quad \text{for } t \in \mathbb{T},$$

where $e_p(.,t_0)$ and $\check{e}_p(.,t_0)$ are the delta and nabla exponential functions defined as in (1.1.5) and (1.2.1), respectively.

Example 1.3.4 ([31]) *Consider the time scale* $\mathbb{T} = \mathbb{Z}$ *and the constant function* $p(t) = 1/2$. *Take* $t_0 = 0$. *Then,* $e_p(t,0) = (3/2)^t$ *is the solution of the initial value problem* $y^\Delta(t) = (1/2)y(t)$, $y(t_0) = 1$. *Moreover* $\check{e}_p(t,0) = 2^t$ *is the unique solution of* $y^\nabla(t) = (1/2)y(t)$, $y(t_0) = 1$. *Now* $E_{p,\alpha}(t;0) = \alpha(3/2)^t + (1-\alpha)(2)^t$, *for* $t \in \mathbb{Z}$.

Remark 1.3.2 *Combined-exponentials cannot be really called an exponential function. Indeed, they seem to fail the most important property of an exponential function, i.e., they are not a solution of an appropriate initial value problem.*

Next we give a direct formulas for the $\Diamond_\alpha$-derivative of exponential functions $e_p(.,t_0)$ and $\check{e}_p(.,t_0)$.

Theorem 1.3.6 *Let* $\mathbb{T}$ *be a regular time scale. Assume that* t, $t_0 \in \mathbb{T}$ *and* $p \in \mathcal{R} \cap \mathcal{R}_\nu$. *Then*

$$e^{\Diamond_\alpha}(t,t_0) = \left[\alpha p(t) + \frac{(1-\alpha)\,p^\rho(t)}{1+\nu(t)p^\rho(t)}\right] e_p(t,t_0),$$

$$\check{e}_p^{\Diamond_\alpha}(t,t_0) = \left[\alpha p(t) + \frac{(1-\alpha)\,p^\sigma(t)}{1+\mu(t)p^\sigma(t)}\right] \check{e}_p(.,t_0),$$

where $e_p(.,t_0)$ *is a solution of the initial value problem* $y^{\Diamond_\alpha}(t) = q(t)y(t)$, $y(t_0) = 1$, *where* $q(t) = \alpha p(t) + \frac{(1-\alpha)p^\rho(t)}{1+\nu(t)p^\rho(t)}$.

1.4 Taylor Monomials and Series

Here we define Taylor monomials and Taylor expansions of functions corresponding to delta and nabla derivatives. To define these functions, we need some basic definitions about calculus of functions of two variables on time scales. Let $\mathbb{T}_1$ and $\mathbb{T}_2$ be two time scales with at least two points and consider the time scale intervals $\Omega_1 = [t_0,\infty) \cap \mathbb{T}_1$ and $\Omega_2 = [s_0,\infty) \cap \mathbb{T}_2$ for $t_0 \in \mathbb{T}_1$ and $s_0 \in \mathbb{T}_2$. Let σ_1, ρ_1, Δ_1 and σ_2, ρ_2, Δ_2 denote the forward jump operators, backward jump operators, and the delta differentiation operator, respectively, on $\mathbb{T}_1$ and $\mathbb{T}_2$. We say that a real valued function f on $\mathbb{T}_1 \times \mathbb{T}_2$

at $(t, s) \in \Omega \equiv \Omega_1 \times \Omega_2$ has a Δ_1 partial derivative $f^{\Delta_1}(t, s)$ with respect to t if for each $\epsilon > 0$ there exists a neighborhood U_t of t such that

$$|f(\sigma_1(t), s) - f(\eta, s) - f^{\Delta_1}(t, s)[\sigma_1(t) - \eta]| \le \epsilon |\sigma(t) - \eta|, \text{ for all } \eta \in U_t.$$

In this case, we say $f^{\Delta_1}(t, s)$ is the (partial delta) derivative of $f(t, s)$ at t. We say that a real valued function f on $\mathbb{T}_1 \times \mathbb{T}_2$ at $(t, s) \in \Omega_1 \times \Omega_2$ has a Δ_2 partial derivative $f^{\Delta_2}(t, s)$ with respect to s if for each $\epsilon > 0$ there exists a neighborhood U_s of s such that

$$|f(t, \sigma_2(s)) - f(t, \xi) - f^{\Delta_2}(t, s)[\sigma_2(t) - \xi]| \le \epsilon |\sigma(t) - \xi|, \text{ for all } \xi \in U_s.$$

In this case, we say $f^{\Delta_2}(t, s)$ is the (partial delta) derivative of $f(t, s)$ at s. The function f is called rd-continuous in t if for every $\alpha_2 \in \mathbb{T}_2$ the function $f(t, \alpha_2)$ is rd-continuous on $\mathbb{T}_1$. The function f is called rd-continuous in s if for every $\alpha_1 \in \mathbb{T}_1$ the function $f(\alpha_1, s)$ is rd-continuous on $\mathbb{T}_2$.

Theorem 1.4.1 *Let $t_0 \in \mathbb{T}^\kappa$ and assume $k : \mathbb{T} \times \mathbb{T}^\kappa \mapsto \mathbb{R}$ is continuous at (t, t), where $t \in \mathbb{T}^\kappa$ with $t > t_0$. Also assume that $k(t, \cdot)$ is rd-continuous on $[t_0, \sigma(t)]$. Suppose for each $\epsilon > 0$ there exists a neighborhood of t, independent U of $\tau \in [t_0, \sigma(t)]$, such that*

$$|k(\sigma(t), \tau) - k(s, \tau) - k^\Delta(t, \tau)(\sigma(t) - s)| \le \epsilon |\sigma(t) - s|, \text{ for all } s \in U,$$

where k^Δ denotes the derivative of k with respect to the first variable. Then

$$g(t) := \int_{t_0}^t k(t, \tau) \Delta \tau, \quad implies \quad g^\Delta(t) = \int_{t_0}^t k^\Delta(t, \tau) \Delta \tau + k(\sigma(t), t).$$

The Taylor monomials $h_k : \mathbb{T} \times \mathbb{T} \to \mathbb{R}$, $k \in \mathbb{N}_0 = \mathbb{N} \cup \{0\}$, are defined recursively as follows. The function h_0 is defined by

$$h_0(t, s) = 1, \text{ for all } s, t \in \mathbb{T},$$

and given h_k for $k \in \mathbb{N}_0$, the function h_{k+1} is defined by

$$h_{k+1}(t, s) = \int_s^t h_k(\tau, s) \Delta \tau, \text{ for all } s, t \in \mathbb{T}.$$

If we let $h_k^\Delta(t, s)$ denote for each fixed $s \in \mathbb{T}$, the derivative of $h(t, s)$ with respect to t, then

$$h_k^\Delta(t, s) = h_{k-1}(t, s), \quad k \in \mathbb{N}, \quad t \in \mathbb{T},$$

for each fixed $s \in \mathbb{T}$. The above definition obviously implies

$$h_1(t, s) = t - s, \text{ for all } s, t \in \mathbb{T}.$$

In the following, we give some formulas of $h_k(t, s)$ as determined in [51]. In the case when $\mathbb{T} = \mathbb{R}$, then

$$h_k(t, s) = \frac{(t - s)^k}{k!}, \quad \text{for all } s, t \in \mathbb{R}. \tag{1.4.1}$$

In the case when $\mathbb{T} = \mathbb{N}$, we see that

$$h_k(n, s) := \frac{(n - s)^{(k)}}{k!}, \quad k = 0, 1, 2, \ldots, \quad t > s, \tag{1.4.2}$$

where $t^{(k)} = t(t - 1) \cdots (t - k + 1)$ is the so-called falling function (see [100]). When $\mathbb{T} = \{t : t = q^n, n \in \mathbb{N}, q > 1\}$, we have that

$$h_k(t, s) = \prod_{m=0}^{k-1} \frac{t - q^m s}{\sum_{j=0}^{m} q^j}, \quad \text{for all } s, t \in \mathbb{T}. \tag{1.4.3}$$

If $\mathbb{T} = h\mathbb{N}$, $h > 0$, we see that

$$h_k(t, s) = \frac{\prod_{i=0}^{k-1}(t - ih - s)}{k!}, \quad \text{for all } s, t \in \mathbb{T}, t > s. \tag{1.4.4}$$

In general for $t \geq s$, we have that $h_k(t, s) \geq 0$, and

$$h_k(t, s) \leq \frac{(t - s)^k}{k!}, \quad \text{for all } t > s, \ k \in \mathbb{N}_0.$$

We also consider the Taylor monomials $g_k : \mathbb{T} \times \mathbb{T} \to \mathbb{R}$, $k \in \mathbb{N}_0 = \mathbb{N} \cup \{0\}$, which are defined recursively. The function g_0 is defined by

$$g_0(t, s) = 1, \quad \text{for all } s, t \in \mathbb{T},$$

and given g_k for $k \in \mathbb{N}_0$, the function g_{k+1} is defined by

$$g_{k+1}(t, s) = \int_s^t g_k(\sigma(\tau), s)\Delta\tau, \quad \text{for all } s, t \in \mathbb{T}.$$

If we let $g_k^\Delta(t, s)$ denote for each fixed $s \in \mathbb{T}$, the derivative of $g(t, s)$ with respect to t, then

$$g_k^\Delta(t, s) = g_{k-1}(\sigma(t), s), \quad k \in \mathbb{N}, \ t \in \mathbb{T},$$

for each fixed $s \in \mathbb{T}$. One can see that

$$h_k(t, s) = (-1)^k g_k(s, t).$$

We denote by $C_{rd}^{(n)}(\mathbb{T})$ the space of all functions $f \in C_{rd}(\mathbb{T})$ such that $f^{\Delta^i} \in C_{rd}(\mathbb{T})$ for $i = 0, 1, 2, \ldots, n$ for $n \in \mathbb{N}$. For the function $f : \mathbb{T} \to \mathbb{R}$, we consider the second derivative f^{Δ_2} provided f^{Δ} is delta differentiable on $\mathbb{T}$ with derivative $f^{\Delta_2} = (f^{\Delta})^{\Delta}$. Similarly, we define the nth order derivative $f^{\Delta_n} = (f^{\Delta_{n-1}})^{\Delta}$. Now, we give the definition of generalized polynomials as follows:

$$h_n(t,s) := \begin{cases} 1, & n = 0 \\ \int_s^t h_{n-1}(\xi,s)\Delta\xi, & n \in \mathbb{N} \end{cases} \tag{1.4.5}$$

and

$$g_n(t,s) := \begin{cases} 1, & n = 0 \\ \int_s^t g_{n-1}(\sigma(\xi),s)\Delta\xi, & n \in \mathbb{N}, \end{cases}$$

for all s, $t \in \mathbb{T}$.

Property. Using induction it is easy to see that $h_n(t,s) \geq 0$ holds for all $n \in \mathbb{N}$ and s, $t \in \mathbb{T}$ with $t \geq s$ and $(-1)^n h_n(t,s) \geq 0$ holds for all $n \in \mathbb{N}$ and s, $t \in \mathbb{T}$ with $t \leq s$. Moreover, $h_n(t,s)$ is increasing with respect to its first component for all $t \geq s$. $\blacksquare$

Recall the following result (see [52]).

Lemma 1.4.1 *For $n \in \mathbb{N}$ and $t \in \mathbb{T}$, we have $g_n(t,s) = 0$ for all $s \in [\rho^{n-1}(t),t]_{\mathbb{T}}$.*

Lemma 1.4.2 *For $n \in \mathbb{N}$, $t \in \mathbb{T}$ and $s \in \mathbb{T}^{\kappa^n}$, we have $h_n(t,s) = (-1)^n g_n(s,t)$.*

From Lemmas 1.4.1 and 1.4.3 we have the following result.

Lemma 1.4.3 *For $n \in \mathbb{N}$ and $t \in \mathbb{T}$, we have $h_n(t,s) = 0$ for all $s \in [\rho^{n-1}(t),t]_{\mathbb{T}}$.*

Theorem 1.4.2 *Let $n \in \mathbb{N}$ and $f \in C_{rd}^n(\mathbb{T},\mathbb{R})$ be an n times differentiable function. For $s \in \mathbb{T}^{\kappa^{n-1}}$, we have*

$$f(t) = \sum_{j=0}^{n-1} h_j(t,s)f^{\Delta^j}(s) + \int_s^{\rho^{n-1}(t)} h_{n-1}(t,\sigma(\xi))f^{\Delta^n}(\xi)\Delta\xi, \text{ for all } t \in \mathbb{T}.$$

Theorem 1.4.3 *Assume that $f \in C_{rd}^{(n)}(\mathbb{T})$ and $s \in \mathbb{T}$. Then*

$$f(t) = \sum_{k=0}^{n-1} f^{\Delta^k}(s)h_k(t,s) + \int_s^t h_{n-1}(t,(\sigma(\tau))f^{\Delta^n}(\tau)\Delta\tau. \tag{1.4.6}$$

As a special case if $m < n$, then

$$f^{\Delta^m}(t) = \sum_{k=0}^{n-m-1} f^{\Delta^{k+m}}(s)h_k(t,s) + \int_s^t h_{n-m-1}(t,(\sigma(\tau))f^{\Delta^n}(\tau)\Delta\tau.$$

Now, we define the Taylor expansions of the functions corresponding to the nabla derivative. The generalized polynomial that will be used in describing these expansions are $\hat{h}_k : \mathbb{T} \times \mathbb{T} \to \mathbb{R}$, $k \in \mathbb{N}_0 = \mathbb{N} \cup \{0\}$, which are defined recursively as follows. The function $\hat{h}_0$ is defined by

$$\hat{h}_0(t, s) = 1, \quad \text{for all } s, \ t \in \mathbb{T},$$

and given $\hat{h}_k$ for $k \in \mathbb{N}_0$, the function $\hat{h}_{k+1}$ is defined by

$$\hat{h}_{k+1}(t, s) = \int_s^t \hat{h}_k(\tau, s) \nabla \tau, \quad \text{for all } s, \ t \in \mathbb{T}. \tag{1.4.7}$$

Note that the functions $\hat{h}_k$ are all well defined. If we let $\hat{h}_k^{\triangle}(t, s)$ denote for each fixed $s \in \mathbb{T}$, the derivative of $\hat{h}(t, s)$ with respect to t, then

$$\hat{h}_k^{\triangle}(t, s) = \hat{h}_{k-1}(t, s), \quad k \in \mathbb{N}, \ t \in \mathbb{T}_k,$$

for each fixed $s \in \mathbb{T}$. The above definition obviously implies

$$\hat{h}_1(t, s) = t - s, \quad \text{for all } s, \ t \in \mathbb{T}.$$

Finding the $\hat{h}_k$ for $k > 1$ is not an easy task in general. However for a particular given time scale it might be easy to find these functions. We will consider some examples first before we present Taylor's formula in general. In the case when $\mathbb{T} = \mathbb{R}$, then $\rho(t) = t$ and

$$\hat{h}_k(t, s) = \frac{(t - s)^k}{k!}, \quad \text{for all } s, \ t \in \mathbb{R}. \tag{1.4.8}$$

In the case when $\mathbb{T} = \mathbb{N}$, we see that $\rho(t) = t - 1$, $\nu(t) = 1$, $y^\nabla(t) = \nabla(t) = y(t) - y(t - 1)$, and

$$\hat{h}_k(t, s) := \frac{(t - s)^{(k)}}{k!}, \quad k = 0, 1, 2, \ldots, \quad t > s, \tag{1.4.9}$$

where $t^{(k)} = t(t - 1) \cdots (t - k + 1)$ is the so-called falling function (see [100]). Noting that $\nabla t^{(k)} = k \, t^{(k-1)}$, we see that

$$\hat{h}_{k+1}(t, s) := \int_s^t \frac{(\tau - s)^{(k)}}{k!} \nabla \tau = \sum_{r=s+1}^t \frac{(\tau - s)^{(k)}}{k!} = \frac{(\tau - s)^{(k+1)}}{(k + 1)!}, \tag{1.4.10}$$

for $k = 0, 1, 2, \ldots, \quad t > s$. In the case when $\mathbb{T} = \{t : t = q^n, \ n \in \mathbb{N}, \ q > 1\}$, we have $\rho(t) = t/q$, $\nu(t) = (q - 1)t/q$, and

$$\hat{h}_k(t, s) = \prod_{m=0}^{k-1} \frac{q^m t - s}{\sum_{j=0}^m q^j}, \quad \text{for all } s, t \in \mathbb{T}. \tag{1.4.11}$$

In general for $t \geq s$, we have that $\hat{h}_k(t,s) \geq 0$, and

$$\hat{h}_k(t,s) \leq \frac{(t-s)^k}{k!}, \quad \text{for all } t > s, \ k \in \mathbb{N}_0.$$

We may also relate the functions $\hat{h}_k$, $\hat{g}_0$ for the nabla derivative to the functions h_k and g_k in the delta derivative.

Definition 1.4.1 *For t, s define the functions*

$$h_0(t,s) = g_0(t,s) = \hat{h}_0(t,s) = \hat{g}_0(t,s) = 1,$$

and given $h_n, g_n, \hat{h}_n$ and $\hat{g}_n$ for $n \in \mathbb{N}_0$,

$$h_{n+1}(t,s) = \int_s^t h_n(\tau,s)\Delta\tau, \ g_{n+1}(t,s) = \int_s^t g_n(\sigma(\tau),s)\Delta\tau,$$

$$\hat{h}_{n+1}(t,s) = \int_s^t \hat{h}_n(\tau,s)\nabla\tau, \ \hat{g}_{n+1} = \int_s^t \hat{g}_n(\rho(\tau),s)\nabla\tau,$$

we have that

$$\hat{h}_n = g_n(t,s) = (-1)^n h_n(s,t) = (-1)^n \hat{g}_n(s,t).$$

We denote by $C_{ld}^{(n)}(\mathbb{T})$ the space of all functions $f \in C_{ld}(\mathbb{T})$ such that $f^{\nabla^i} \in C_{ld}(\mathbb{T})$ for $i = 0,1,2,\ldots,n$ for $n \in \mathbb{N}$. For the function $f : \mathbb{T} \to \mathbb{R}$, we consider the second derivative f^{∇^2} provided f^∇ is nabla differentiable on $\mathbb{T}$ with derivative $f^{\nabla^2} = (f^\nabla)^\nabla$. Similarly, we define the nth order nabla derivative $f^{\nabla^n} = (f^{\nabla^{n-1}})^\nabla$.

Theorem 1.4.4 *Let $n \in \mathbb{N}$. Suppose that the function f is such that $f^{\nabla^{n+1}}$ is ld-continuous on $\mathbb{T}_{\kappa^{n+1}}$. Let $s \in \mathbb{T}_{\kappa^n}$, $t \in \mathbb{T}$, and define*

$$\hat{h}_0(t,s) = 1, \ \hat{h}_{k+1}(t,s) = \int_s^t \hat{h}_k(\tau,s)\nabla\tau, \ \text{for all } s, \ t \in \mathbb{T} \text{ and } k \in \mathbb{N}_0.$$

Then, we have

$$f(t) = \sum_{k=0}^n \hat{h}_k(t,s) f^{\nabla^k}(s) + \int_s^t \hat{h}_n(t,\rho(\xi)) f^{\nabla^{n+1}}(\xi)\nabla\xi.$$

We end this section with the time scale version of L'Hôpital's rule. We present the rule for delta and nabla derivatives.

Theorem 1.4.5 *Assume that f and g are Δ-differentiable on $\mathbb{T}$ and let $t_0 \in \mathbb{T} \cup \{\infty\}$. If $t_0 \in \mathbb{T}$, assume that t_0 is right-dense. Furthermore, assume that $\lim_{t \to t_0^-} f(t) = \lim_{t \to t_0^-} g(t) = 0$, and suppose that there exists $\varepsilon > 0$ with $g(t)g^\Delta(t) > 0$ for all $t \in L_\varepsilon(t_0) = \{t \in \mathbb{T} : 0 < t_0 - t < \varepsilon\}$. Then*

$$\liminf_{t \to t_0^-} \frac{f^\Delta(t)}{g^\Delta(t)} \leq \liminf_{t \to t_0^-} \frac{f(t)}{g(t)} \leq \limsup_{t \to t_0^-} \frac{f(t)}{g(t)} \leq \limsup_{t \to t_0^-} \frac{f^\Delta(t)}{g^\Delta(t)}.$$

Theorem 1.4.6 *Assume that f and g are ∇-differentiable on $\mathbb{T}$ and let $t_0 \in \mathbb{T} \cup \{-\infty\}$. If $t_0 \in \mathbb{T}$, assume that t_0 is right-dense. Furthermore, assume that $\lim_{t \to t_0^+} f(t) = \lim_{t \to t_0^+} g(t) = 0$, and suppose that there exists $\varepsilon > 0$ with $g(t)g^{\nabla}(t) > 0$ for all $t \in R_{\varepsilon}(t_0) = \{t \in \mathbb{T} : 0 < t - t_0 < \varepsilon\}$. Then*

$$\liminf_{t \to t_0^-} \frac{f^{\nabla}(t)}{g^{\nabla}(t)} \leq \liminf_{t \to t_0^-} \frac{f(t)}{g(t)} \leq \limsup_{t \to t_0^-} \frac{f(t)}{g(t)} \leq \limsup_{t \to t_0^-} \frac{f^{\nabla}(t)}{g^{\nabla}(t)}.$$

Chapter 2

Basic Inequalities

In so far as the theorems of mathematics relate to reality, they are not certain, and in so far as they are certain they do not relate to reality. Every thing should be made as simple as possible but not simpler.
Albert Einstein (1879–1955).

This chapter deals with the basic inequalities used in the rest of the book. The chapter is divided into seven sections and is organized as follows. In Sect. 2.1 we consider Young type inequalities which will be used in the proof of the Hölder and Minkowski inequalities. Section 2.2 discusses Jensen's inequality on time scales and Sect. 2.3 considers Hölder type inequalities. In Sect. 2.4 we consider the Minkowski inequality and Sect. 2.5 is devoted to Steffensen type inequalities on time scales. Section 2.6 considers Hermite–Hadamard type inequalities and finally Sect. 2.7 discusses Čebyšev type inequalities on time scales.

2.1 Young Inequalities

In 1912, Young [157] presented the following highly intuitive integral inequality

$$ab \leq \int_0^a f(t)dt + \int_0^b (f^{-1})(s)ds, \qquad (2.1.1)$$

for any real-valued continuous function $f : [0, \infty) \to [0, \infty)$ satisfying $f(0) = 0$ with f strictly increasing on $[0, \infty)$ and $a, b \in [0, \infty)$. The equality holds if

© Springer International Publishing Switzerland 2014
R. Agarwal et al., *Dynamic Inequalities On Time Scales*,
DOI 10.1007/978-3-319-11002-8_2

and only if $b = f(a)$. A useful consequence of this inequality, by taking $f(t) = t^{p-1}$ and $q = \frac{p}{p-1}$, is the classical Young inequality

$$ab \leq \frac{a^p}{p} + \frac{b^q}{q}, \qquad \frac{1}{p} + \frac{1}{q} = 1. \tag{2.1.2}$$

Hardy, Littlewood, and Pólya included (2.1.1) in their classical book [72]. The purpose of this section is to establish this inequality and its extensions on time scales. These will be used in the next sections to prove Hölder and Minkowski inequalities on time scales. The results are adapted from [25, 29, 151].

Theorem 2.1.1 *Let* $g \in C_{rd}([0, c]_{\mathbb{T}}, \mathbb{R})$ *be a strictly increasing function with* $c > 0$. *If* $g(0) = 0$, $a \in [0, c]_{\mathbb{T}}$ *and* $b \in [0, g(c)]_{g(\mathbb{T})}$, *then*

$$ab \leq \int_0^a g^\sigma(x)\Delta x + \int_0^b (g^{-1})^\sigma(y)\Delta y.$$

Proof. Since $g^{-1}(x)$ is strictly increasing and $\sigma(s) \geq s$, we see that

$$\int_0^b (g^{-1})^\sigma(x)\Delta x = \int_0^b (g^{-1})(\sigma(x))\Delta x \geq \int_0^b (g^{-1}(x))\Delta x. \tag{2.1.3}$$

Letting $v(x) = g(x)$ and $f(x) = x$ in Lemma 1.1.2, we see that

$$\int_0^{g^{-1}(b)} g^\Delta(x)x\Delta x = \int_{g(0)}^{g(g^{-1}(b))} g^{-1}(y)\Delta y = \int_0^b g^{-1}(y)\Delta y. \tag{2.1.4}$$

Integration by parts yields

$$\begin{aligned} \int_0^{g^{-1}(b)} g^\Delta(x)x\Delta x &= g(x)x\big|_0^{g^{-1}(b)} - \int_0^{g^{-1}(b)} g^\sigma(x)\Delta x \\ &= bg^{-1}(b) - \int_0^{g^{-1}(b)} g^\sigma(x)\Delta x. \end{aligned}$$

Thus, (2.1.3) and (2.1.4) imply that

$$\int_0^a g^\sigma(x)\Delta x + \int_0^b (g^{-1})^\sigma(y)\Delta y \geq bg^{-1}(b) + \int_{g^{-1}(b)}^0 g^\sigma(x)\Delta x. \tag{2.1.5}$$

Case (a). $a > g^{-1}(b)$.

It follows from the strictly increasing property of g that

$$\begin{aligned} \int_{g^{-1}(b)}^a g^\sigma(x)\Delta x &\geq \int_{g^{-1}(b)}^a g(\sigma(g^{-1}(b)))\Delta x \geq \int_{g^{-1}(b)}^a g(g^{-1}(b))\Delta x \\ &= b(a - g^{-1}(b)) = ab - bg^{-1}(b). \end{aligned}$$

This and (2.1.5) imply

$$\int_0^a g^\sigma(x)\Delta x + \int_0^b (g^{-1})^\sigma(y)\Delta y \geq ab.$$

Case (b). $a < g^{-1}(b)$.

Let $h = g^{-1}$. Then $a < h(b)$. Applying case (a) yields

$$ab \leq \int_0^b h^\sigma(x)\Delta x + \int_0^a (h^{-1})^\sigma(y)\Delta y = \int_0^b \left(g^{-1}\right)^\sigma(x)\Delta x + \int_0^a (g)^\sigma(y)\Delta y.$$

Combining Case (a) and Case (b), we get the desired inequality. The proof is complete. ∎

As an application of Theorem 2.1.1 by taking $g(x) = x^{p-1}$ on $[0,\infty)_\mathbb{T}$ and $g^{-1}(y) = y^{q-1}$ on $[0,\infty)_\mathbb{T}$, we get the following result.

Corollary 2.1.1 *Let $p > 1$ and $q > 1$ with $1/p + 1/q = 1$. If $a \geq 0$ and $b \geq 0$, then*

$$ab \leq \int_0^a (\sigma(x))^{p-1}\Delta x + \int_0^b (\sigma(y))^{q-1}\Delta y.$$

Example 2.1.1 *Let $\mathbb{T} = \mathbb{R}$, then Corollary 2.1.1 says, note that in $\mathbb{R}$ $\sigma(x) = x$, that*

$$ab \leq \frac{a^p}{p} + \frac{b^q}{q}, \qquad \frac{1}{p} + \frac{1}{q} = 1, \tag{2.1.6}$$

which is the classical Young inequality.

Example 2.1.2 *Let $\mathbb{T} = \mathbb{Z}$ and $g(t) = t$, then Theorem 2.1.1 says that*

$$ab \leq \sum_{t=0}^{a-1}(t+1) + \sum_{y=0}^{b-1}(y+1) = \frac{1}{2}a(a+1) + \frac{1}{2}b(b+1). \tag{2.1.7}$$

Theorem 2.1.2 *Let $\mathbb{T}$ be any time scale (unbounded above) with $0 \in \mathbb{T}$. Further suppose that $f : [0,\infty)_\mathbb{T} \to \mathbb{R}$ is a real-valued function satisfying*

(1). $f(0) = 0$;

(2). *f is continuous on $[0,\infty)_\mathbb{T}$, right-dense continuous at 0;*

(3). *f is strictly increasing on $[0,\infty)_\mathbb{T}$ such that $\tilde{\mathbb{T}} = f(\mathbb{T})$ is also a time scale.*

Then for any $a \in [0,\infty)_\mathbb{T}$ and $b \in [0,\infty)_{\tilde{\mathbb{T}}}$, we have

$$\int_0^a f(t)\Delta t + \int_0^a f(t)\nabla t + \int_0^b f^{-1}(y)\Delta y + \int_0^b f^{-1}(y)\nabla y \geq 2ab, \tag{2.1.8}$$

with equality if and only if $b = f(a)$.

Proof. From the continuity assumption (2), we see that f is both delta and nabla integrable. For simplicity, define

$$F(a,b) := \int_0^a f(t)\Delta t + \int_0^a f(t)\nabla t + \int_0^b f^{-1}(y)\Delta y + \int_0^b f^{-1}(y)\nabla y - 2ab.$$

Then it is enough to prove that $F(a,b) \geq 0$.

(I). We will first show that

$$F(a,b) \geq F(a,f(a)), \quad a \in [0,\infty)_{\mathbb{T}} \text{ and } b \in [0,\infty)_{\widetilde{\mathbb{T}}},$$

with equality if and only if $b = f(a)$. For any such a and b, we have

$$
\begin{aligned}
F(a,b) - F(a,f(a)) &= \int_{f(a)}^b [f^{-1}(y) - a]\Delta y + \int_{f(a)}^b [f^{-1}(y) - a]\nabla y \\
&= \int_b^{f(a)} [a - f^{-1}(y)]\Delta y + \int_b^{f(a)} [a - f^{-1}(y)]\nabla y.
\end{aligned}
$$

There are two cases to consider. The first case is $b > f(a)$. Here, whenever $y \in [f(a),b]_{\widetilde{\mathbb{T}}}$, we have $f^{-1}(b) \geq f^{-1}(y) \geq f^{-1}(f(a)){=}a$. Consequently,

$$F(a,b) - F(a,f(a)) = \int_b^{f(a)} [a - f^{-1}(y)]\Delta y + \int_b^{f(a)} [a - f^{-1}(y)]\nabla y \geq 0.$$

Since $f^{-1}(y) - a$ is continuous and strictly increasing for $y \in [f(a),b]_{\widetilde{\mathbb{T}}}$, equality will hold if and only if $b = f(a)$. The second case is $b \leq f(a)$. Here whenever $y \in [f(a),b] \cap f(\mathbb{T})$, we have $f^{-1}(b) \leq f^{-1}(y) \leq f^{-1}(f(a)) = a$. Consequently,

$$F(a,b) - F(a,f(a)) = \int_b^{f(a)} [a - f^{-1}(y)]\Delta y + \int_b^{f(a)} [a - f^{-1}(y)]\nabla y \geq 0.$$

Since $a - f^{-1}(y)$ is continuous and strictly decreasing for $y \in [b,f(a)]_{\widetilde{\mathbb{T}}}$, equality will hold if and only if $b = f(a)$.

(II). We will next show that $F(a,f(a)) = 0$.

Now, for brevity, we put $\delta(a) = F(a,f(a))$, that is

$$\delta(a) = \int_0^a f(t)\Delta t + \int_0^a f(t)\nabla t + \int_0^{f(a)} f^{-1}(y)\Delta y + \int_0^{f(a)} f^{-1}(y)\nabla y - 2af(a).$$

First, assume a is right scattered point. Then

$$
\begin{aligned}
\delta^\sigma(a) - \delta(a) &= [\sigma(a) - a]f(a) + [\sigma(a) - a]f^\sigma(a) \\
&\quad + [f^\sigma(a) - f(a)]f^{-1}(f(a)) + [f^\sigma(a) - f(a)]f^{-1}(f^\sigma(a)) \\
&\quad - 2[\sigma(a)f^\sigma(a) - af(a)] \\
&= [\sigma(a) - a][f(a) + f^\sigma(a)] + [f^\sigma(a) - f(a)][\sigma(a) + a] \\
&\quad - 2[\sigma(a)f^\sigma(a) - af(a)] = 0.
\end{aligned}
$$

Therefore if a is right-scattered point, then $\delta^\Delta(a) = 0$. Next, assume a is a right-dense point. Let $\{a_n\}_{n\in\mathbb{N}} \subset [a,\infty)_\mathbb{T}$ be a decreasing sequence converging to a. Then

$$\delta(a_n) - \delta(a)$$
$$= \int_a^{a_n} f(t)\Delta t + \int_a^{a_n} f(t)\nabla t + \int_{f(a)}^{f(a_n)} f^{-1}(y)\Delta y + \int_{f(a)}^{f(a_n)} f^{-1}(y)\nabla y$$
$$-2a_n f(a_n) + 2af(a).$$
$$= \int_a^{a_n} [f(t) - f(a_n)]\Delta t + \int_a^{a_n} [f(t) - f(a_n)]\nabla t + \int_{f(a)}^{f(a_n)} [f^{-1}(y) - a]\Delta y$$
$$+ \int_{f(a)}^{f(a_n)} [f^{-1}(y) - a]\nabla y.$$

Since the functions f and f^{-1} are strictly increasing, we get that

$$\delta(a_n) - \delta(a) \geq \int_a^{a_n} [f(a) - f(a_n)]\Delta t + \int_a^{a_n} [f(a) - f(a_n)]\nabla t$$
$$+ \int_{f(a)}^{f(a_n)} [f^{-1}(f(a)) - a]\Delta y + \int_{f(a)}^{f(a_n)} [f^{-1}(f(a)) - a]\nabla y$$
$$= 2(a_n - a)[f(a) - f(a_n)].$$

Similarly,

$$\delta(a_n) - \delta(a) \leq \int_a^{a_n} [f(a_n) - f(a_n)]\Delta t + \int_a^{a_n} [f(a_n) - f(a_n)]\nabla t$$
$$+ \int_{f(a)}^{f(a_n)} [f^{-1}(f(a_n)) - a_n]\Delta y + \int_{f(a)}^{f(a_n)} [f^{-1}(f(a)) - a]\nabla y$$
$$= 2(a_n - a)[f(a_n) - f(a)].$$

Therefore

$$0 = \lim_{n\to\infty} 2[f(a_n) - f(a)] \leq \lim_{n\to\infty} \frac{\delta(a_n) - \delta(a)}{(a_n - a)}$$
$$\leq \lim_{n\to\infty} 2[f(a_n) - f(a)] = 0.$$

It follows that $\delta^\Delta(a)$ exists, and $\delta^\Delta(a) = 0$ for right-dense a as well. As $\delta(0) = 0$, by a uniqueness theorem for initial value problems, we have that $\delta(a) = 0$ for all $a \in [0,\infty)_\mathbb{T}$. This implies that $F(a,b) \geq F(a,f(a)) = 0$, with equality if and only if $b = f(a)$. The proof is complete. ∎

As an application of Theorem 2.1.2 when $f(t) = t^{p-1}$ and $f^{-1}(y) = y^{q-1}$, we have the following result.

Corollary 2.1.2 *Let* $\mathbb{T}$ *be any time scale (unbounded above) with* $0 \in \mathbb{T}$. *Let* $p, q > 1$ *be real numbers with* $1/p + 1/q = 1$. *Then for any* $a \in [0, \infty)_{\mathbb{T}}$ *and* $b \in [0, \infty)_{\mathbb{T}^*}$ *where* $\mathbb{T}^* = \{t^{p-1} : t \in \mathbb{T}\}$, *we have*

$$\int_0^a t^{p-1} \Delta t + \int_0^a t^{p-1} \nabla t + \int_0^b y^{q-1} \Delta y + \int_0^b y^{q-1} \nabla y \geq 2ab,$$

with equality if and only if $b = a^{p-1}$.

Example 2.1.3 *If* $\mathbb{T} = \mathbb{R}$, *we see that* $\sigma(t) = t$ *and then Theorem 2.1.2 yields the classical Young inequality (2.1.1).*

Example 2.1.4 *If* $\mathbb{T} = \mathbb{Z}$, *we see that* $\sigma(t) = t + 1$ *and then Theorem 2.1.2 yields Young's discrete inequality*

$$2ab \leq \sum_{t=0}^{a-1}[f(t) + f(t+1)] + \sum_{y \in [0,b) \cap f(\mathbb{Z})}^{b-1} \mu(y)[2f^{-1}(y) + 1],$$

since here $f^{-1}(\sigma(y)) = \sigma(f^{-1}(y)) = f^{-1}(y) + 1$.

Theorem 2.1.3 *Let* $\mathbb{T}$ *be any time scale (unbounded above) with* $0 \in \mathbb{T}$. *Further suppose that* $f : [0, \infty)_{\mathbb{T}} \to \mathbb{R}$ *is a real-valued function satisfying:*

(1). $f(0) = 0$;

(2). f *is continuous on* $[0, \infty)_{\mathbb{T}}$, *right-dense continuous at 0;*

(3). f *is strictly increasing on* $[0, \infty)_{\mathbb{T}}$ *such that* $\tilde{\mathbb{T}} = f(\mathbb{T})$ *is also a time scale.*

Then for any $a \in [0, \infty)_{\mathbb{T}}$ *and* $b \in [0, \infty)_{\tilde{\mathbb{T}}}$, *we have*

$$\int_0^a [f(t) + f^\sigma(t)] \, \Delta t + \int_0^b [f^{-1}(y) + f^{-1}(\sigma(y))] \, \Delta y \geq 2ab, \qquad (2.1.9)$$

with equality if and only if $b = f(a)$.

Proof. For a continuous function g and $a \in [0, \infty)_{\mathbb{T}}$, define the function

$$G(a) = \int_0^a g(t) \Delta t + \int_0^a g(t) \nabla t - \int_0^a [g(t) + g^\sigma(t)] \, \Delta t.$$

Then $G(0) = 0$, and

$$G^\Delta(a) = g(a) + g^\sigma(a) - [g(a) + g^\sigma(a)] = 0.$$

Therefore $G \equiv 0$, and Theorem 2.1.3 follows from Theorem 2.1.2. The proof is complete. ∎

Next we establish Young integral inequalities with upper and lower bounds for the remainder.

Theorem 2.1.4 *Let $\mathbb{T}$ be any time scale (unbounded above) with $\alpha_1 \in \mathbb{T}$ and $\sup \mathbb{T} = \infty$. Further suppose that $f : [\alpha_1, \infty)_{\mathbb{T}} \to \mathbb{R}$ is a real-valued function satisfying*

(i). $f(\alpha_1) = \beta_1$;

(ii). *f is continuous on $[\alpha_1, \infty)_{\mathbb{T}}$, right-dense continuous at α_1;*

(iii). *f is strictly increasing on $[\alpha_1, \infty)_{\mathbb{T}}$ such that $\tilde{\mathbb{T}} = f(\mathbb{T})$ is also a time scale.*

Then for any $a \in [\alpha_1, \infty)_{\mathbb{T}}$ and $b \in [\beta_1, \infty)_{\tilde{\mathbb{T}}}$, we have

$$ab \leq \int_{\alpha_1}^{a} f(t)\Delta t + \int_{\beta_1}^{b} f^{-1}(y)\tilde{\nabla} y + \alpha_1 \beta_1, \qquad (2.1.10)$$

with equality if and only if $b \in \{f^{\rho}(a), f(a)\}$ for fixed a or with equality if and only if $a \in \{f^{-1}(b), \sigma(f^{-1}(b))\}$ for fixed b. The inequality (2.1.10) is reversed if f is strictly decreasing.

Proof. By the continuity assumption (ii), we see that the function f is delta integrable and the function f^{-1} is nabla integrable. For simplicity, we define

$$F(a, b) = \int_{\alpha_1}^{a} f(t)\Delta t + \int_{\beta_1}^{b} f^{-1}(y)\tilde{\nabla} y + \alpha_1 \beta_1 - ab. \qquad (2.1.11)$$

To prove (2.1.10), we need to show that $F(a, b) \geq 0$.

(I). We will first show that

$$F(a, b) \geq F(a, f(a)), \text{ for } a \in [\alpha_1, \infty)_{\mathbb{T}} \text{ and } b \in [\beta_1, \infty)_{\tilde{\mathbb{T}}},$$

with equality if and only if $b \in \{f^{\rho}(a), f(a)\}$. For any such a and b, we have

$$F(a, b) - F(a, f(a)) = \int_{f(a)}^{b} [f^{-1}(y) - a]\tilde{\nabla} y. \qquad (2.1.12)$$

Clearly if $b = f(a)$, then the integral equals to zero and if $b = f^{\rho}(a)$, then

$$
\begin{aligned}
F(a, f^{\rho}(a)) - F(a, f(a)) &= \int_{f^{\rho}(a)}^{f(a)} [a - f^{-1}(y)]\tilde{\nabla} y \\
&= [f(a) - f^{\rho}(a)][a - f^{-1}(f(a))] = 0.
\end{aligned}
$$

Otherwise, since $f^{-1}(y)$ is continuous and strictly increasing for $y \in \tilde{\mathbb{T}}$, the integrals in (2.1.12) are strictly positive for $b < f^{\rho}(a)$ and $b > f(a)$.

(II). We will next show that $F(a, f(a)) = F(a, f^{\rho}(a)) = 0$.

Now, for brevity, we put $\varphi(a) = F(a, f(a))$, that is

$$\varphi(a) = \int_{\alpha_1}^{a} f(t)\Delta t + \int_{\beta_1}^{f(a)} f^{-1}(y)\tilde{\nabla} y - af(a) + \alpha_1 \beta_1.$$

First, assume that a is right scattered point. Then

$$\varphi^\sigma(a) - \varphi(a)$$
$$= \int_{\alpha_1}^{\sigma(a)} f(t)\Delta t + \int_{f(a)}^{f^\sigma(a)} f^{-1}(y)\tilde{\nabla}y - \sigma(a)f^\sigma(a) + af(a)$$
$$= [\sigma(a) - a]f(a) + [f^\sigma(a) - f(a)]f^{-1}(f^\sigma(a)) - \sigma(a)f^\sigma(a) + af(a)$$
$$= 0.$$

Therefore if a is right-scattered point, then $\varphi^\Delta(a) = 0$. Next, assume a is a right-dense point. Let $\{a_n\}_{n\in\mathbb{N}} \subset [a, \infty)_\mathbb{T}$ be a decreasing sequence converging to a. Then

$$\varphi(a_n) - \varphi(a)$$
$$= \int_a^{a_n} f(t)\Delta t + \int_{f(a)}^{f(a_n)} f^{-1}(y)\tilde{\nabla}y - a_n f(a_n) + af(a)$$
$$\geq (a_n - a)f)a) + [f(a) - f(a_n)]a - a_n f(a_n) + af(a)$$
$$= (a_n - a)[f(a) - f(a_n)],$$

since the functions f and f^{-1} are strictly increasing. Similarly,

$$\varphi(a_n) - \varphi(a) \leq (a_n - a)[f(a_n) - f(a)].$$

Therefore

$$0 = \lim_{n\to\infty}[f(a_n) - f(a)] \leq \lim_{n\to\infty}\frac{\varphi(a_n) - \varphi(a)}{(a_n - a)} \leq \lim_{n\to\infty}[f(a_n) - f(a)] = 0.$$

It follows that $\varphi^\Delta(a)$ exists, and $\varphi^\Delta(a) = 0$ for right-dense a as well. In other words, in either case $\varphi^\Delta(a) = 0$ for $a \in [\alpha_1, \infty)_\mathbb{T}$. As $\varphi(\alpha_1) = 0$, by a uniqueness theorem for initial value problems, we have that $\varphi(a) = 0$ for all $a \in [\alpha_1, \infty)_\mathbb{T}$. As $F(a, f(a)) = F(a, f^\rho(a)) = 0$, we have that $F(a, b) \geq F(a, f(a)) = 0$, with equality if and only if $b = f(a)$ or $b = f^\rho(a)$. The case with $a \in \{f^{-1}(b), \sigma(f^{-1}(b))\}$ for fixed b is similar and thus omitted. If f is strictly decreasing, it is straightforward to see that the inequality (2.1.10) is reversed. The proof is complete. ∎

Now to establish upper bounds for Young's integral inequality we need the following result.

Lemma 2.1.1 *Let f satisfy the hypotheses of Theorem 2.1.4, and let $F(a, b)$ be given as in (2.1.11). Then for any a, $\alpha \in \mathbb{T}$ and $b, \beta \in \tilde{\mathbb{T}}$, we have*

$$F(a, b) + F(\alpha, \beta) \geq -(\alpha - a)(\beta - b), \qquad (2.1.13)$$

with equality if and only if $\alpha \in \{f^{-1}(b), \sigma(f^{-1}(b))\}$ and $\beta \in \{f^\rho(a), f(a)\}$.

Proof. Fix $a \in \mathbb{T}$ and $b \in \tilde{\mathbb{T}}$. By Young's integral inequality (2.1.10), we see that

$$\int_{\alpha_1}^{a} f(t)\Delta t + \int_{\beta_1}^{\beta} f^{-1}(y)\tilde{\nabla}y + \alpha_1\beta_1 \geq a\beta, \qquad (2.1.14)$$

and

$$\int_{\alpha_1}^{\alpha} f(t)\Delta t + \int_{\beta_1}^{\beta} f^{-1}(y)\tilde{\nabla}y + \alpha_1\beta_1 \geq \alpha b, \qquad (2.1.15)$$

with equality if and only if $\beta \in \{f^\rho(a), f(a)\}$ and $\alpha \in \{f^{-1}(b), \sigma(f^{-1}(b))\}$, respectively. By rearranging it follows that

$$\begin{aligned}
&\int_{\alpha_1}^{a} f(t)\Delta t + \int_{\beta_1}^{b} f^{-1}(y)\tilde{\nabla}y + \alpha_1\beta_1 - ab \\
&+ \int_{\alpha_1}^{\alpha} f(t)\Delta t + \int_{\beta_1}^{\beta} f^{-1}(y)\tilde{\nabla}y + \alpha_1\beta_1 - \alpha b \\
=~&\int_{\alpha_1}^{a} f(t)\Delta t + \int_{\beta_1}^{\beta} f^{-1}(y)\tilde{\nabla}y + \alpha_1\beta_1 \\
&+ \int_{\alpha_1}^{\alpha} f(t)\Delta t + \int_{\beta_1}^{b} f^{-1}(y)\tilde{\nabla}y + \alpha_1\beta_1 - ab - \alpha\beta \\
\geq~&a\beta + \alpha b - ab - \alpha\beta = -(\alpha - a)(\beta - b).
\end{aligned}$$

Note that equality holds here if and only if it holds in (2.1.14) and (2.1.15), and this happens if and only if $\beta \in \{f^\rho(a), f(a)\}$ and $\alpha \in \{f^{-1}(b), \sigma(f^{-1}(b))\}$. The proof is complete. ∎

Theorem 2.1.5 *Let $\mathbb{T}$ be any time scale and $f : [\alpha_1, \alpha_2]_\mathbb{T} \to [\beta_1, \beta_2]_{\tilde{\mathbb{T}}}$ be a continuous strictly increasing function such that $\tilde{\mathbb{T}} = f(\mathbb{T})$ is also a time scale. Then for every $a, A \in [\alpha_1, \alpha_2]_\mathbb{T}$ and $b, B \in [\beta_1, \beta_2]_{\tilde{\mathbb{T}}}$, we have*

$$\begin{aligned}
(f^{-1}(B) - A)(f^\rho(A) - B)ab ~&\leq~ \int_A^a f(t)\Delta t + \int_B^b f^{-1}(y)\tilde{\nabla}y - ab + AB \\
&\leq~ -(f^{-1}(b) - a)(f^\rho(a) - b), \qquad (2.1.16)
\end{aligned}$$

with equality if and only if $B \in \{f^\rho(A), f(A)\}$ and $b \in \{f^\rho(a), f(a)\}$. The inequalities are reversed if f is strictly decreasing.

Proof. Considering F as in (2.1.11) and (2.1.13) with $\alpha = f^{-1}(b)$ and $\beta = f(a)$, we have the equality

$$F(a, b) + F(f^{-1}(b), f^\rho(a)) = -(f^{-1}(b) - a)(f^\rho(a) - b).$$

As $f^{-1} \in [\alpha_1, \alpha_2]_{\mathbb{T}}$ and $f^\rho \in [\beta_1, \beta_2]_{\widetilde{\mathbb{T}}}$, via Young's inequality (2.1.10), we see that $F(f^{-1}(b), f^\rho(a)) \geq 0$. Consequently, we have that

$$0 \leq F(a, b) \leq -(f^{-1}(b) - a)(f^\rho(a) - b), \qquad (2.1.17)$$

and inequality holds if and only if $b \in \{f^\rho(a), f(a)\}$. Thus for any $A \in [\alpha_1, \alpha_2]_{\mathbb{T}}$ and $B \in [\beta_1, \beta_2]_{\widetilde{\mathbb{T}}}$, we have from (2.1.17) that

$$0 \leq -(f^{-1}(B) - A)(f^\rho(A) - B) - F(A, B), \qquad (2.1.18)$$

with equality if and only if $B \in \{f^\rho(A), f(A)\}$. Combining (2.1.17) and (2.1.18), we get

$$
\begin{aligned}
0 \ &\leq \ F(a, b) - (f^{-1}(B) - A)(f^\rho(A) - B) - F(A, B) \\
&\leq \ -(f^{-1}(b) - a)(f^\rho(a) - b) - (f^{-1}(B) - A)(f^\rho(A) - B) - F(A, B),
\end{aligned}
$$

which can be rewritten to obtain (2.1.16). If f strictly decreasing the proof is similar and omitted. The proof is complete. ∎

In the following, we establish a theorem which can be considered as a modification of Theorem 2.1.5 above. This theorem allows us to get a Young type integral inequality without having to find f^{-1}.

Theorem 2.1.6 *Let the hypotheses of Theorem 2.1.5 hold. Then for any $a, \alpha,\ A,\ \Lambda \in [\alpha_1, \alpha_2]_{\mathbb{T}}$, we have*

$$
\begin{aligned}
(\Lambda - A)(f^\rho(A) - f(\Lambda)) \ &\leq \ \int_A^a f(t)\Delta t - \int_\Lambda^\alpha f(t)\Delta t \\
&\quad + (\alpha - a)f(\alpha) + (A - \Lambda)f(\Lambda) \\
&\leq \ -(\alpha - a)(f^\rho(a) - f(\alpha)), \qquad (2.1.19)
\end{aligned}
$$

where equalities hold if and only if $\Lambda \in \{\rho(A), A\}$ and $\alpha \in \{\rho(a), a\}$.

Proof. By Theorem 2.1.5 with $A = \Lambda$, $B = f(\Lambda)$, $a = \alpha$ and $b = f(\alpha)$, we have

$$\int_{f(\Lambda)}^{f(\alpha)} f^{-1}(y)\widetilde{\nabla}y = \alpha f(\alpha) - \Lambda f(\Lambda) - \int_\Lambda^\alpha f(t)\Delta t, \qquad (2.1.20)$$

for any $\alpha,\ \Lambda \in [\alpha_1, \alpha_2]_{\mathbb{T}}$. Since $\alpha,\ \Lambda \in [\alpha_1, \alpha_2]_{\mathbb{T}}$ are arbitrary, we substitute (2.1.20) into (2.1.16) to obtain (2.1.19). The proof is complete. ∎

In the following, we apply the results when $\mathbb{T} = \mathbb{Z}$ and derive some discrete inequalities. Recall that $[\alpha_1, \alpha_2]_{\mathbb{Z}} = \{\alpha_1,\ \alpha_1 + 1, \ldots, \alpha_2 - 1, \alpha_2\}$. The first two theorems are direct translations to $\mathbb{T} = \mathbb{Z}$ of Theorem 2.1.5 and Theorem 2.1.6, respectively.

Theorem 2.1.7 *Let* $f : [\alpha_1, \alpha_2]_{\mathbb{Z}} \to [\beta_1, \beta_2]_{\widetilde{\mathbb{Z}}}$ *be strictly increasing, where* $\widetilde{\mathbb{Z}} = f(\mathbb{Z})$. *Then for every* a, $A \in [\alpha_1, \alpha_2]_{\mathbb{Z}}$ *and* b, $B \in [\beta_1, \beta_2]_{\widetilde{\mathbb{Z}}}$, *we have*

$$\left[f^{-1}(B) - A) \right] \left(f(A-1) - B \right)$$

$$\leq \sum_{n=A}^{a-1} f(n) + \sum_{m \in (B,b) \cap \widetilde{\mathbb{Z}}}^{a-1} f^{-1}(m) \widetilde{\nu}(m) - ab + AB$$

$$\leq -(f^{-1}(b) - a)(f(a-1) - b),$$

where equalities hold if and only if $B \in \{f(A-1), f(A)\}$ *and* $b \in \{f(a-1), f(a)\}$.

Theorem 2.1.8 *Let* $f : \mathbb{Z} \to \mathbb{R}$ *be strictly increasing. Then for every* a, A, α, Λ, *we have*

$$[\Lambda - A)] \left(f(A-1) - f(\Lambda) \right)$$

$$\leq \sum_{n=A}^{a-1} f(n) - \sum_{m=\Lambda}^{\alpha-1} f(m) + (\alpha - a)f(\alpha) + (A - \Lambda)$$

$$\leq -(\alpha - a)(f(a-1) - f(\alpha)),$$

where equalities holds if and only if $\Lambda \in \{(A-1), (A)\}$ *and* $\alpha \in \{(a-1), a\}$.

Example 2.1.5 *Consider the factorial function*

$$f_k(t) = t^{(k)} = t(t-1)\ldots(t-k+1), \text{ for } t, \; k \in \mathbb{Z}.$$

It is clear that f_k *is increasing on the interval* $[k-1, \infty)_{\mathbb{Z}}$. *By Theorem 2.1.8, we have*

$$(a - \alpha)f_k(\alpha) \leq \frac{1}{k+1}[f_{k+1}(a) - f_{k+1}(\alpha)] \leq (a - \alpha)f_k(a-1),$$

for a, $\alpha \in \{k-1, k, k+1, \ldots\}$, *where equalities hold if and only if* $\alpha \in \{a-1, a\}$.

Example 2.1.6 *Let* $f(t) = \sin[\pi t/2k]$ *for* $k \in \mathbb{N}$. *Then* f *is increasing on* $[-k, k]$, *so that for any* $a \geq \alpha \in [-k, k]_{\mathbb{Z}}$, *we have by Theorem 2.1.8 that*

$$\sin\frac{\alpha\pi}{2k} \leq \frac{1}{2(a-\alpha)} \left(\cos\left[\frac{(2\alpha - 1)\pi}{4k}\right] - \cos\left[\frac{(2a-1)\pi}{4k}\right] \right) \csc\frac{\pi}{4k}$$

$$\leq \sin\frac{(a-1)\pi}{2k},$$

with equalities if and only if $\alpha \in \{a-1, a\}$.

2.2   Jensen Inequalities

The original Jensen inequality proved by Jensen states that if $g \in C([a, b], (c, d))$ and $F \in C([a, b], \mathbb{R})$ is convex, then

$$F\left(\frac{\int_a^b g(s)ds}{b-a}\right) \leq \frac{1}{b-a}\int_a^b F(g(s))ds. \tag{2.2.1}$$

In this section we give extensions of this inequality on time scales. The inequalities will be proved for delta derivative, nabla derivative as well as for diamond-α derivative. The results are adapted from [11, 23, 30, 39, 115, 150].

We begin with a lemma adapted from [67].

Lemma 2.2.1 *Let $f \in C((c, d), \mathbb{R})$ be convex. Then for each $t \in (c, d)$, there exits $\beta_t \in \mathbb{R}$ such that*

$$f(x) - f(t) \geq \beta_t(x - t), \quad \text{for all } x \in (c, d). \tag{2.2.2}$$

If f is strictly convex, then the inequality sign $\geq$ in (2.2.2) should be replaced by $>$.

Theorem 2.2.1 *Let $a, b \in \mathbb{T}$ and $c, d \in \mathbb{R}$. Let $g \in C_{rd}([a, b], (c, d))$ and $F \in C((c, d), \mathbb{R})$ is convex. Then*

$$F\left(\frac{\int_a^b g(s)\Delta s}{b-a}\right) \leq \frac{1}{b-a}\int_a^b F(g(s))\Delta s. \tag{2.2.3}$$

If F is strictly convex, then the inequality $\leq$ can be replaced by $<$.

Proof. Since F is convex, it follows from Lemma 2.2.1 that for each $t \in (c, d)$, there exists $\beta_t \in \mathbb{R}$ such that (2.2.2) holds. Let

$$t = \frac{1}{b-a}\int_a^b g(s)\Delta s.$$

Now

$$\int_a^b F(g(s))\Delta s - (b-a)F\left(\frac{\int_a^b g(s)\Delta s}{b-a}\right)$$

$$= \int_a^b F(g(s))\Delta s - (b-a)F(t)$$

$$\geq \beta_t\int_a^b [g(s) - t]\Delta s = \beta\left[\int_a^b g(s)\Delta s - t(b-a)\right] = 0.$$

The proof is complete.   ■

Example 2.2.1 *As a special case let* $\mathbb{T} = \mathbb{R}$ *and* $F = -\log$. *Note* F *is convex and continuous on* $(0, \infty)$. *Apply Theorem 2.2.1 with* $a = 0$ *and* $b = 1$ *to obtain* $\log \int_0^1 g(t) dt \geq \int_0^1 \log(g(t)) dt$, *and hence* $\int_0^1 g(t) dt \geq \exp\left(\int_0^1 \log(g(t)) dt\right)$, *whenever* $g \in C([0, 1), (0, \infty))$ *is continuous.*

Example 2.2.2 *Let* $\mathbb{T} = \mathbb{N}$ *and* $N \in \mathbb{N}$. *Apply Jensen's inequality (Theorem 2.2.1) with* $a = 1$ *and* $b = N + 1$ *and* $g : [1, N + 1]_{\mathbb{N}} \to (0, \infty)$ *to find*

$$\log\left[\frac{1}{N} \sum_{n=1}^{N} g(n)\right] \geq \log\left[\frac{1}{N} \int_1^{N+1} g(t) \Delta t\right]$$

$$\geq \frac{1}{N} \int_1^{N+1} \log(g(t)) \Delta t$$

$$= \frac{1}{N} \sum_{n=1}^{N} \log(g(n)) = \log\left(\prod_{n=1}^{N} g(n)\right)^{1/N},$$

and hence

$$\frac{1}{N} \sum_{n=1}^{N} g(n) \geq \left(\prod_{n=1}^{N} g(n)\right)^{1/N}.$$

This is the well-known arithmetic-mean geometric-mean inequality.

Example 2.2.3 *Let* $\mathbb{T} = 2^{\mathbb{N}_0}$ *and* $N \in \mathbb{N}$. *Apply Jensen's inequality (Theorem 2.2.1) with* $a = 1$ *and* $b = 2^N$ *and* $g : [1, 2^N]_{2^{\mathbb{N}_0}} \to (0, \infty)$ *to find*

$$\log\left[\frac{1}{2^N - 1} \sum_{n=0}^{N-1} 2^n g(2^n)\right]$$

$$\geq \log\left[\frac{1}{2^N - 1} \int_1^{2^N} g(t) \Delta t\right]$$

$$\geq \frac{1}{2^N - 1} \int_1^{2^N} \log(g(t)) \Delta t = \frac{1}{2^N - 1} \sum_{n=0}^{N-1} 2^n \log(g(2^n))$$

$$= \frac{1}{2^N - 1} \sum_{n=0}^{N-1} \log((g(2^n))^{2^n}) = \log\left(\prod_{n=1}^{N} ((g(2^n))^{2^n}\right)^{1/(2^N - 1)},$$

and hence

$$\frac{1}{2^N - 1} \sum_{n=0}^{N-1} 2^n g(2^n) \geq \left(\prod_{n=1}^{N} ((g(2^n))^{2^n}\right)^{1/(2^N - 1)}.$$

Theorem 2.2.2 *Let $a, b \in \mathbb{T}$ and $c, d \in \mathbb{R}$. Suppose that $g \in C_{rd}([a, b], (c, d))$ and $h \in C_{rd}([a, b]_\mathbb{T}, \mathbb{R})$ with*

$$\int_a^b |h(s)|\, \Delta s > 0.$$

If $F \in C((c, d), \mathbb{R})$ is convex, then

$$F\left(\frac{\int_a^b |h(s)|\, g(s)\Delta s}{\int_a^b |h(s)|\, \Delta s}\right) \leq \int_a^b \frac{|h(s)|\, F(g(s))\Delta s}{\int_a^b |h(s)|\, \Delta s}. \tag{2.2.4}$$

If F is strictly convex, then the inequality $\leq$ can be replaced by $<$.

Proof. Since F is convex it follows from Lemma 2.2.1 that for each $t \in (c, d)$, there exists $\beta_t \in \mathbb{R}$ such that (2.2.2) holds. Let

$$t = \frac{\int_a^b |h(s)|\, g(s)\Delta s}{\int_a^b |h(s)|\, \Delta s}.$$

Thus

$$\int_a^b |h(s)|\, F(g(s))\Delta s - \left(\int_a^b |h(s)|\, \Delta s\right) F\left(\frac{\int_a^b |h(s)|\, g(s)\Delta s}{\int_a^b |h(s)|\, \Delta s}\right)$$

$$= \int_a^b |h(s)|\, F(g(s))\Delta s - \left(\int_a^b |h(s)|\, \Delta s\right) F(t)$$

$$= \int_a^b |h(s)|\, [F(g(s)) - F(t)]\, \Delta s \geq \beta_t \int_a^b |h(s)|\, [g(s) - t]\, \Delta s$$

$$= \beta_t \left[\int_a^b |h(s)|\, g(s)\Delta s - t \int_a^b |h(s)|\, \Delta s\right]$$

$$= \beta_t \left[\int_a^b |h(s)|\, g(s)\Delta s - \frac{\int_a^b |h(s)|\, g(s)\Delta s}{\int_a^b |h(s)|\, \Delta s} \int_a^b |h(s)|\, \Delta s\right] = 0.$$

The proof is complete. ∎

Remark 2.2.1 *If the condition of convexity of the function F is changed to concavity, then the inequality sign of the inequality (2.2.4) is reversed.*

As a special case of Theorem 2.2.2, when $g(t) \geq 0$ on $[a, b]$ and $F(t) = t^\gamma$ on $[0, \infty)$, we see that F is convex on $[0, \infty)$ for $\alpha < 0$ or $\alpha > 1$ and F is concave on $[0, \infty)$ for $\alpha \in (0, 1)$.

Corollary 2.2.1 *Let $g \in C_{rd}([a, b], (c, d))$ such that $g(t) \geq 0$ on $[a, b]$ and $h \in C_{rd}([a, b], \mathbb{R})$ with*

$$\int_a^b |h(s)|\, \Delta s > 0,$$

where $a, b \in \mathbb{T}$ and $(c, d) \subset \mathbb{R}$. Then

$$\left(\frac{\int_a^b |h(s)|\, g(s)\Delta s}{\int_a^b |h(s)|\, \Delta s} \right)^\alpha \leq \frac{\int_a^b |h(s)|\, g^\alpha(s)\Delta s}{\int_a^b |h(s)|\, \Delta s}, \quad \text{for } \alpha < 0 \text{ or } \alpha > 1,$$

and

$$\left(\frac{\int_a^b |h(s)|\, g(s)\Delta s}{\int_a^b |h(s)|\, \Delta s} \right)^\alpha \geq \frac{\int_a^b |h(s)|\, g^\alpha(s)\Delta s}{\int_a^b |h(s)|\, \Delta s}, \quad \text{for } \alpha \in (0, 1).$$

We now present nabla Jensen inequalities.

Theorem 2.2.3 *Let $a, b \in \mathbb{T}$ and c, $d \in \mathbb{R}$, and $h \in C_{ld}([a, b]_\mathbb{T}, \mathbb{R})$ and $g \in C_{ld}([a, b], (c, d))$ with $\int_a^b |h(\tau)|\, \nabla\tau > 0$, and $\phi \in C((c, d), \mathbb{R})$ is convex, then*

$$\phi\left(\frac{\int_a^b |h(\tau)|\, g(\tau)\nabla\tau}{\int_a^b |h(\tau)|\, \nabla\tau} \right) \leq \frac{\int_a^b |h(\tau)|\, \phi(g(\tau))\nabla\tau}{\int_a^b |h(\tau)|\, \nabla\tau}. \qquad (2.2.5)$$

If ϕ is strictly convex, then the inequality $\leq$ can be replaced by $<$.

Proof. Since ϕ is convex, it follows from Lemma 2.2.1 that for each $t \in (c, d)$, there exists $\beta_t \in \mathbb{R}$ such that (2.2.2) holds. Let

$$t = \frac{\int_a^b |h(s)|\, g(s)\nabla s}{\int_a^b |h(s)|\, \nabla s}.$$

Thus

$$\int_a^b |h(s)|\, \phi(g(s))\nabla s - \left(\int_a^b |h(s)|\, \nabla s \right) \phi\left(\frac{\int_a^b |h(s)|\, g(s)\nabla s}{\int_a^b |h(s)|\, \nabla s} \right)$$

$$= \int_a^b |h(s)|\, \phi(g(s))\Delta s - \left(\int_a^b |h(s)|\, \nabla s \right) \phi(t)$$

$$= \int_a^b |h(s)|\, [\phi(g(s)) - \phi(t)]\, \Delta s \geq \beta_t \int_a^b |h(s)|\, [g(s) - t]\, \nabla s$$

$$= \beta_t \left[\int_a^b |h(s)|\, g(s)\nabla s - t \int_a^b |h(s)|\, \nabla s \right]$$

$$= \beta_t \left[\int_a^b |h(s)|\, g(s)\nabla s - \frac{\int_a^b |h(s)|\, g(s)\nabla s}{\int_a^b |h(s)|\, \Delta s} \int_a^b |h(s)|\, \nabla s \right] = 0.$$

The proof is complete. ∎

As a consequence of Theorem 2.2.3, we have the following result.

Theorem 2.2.4 *Let* $a, b \in \mathbb{T}$ *and* $c, d \in \mathbb{R}$*. If* $h \in C_{ld}([a, b]_\mathbb{T}, \mathbb{R})$ *and* $g \in C_{ld}([a, b], (c, d))$ *are nonnegative, with* $\int_a^b h(t)\nabla t > 0$*, and* $\phi : (c, d) \to \mathbb{R}$ *is continuous and convex, then*

$$\phi\left(\frac{\int_a^b h(t)g(t)\nabla t}{\int_a^b h(t)\nabla t}\right) \leq \frac{\int_a^b h(t)\phi(g(t))\nabla t}{\int_a^b h(t)\nabla t}.$$

If ϕ *is strictly convex, then the inequality* $\leq$ *can be replaced by* $<$*.*

Now, we give some generalized versions of Jensen's inequality on time scales via the diamond-α integral.

Theorem 2.2.5 *Let* $\mathbb{T}$ *be a time scale,* $a, b \in \mathbb{T}$ *and* $c, d \in \mathbb{R}$*. Suppose that* $g \in C([a, b]_\mathbb{T}, (c, d))$ *and* $F \in C((c, d), \mathbb{R})$ *is convex. Then*

$$F\left(\frac{\int_a^b g(s)\Diamond_\alpha s}{b - a}\right) \leq \frac{1}{b - a} \int_a^b F(g(s))\Diamond_\alpha s. \qquad (2.2.6)$$

If F *is strictly convex, then the inequality* $\leq$ *can be replaced by* $<$*.*

Proof. Since F is convex, we have

$$F\left(\frac{\int_a^b g(s)\Diamond_\alpha \Delta s}{b - a}\right) = F\left(\frac{\alpha}{b - a}\int_a^b g(s)\Delta s + \frac{(1 - \alpha)}{b - a}\int_a^b g(s)\nabla s\right)$$

$$\leq \alpha F\left(\frac{1}{b - a}\int_a^b g(s)\Delta s\right) + (1 - \alpha)F\left(\frac{1}{b - a}\int_a^b g(s)\nabla s\right).$$

Now, using delta and nabla Jensen inequalities, we get that

$$F\left(\frac{\int_a^b g(s)\Diamond_\alpha \Delta s}{b - a}\right) \leq \frac{\alpha}{b - a}\left(\int_a^b F(g(s))\Delta s\right) + \frac{(1 - \alpha)}{b - a}\left(\int_a^b F(g(s))\nabla s\right)$$

$$= \frac{1}{b - a}\left[\left(\int_a^b F(g(s))\Delta s\right) + \left(\int_a^b F(g(s))\nabla s\right)\right]$$

$$= \frac{1}{b - a}\left(\int_a^b F(g(s))\Diamond_\alpha \Delta s\right).$$

The proof is complete. ■

In the following, we give a generalization of (2.2.6) on time scales.

Theorem 2.2.6 *Let* $\mathbb{T}$ *be a time scale,* $a, b \in \mathbb{T}$ *and* $c, d \in \mathbb{R}$*. Suppose that* $g \in C([a, b], (c, d))$ *and* $h \in C([a, b]_\mathbb{T}, \mathbb{R})$ *with* $\int_a^b |h(s)| \Diamond_\alpha s > 0$*. If* $F \in C((c, d), \mathbb{R})$ *is convex, then*

$$F\left(\frac{\int_a^b |h(s)| g(s)\Diamond_\alpha s}{\int_a^b |h(s)| \Diamond_\alpha s}\right) \leq \frac{\int_a^b |h(s)| F(g(s))\Diamond_\alpha s}{\int_a^b |h(s)| \Diamond_\alpha s}. \qquad (2.2.7)$$

If F *is strictly convex, then the inequality* $\leq$ *can be replaced by* $<$*.*

Proof. Since F is convex, it follows from Lemma 2.2.1 that for each $t \in (c,d)$, there exists $\beta_t \in \mathbb{R}$ such that (2.2.2) holds. Setting

$$t = \frac{\int_a^b |h(s)|\, g(s) \Diamond_\alpha s}{\int_a^b |h(s)|\, \Diamond_\alpha s},$$

we get that

$$\int_a^b |h(s)|\, F(g(s)) \Diamond_\alpha s - \left(\int_a^b |h(s)|\, \Diamond_\alpha s \right) F\left(\frac{\int_a^b |h(s)|\, g(s) \Diamond_\alpha s}{\int_a^b |h(s)|\, \Diamond_\alpha s} \right)$$

$$= \int_a^b |h(s)|\, F(g(s)) \Diamond_\alpha s - \left(\int_a^b |h(s)|\, \Delta s \right) F(t)$$

$$= \int_a^b |h(s)|\, [F(g(s)) - F(t)]\, \Diamond_\alpha s \geq \beta_t \int_a^b |h(s)|\, [g(s) - t]\, \Diamond_\alpha s$$

$$= \beta_t \left[\int_a^b |h(s)|\, g(s) \Diamond_\alpha s - t \int_a^b |h(s)|\, \Diamond_\alpha s \right]$$

$$= \beta_t \left[\int_a^b |h(s)|\, g(s) \Diamond_\alpha s - \frac{\int_a^b |h(s)|\, g(s) \Diamond_\alpha s}{\int_a^b |h(s)|\, \Diamond_\alpha s} \int_a^b |h(s)|\, \Diamond_\alpha s \right] = 0.$$

The proof is complete. ∎

Remark 2.2.2 *If the convexity condition of the function F is changed to concavity, then the inequality sign of the inequality (2.2.7) is reversed.*

As a special case of Theorem 2.2.6, when $F(t) = t^\gamma$ on $[0, \infty)$, we see that F is convex on $[0, \infty)$ for $\gamma < 0$ or $\gamma > 1$ and F is concave on $[0, \infty)$ for $\gamma \in (0,1)$. This gives us the following result.

Corollary 2.2.2 *Let $g \in C([a,b],(c,d))$ such that $g(t) > 0$ on $[a,b]_{\mathbb{T}}$ and $h \in C([a,b]_{\mathbb{T}}, \mathbb{R})$ with*

$$\int_a^b |h(s)|\, \Diamond_\alpha s > 0,$$

where $a, b \in \mathbb{T}$ and $(c,d) \subset \mathbb{R}$. Then

$$\left(\frac{\int_a^b |h(s)|\, g(s) \Diamond_\alpha s}{\int_a^b |h(s)|\, \Diamond_\alpha s} \right)^\gamma \leq \frac{\int_a^b |h(s)|\, g^\gamma(s) \Diamond_\alpha s}{\int_a^b |h(s)|\, \Diamond_\alpha s}, \ \text{for } \gamma < 0 \text{ or } \gamma > 1,$$

and

$$\left(\frac{\int_a^b |h(s)|\, g(s) \Delta s}{\int_a^b |h(s)|\, \Diamond_\alpha s} \right)^\gamma \geq \frac{\int_a^b |h(s)|\, g^\gamma(s) \Diamond_\alpha s}{\int_a^b |h(s)|\, \Diamond_\alpha s}, \ \text{for } \gamma \in (0,1).$$

Example 2.2.4 *Let* $g(t) > 0$ *on* $[a,b]_{\mathbb{T}}$ *and* $F(t) = \ln(t)$ *on* $(0,\infty)$. *Now, since* F *is concave on* $(0,\infty)$, *it follows from Theorem 2.2.6 that*

$$\ln\left(\frac{\int_a^b |h(s)|\, g(s)\Delta s}{\int_a^b |h(s)|\, \Diamond_\alpha s}\right) \geq \frac{\int_a^b |h(s)|\ln(g(s))\, \Diamond_\alpha s}{\int_a^b |h(s)|\, \Diamond_\alpha s}.$$

Example 2.2.5 *Let* $\mathbb{T} = \mathbb{Z}$ *and* $n \in \mathbb{N}$. *Fix* $a = 1$ *and* $b = N+1$ *and consider* $g : [1, N+1]_{\mathbb{N}} \to (0,\infty)$ *and let* $F(t) = -\ln t$. *Now* F *is convex and continuous on* $(0,\infty)$. *Apply the Jensen inequality (2.2.7) to obtain*

$$\ln\left[\frac{\alpha}{N}\sum_{n=1}^N g(n) + \frac{1-\alpha}{N}\sum_{n=2}^{N+1} g(n)\right]$$

$$= \ln\left(\int_1^{N+1} \frac{1}{N}g(t)\Diamond_\alpha t\right)$$

$$\geq \frac{1}{N}\int_1^{N+1}\ln(g(t))\Diamond_\alpha t = \frac{\alpha}{N}\sum_{n=1}^N \ln g(n) + \frac{1-\alpha}{N}\sum_{n=2}^{N+1}\ln g(n)$$

$$= \ln\left(\prod_{n=1}^N g(n)\right)^{\frac{\alpha}{N}} + \ln\left(\prod_{n=2}^{N+1} g(n)\right)^{\frac{1-\alpha}{N}},$$

and hence

$$\frac{1}{N}\left[\alpha\sum_{n=1}^N g(n) + (1-\alpha)\sum_{n=2}^{N+1} g(n)\right] \geq \left(\prod_{n=1}^N g(n)\right)^{\frac{\alpha}{N}}\left(\prod_{n=2}^{N+1} g(n)\right)^{\frac{1-\alpha}{N}}.$$

When $\alpha = 1$, *we obtain the well-known arithmetic-mean geometric-mean inequality*

$$\frac{1}{N}\sum_{n=1}^N g(n) \geq \left(\prod_{n=1}^N g(n)\right)^{\frac{1}{N}},$$

and when $\alpha = 0$, *we obtain*

$$\frac{1}{N}\sum_{n=2}^{N+1} g(n) \geq \left(\prod_{n=2}^{N+1} g(n)\right)^{\frac{1}{N}}.$$

Example 2.2.6 *Let* $\mathbb{T} = 2^{\mathbb{N}_0}$ *and* $F(t) = -\ln t$. *Apply the Jensen inequality (Theorem 2.2.6) with* $a = 1$ *and* $b = 2^N$ *and* $g : [1, 2^N]_{2^{\mathbb{N}_0}} \to (0,\infty)$, *we find that*

$$\ln\left[\frac{1}{2^N-1}\int_1^{2^N}g(t)\Diamond_\alpha t\right]$$

$$=\ln\left[\frac{\alpha}{2^N-1}\int_1^{2^N}g(t)\Delta t+\frac{1-\alpha}{2^N-1}\int_1^{2^N}g(t)\nabla t\right]$$

$$=\ln\left[\frac{\alpha}{2^N-1}\sum_{n=0}^{N-1}2^n\log(g(2^n))+\frac{1-\alpha}{2^N-1}\sum_{n=1}^{N}2^n\log(g(2^n))\right]$$

$$\geq\left[\frac{1}{2^N-1}\int_1^{2^N}\ln g(t)\Diamond_\alpha t\right]$$

$$=\frac{\alpha}{2^N-1}\sum_{n=0}^{N-1}2^n\log((g(2^n))+\frac{1-\alpha}{2^N-1}\sum_{n=1}^{N}2^n\log((g(2^n))$$

$$=\frac{\alpha}{2^N-1}\sum_{n=0}^{N-1}\log((g(2^n))^{2^n}+\frac{1-\alpha}{2^N-1}\sum_{n=1}^{N}\log((g(2^n))^{2^n}$$

$$=\frac{1}{2^N-1}\ln\prod_{n=0}^{N-1}((g(2^n))^{\alpha2^n}+\frac{1}{2^N-1}\ln\prod_{n=1}^{N}((g(2^n))^{(1-\alpha)2^n}$$

$$=\ln\left(\prod_{n=1}^{N}((g(2^n))^{2^n}\right)^{\frac{1}{2^N-1}}+\ln\left(\prod_{n=1}^{N}((g(2^n))^{(1-\alpha)2^n}\right)^{\frac{1}{2^N-1}}.$$

From this we conclude that

$$\ln\left[\frac{\alpha}{2^N-1}\sum_{n=0}^{N-1}2^n\log(g(2^n))+\frac{1-\alpha}{2^N-1}\sum_{n=1}^{N}2^n\log(g(2^n))\right]$$

$$\geq\ln\left[\left(\prod_{n=1}^{N}((g(2^n))^{2^n}\right)^{\frac{1}{2^N-1}}\left(\prod_{n=1}^{N}((g(2^n))^{(1-\alpha)2^n}\right)^{\frac{1}{2^N-1}}\right],$$

and hence

$$\frac{\alpha}{2^N-1}\sum_{n=0}^{N-1}2^n\log(g(2^n))+\frac{1-\alpha}{2^N-1}\sum_{n=1}^{N}2^n\log(g(2^n))$$

$$\geq\left(\prod_{n=1}^{N}((g(2^n))^{2^n}\right)^{\frac{1}{2^N-1}}\left(\prod_{n=1}^{N}((g(2^n))^{(1-\alpha)2^n}\right)^{\frac{1}{2^N-1}}.$$

Since

$$\frac{1}{2^N-1}\left[\alpha\sum_{n=0}^{N-1}2^n\log(g(2^n))+(1-\alpha)\sum_{n=1}^{N}2^n\log(g(2^n))\right]$$

$$=\frac{1}{2^N-1}\sum_{n=1}^{N-1}2^n\log(g(2^n))+\alpha g(1)+(1-\alpha)2^N g(2^N),$$

we get that

$$\frac{1}{2^N - 1} \sum_{n=1}^{N-1} 2^n \log(g(2^n)) + \alpha g(1) + (1 - \alpha) 2^N g(2^N)$$

$$\geq \left(\prod_{n=1}^{N} ((g(2^n))^{2^n} \right)^{\frac{1}{2^N - 1}} \left(\prod_{n=1}^{N} ((g(2^n))^{(1-\alpha)2^n} \right)^{\frac{1}{2^N - 1}}.$$

As an application of Theorem 2.2.6, we have the following result.

Theorem 2.2.7 *Let* $\mathbb{T}$ *be a time scale,* a, $b \in \mathbb{T}$ *with* $a < b$ *and* f, g, $h \in C([a, b]_{\mathbb{T}}, (0, \infty))$.

(i) If $p > 1$, *then*

$$\left[\left(\int_a^b h(s) f(s) \Diamond_\alpha s \right)^p + \left(\int_a^b h(s) g(s) \Diamond_\alpha s \right)^p \right]^{1/p}$$

$$\leq \int_a^b h(s) \left[f^p(s) + g^p(s) \right]^{1/p} \Diamond_\alpha s. \tag{2.2.8}$$

(ii) If $0 < p < 1$, *then*

$$\left[\left(\int_a^b h(s) f(s) \Diamond_\alpha s \right)^p + \left(\int_a^b h(s) g(s) \Diamond_\alpha s \right)^p \right]^{1/p}$$

$$\geq \int_a^b h(s) \left[f^p(s) + g^p(s) \right]^{1/p} \Diamond_\alpha s. \tag{2.2.9}$$

Proof. We prove only (i), since the proof of (ii) is similar. Inequality (2.2.8) is trivially true when f is zero. Otherwise, applying Theorem 2.2.6 with $F(x) = (1 + x^p)^{1/p}$, which is clearly convex on $(0, \infty)$, we obtain

$$\left(1 + \frac{\int_a^b h(s) f(s) \Diamond_\alpha s}{\int_a^b h(s) \Diamond_\alpha s} \right)^{1/p} \leq \frac{\int_a^b h(s) (1 + f^p(s))^{1/p} \Diamond_\alpha s}{\int_a^b h(s) \Diamond_\alpha s}.$$

In other words

$$\left(\int_a^b h(s) \Diamond_\alpha s + \int_a^b h(s) f(s) \Diamond_\alpha s \right)^{1/p} \leq \int_a^b h(s) (1 + f^p(s))^{1/p} \Diamond_\alpha s.$$

Changing h and f with $hf / \int_a^b h(s) f(s) \Diamond_\alpha s$ and g/f in the last inequality we obtain (2.2.8). The proof is complete. ∎

Using the fact that the time scale integral is an isotonic linear functional, we prove some Jensen type inequalities on time scales.

Definition 2.2.1 *Let E be a nonempty set and L be a linear class of real-valued functions $f : E \to \mathbb{R}$, having the following properties:*

(L_1). *If f, $g \in L$ and a, $b \in \mathbb{R}$, then $(af + bg) \in L$.*

(L_2). *If $f(t) = 1$ for all $t \in E$, then $f \in L$.*

An isotonic linear functional is a functional $A : L \to \mathbb{R}$ having the following properties:

(A_1). *If f, $g \in L$ and a, $b \in \mathbb{R}$, then $A(af + bg) = aA(f) + bA(g)$.*

(A_2). *If $f \in L$ and $f(t) \geq 0$ for all $t \in E$, then $A(f) \geq 0$.*

Furthermore, if the functional A has a property

(A_3). *$A(\mathbf{1}) = 1$, where $\mathbf{1}(t) = 1$ for all $t \in E$, then we will say that A is normalized.*

Our next theorem proves that the Cauchy integral on time scales is an isotonic functional. The proof is straightforward from its definition and properties presented in [51, Defintion 1.58 and Theorem 1.77].

Theorem 2.2.8 *Let $\mathbb{T}$ be a time scale, a, $b \in \mathbb{T}$ with $a < b$ and let*

$$E = [a, b) \cap \mathbb{T}, \quad L = C_{rd}([a, b), \mathbb{R}). \tag{2.2.10}$$

Then (L_1) and (L_2) are satisfied. Moreover, let

$$A(f) = \int_a^b f(t)\Delta t, \tag{2.2.11}$$

where the integral is the Cauchy delta time-scale integral. Then (A_1) and (A_2) are satisfied.

Example 2.2.7 *If $\mathbb{T} = \mathbb{R}$ in Theorem 2.2.8, then $L = C([a, b], \mathbb{R})$ and $A(f) = \int_a^b f(t)dt$. If $\mathbb{T} = \mathbb{Z}$ in Theorem 2.2.8, then L consists of real-valued functions on $[a, b - 1] \cap \mathbb{Z}$ and $A(f) = \sum_{n=a}^{b-1} f(n)$. If $\mathbb{T} = q^{\mathbb{N}_0}$, where $q > 1$, in Theorem 2.2.8, then L consists of real-valued functions on $[a, b/q] \cap q^{\mathbb{N}_0}$ and $A(f) = (q - 1) \sum_{n=\log_q(a)}^{\log_q(b)-1} q^n f(q^n)$.*

Theorem 2.2.8 also has corresponding versions for the nabla and the α-diamond integral.

Theorem 2.2.9 *Let $\mathbb{T}$ be a time scale, a, $b \in \mathbb{T}$ with $a < b$ and let*

$$E = (a, b] \cap \mathbb{T}, \quad L = C_{ld}((a, b], \mathbb{R}).$$

Then (L_1) and (L_2) are satisfied. Moreover, let

$$A(f) = \int_a^b f(t)\nabla t,$$

where the integral is the Cauchy nabla time-scale integral. Then (A_1) and (A_2) are satisfied.

Theorem 2.2.10 *Let $\mathbb{T}$ a time scale, a, $b \in \mathbb{T}$ with $a < b$ and let*

$$E = [a, b] \cap \mathbb{T}, \quad L = C([a, b], \mathbb{R}).$$

Then (L_1) and (L_2) are satisfied. Moreover, let

$$A(f) = \int_a^b f(t)\Diamond_\alpha t,$$

where the integral is the Cauchy α-diamond time-scale integral. Then (A_1) and (A_2) are satisfied.

The Riemann multiple integral is also an isotonic linear functional.

Theorem 2.2.11 *Let $\mathbb{T}_1, \ldots, \mathbb{T}_n$ a time scales. For a_i, $b_i \in \mathbb{T}_i$ with $a_i < b_i$, $1 \le i \le n$, let*

$$E \subset ([a_1, b_1) \cap \mathbb{T}_1 \times \ldots \times [a_n, b_n) \cap \mathbb{T}_n,$$

be Jordan Δ-measurable and let L be the set of all bounded Δ-integrable functions from E to $\mathbb{R}$. Then (L_1) and (L_2) are satisfied. Moreover, let

$$A(f) = \int_E f(t)\Delta t,$$

where the integral is the multiple Riemann delta-time scale integral. Then (A_1) and (A_2) are satisfied.

Theorem 2.2.12 *Let $\mathbb{T}_1, \ldots, \mathbb{T}_n$ be time scales. For a_i, $b_i \in \mathbb{T}_i$ with $a_i < b_i$, $1 \le i \le n$, let*

$$E \subset ([a_1, b_1) \cap \mathbb{T}_1 \times \ldots \times [a_n, b_n) \cap \mathbb{T}_n,$$

be Lebesgue Δ-measurable and let L be the set of all bounded Δ-integrable functions from E to $\mathbb{R}$. Then (L_1) and (L_2) are satisfied. Moreover, let

$$A(f) = \int_E f(t)\Delta t,$$

where the integral is the multiple Lebesgue delta-time scale integral. Then (A_1) and (A_2) are satisfied.

Theorem 2.2.13 *Let the assumptions of Theorem 2.2.12 be satisfied. Let $A(f)$ be replaced by*

$$A(f) = \frac{\int_E |h(t)| \, f(t) \Delta t}{\int_E |h(t)| \, \Delta t},$$

where $h : \mathbb{E} \to \mathbb{R}$ is Δ-integrable such that $\int_E |h(t)| \, \Delta t > 0$. Then A is an isotonic linear functional satisfying $A(1) = 1$.

We next note the following theorem that has been proved by Jessen [87] (see also [117]).

Theorem 2.2.14 *Let L satisfy properties (L_1) and (L_2). Assume $\Phi \in C(\mathbb{I}, \mathbb{R})$ is convex where $\mathbb{I} \subset \mathbb{R}$ is an interval. If A satisfies (A_1) and (A_2) such that $A(1) = 1$, then for all $f \in L$ such that $\Phi(f) \in L$, one has $A(f) \in \mathbb{I}$ and*

$$\Phi(A(f)) \leq A(\Phi(f)).$$

Now, the application of Theorems 2.2.13 and 2.2.14 gives the following result.

Theorem 2.2.15 *Assume that $\Phi \in C(\mathbb{I}, \mathbb{R})$ is convex where $\mathbb{I} \subset \mathbb{R}$ is an interval. Let $E \subset \mathbb{R}^n$ be as in Theorem 2.2.12 and suppose that f is Δ-integrable on E such that $f(E) = \mathbb{I}$. Moreover, let $h : E \to \mathbb{R}$ be Δ-integrable such that $\int_E |h(t)| \, \Delta t > 0$. Then*

$$\Phi\left(\frac{\int_E |h(t)| \, f(t) \Delta t}{\int_E |h(t)| \, \Delta t}\right) \leq \frac{\int_E |h(t)| \, \Phi(f(t)) \Delta t}{\int_E |h(t)| \, \Delta t}.$$

The concept of superquadratic functions in one variable, as a generalization of the class of convex functions was introduced by S. Abramovich, G. Jameson, and G. Sinnamon in [1, 2].

Definition 2.2.2 *A function $\varphi : [0, \infty) \to \mathbb{R}$ is called superquadratic if there exists a function $C : [0, \infty) \to \mathbb{R}$ such that*

$$\varphi(y) - \varphi(x) - \varphi(|y - x|) \geq C(x)(y - x), \quad \text{for all } x, y > 0.$$

We say that φ is subquadratic if $-\varphi$ is superquadratic.

For example, the function $\varphi(x) = x^p$ is superquadratic for $p \geq 2$ and subquadratic for $p \in (0, 2]$.

Lemma 2.2.2 *Let φ be a superquadratic function with C as in Definition 2.2.2. Then*

(i) $\varphi(0) \leq 0$,

(ii) if $\varphi(0) = \varphi'(0)$, then $C(x) = \varphi'(x)$ whenever φ is differentiable at $x > 0$,

(iii) if $\varphi \geq 0$, then φ is convex and $\varphi(0) = \varphi'(0) = 0$.

In the following, we prove a Jensen type inequality on time scales for superquadratic functions.

Theorem 2.2.16 *Let a, $b \in \mathbb{T}$. Suppose $f \in C_{rd}([a,b]_{\mathbb{T}}, [0, \infty))$ and $\varphi : [0, \infty) \to \mathbb{R}$ is continuous and superquadratic. Then*

$$\varphi \left(\frac{\int_a^b f(t)\Delta t}{b-a} \right) \leq \frac{1}{b-a} \int_a^b \left[\varphi(f(s)) - \varphi \left(\left| f(s) - \frac{\int_a^b f(t)\Delta t}{b-a} \right| \right) \right] \Delta s.$$
(2.2.12)

Proof. Since $\varphi : [0, \infty) \to \mathbb{R}$ is a superquadratic function, then there exists a function $C : [0, \infty) \to \mathbb{R}$ such that

$$\varphi(y) \geq \varphi(x_0) + \varphi(|y - x_0|) + C(x_0)(y - x_0), \text{ for all } x_0, y > 0. \quad (2.2.13)$$

Let

$$x_0 = \frac{1}{(b-a)} \int_a^b f(t)\Delta t.$$

Applying (2.2.13) with $y = f(s)$, we see that

$$\varphi(f(s)) \geq \varphi \left(\frac{\int_a^b f(t)\Delta t}{b-a} \right) + \varphi \left(\left| f(s) - \frac{\int_a^b f(t)\Delta t}{b-a} \right| \right)$$
$$+ C(x_0)(f(s) - x_0).$$

Integrating from a to b, we see that

$$\int_a^b \left[\varphi(f(s)) - \varphi \left(\left| f(s) - \frac{\int_a^b f(t)\Delta t}{b-a} \right| \right) - \varphi \left(\frac{\int_a^b f(t)\Delta t}{b-a} \right) \right] \Delta s$$
$$\geq C(x_0) \int_a^b (f(s) - x_0)\Delta s = C(x_0) \left[\int_a^b f(s)\Delta s - (b-a)x_0 \right] = 0.$$

This implies that

$$\varphi \left(\frac{\int_a^b f(t)\Delta t}{b-a} \right) \leq \frac{1}{b-a} \int_a^b \left[\varphi(f(s)) - \varphi \left(\left| f(s) - \frac{\int_a^b f(t)\Delta t}{b-a} \right| \right) \right] \Delta s,$$

which is the desired inequality (2.2.12). The proof is complete. ∎

2.3 Hölder Inequalities

In 1889 Hölder [84] proved that

$$\sum_{k=1}^{n} x_k y_k \leq \left(\sum_{k=1}^{n} x_k^p\right)^{1/p} \left(\sum_{k=1}^{n} y_k^q\right)^{1/q}, \tag{2.3.1}$$

where x_n and y_n are positive sequences and p and q are two positive numbers such that $1/p + 1/q = 1$. The inequality reverses if either p or q is negative. The integral form of this inequality is

$$\int_a^b |f(t)g(t)|\, dt \leq \left[\int_a^b |f(t)|^p\, dt\right]^{\frac{1}{p}} \left[\int_a^b |g(t)|^q\, dt\right]^{\frac{1}{q}}, \tag{2.3.2}$$

where a, $b \in \mathbb{R}$ and $f, g \in C([a,b], \mathbb{R})$. In this section, we discuss various versions of the Hölder inequality on time scales which not only give a unification of (2.3.1) and (2.3.2) but can be applied on different types of time scales. The results in this section are adapted from [11, 24, 30, 39, 145, 155]. We begin with the proof of the classical Hölder inequality on time scales.

Theorem 2.3.1 *Let a, $b \in \mathbb{T}$. For $f, g \in C_{rd}(\mathbb{I}, \mathbb{R})$, we have*

$$\int_a^b |f(t)g(t)|\, \Delta t \leq \left[\int_a^b |f(t)|^p\, \Delta t\right]^{\frac{1}{p}} \left[\int_a^b |g(t)|^q\, \Delta t\right]^{\frac{1}{q}}, \tag{2.3.3}$$

where $p > 1$ and $\frac{1}{p} + \frac{1}{q} = 1$.

Proof. For nonnegative real numbers α and β, the classical Young inequality

$$\alpha^{1/p}\beta^{1/q} \leq \frac{\alpha}{p} + \frac{\beta}{q}, \tag{2.3.4}$$

holds. Now suppose without loss of generality that

$$\left(\int_a^b |f(t)|^p\, \Delta t\right) \left(\int_a^b |g(t)|^q\, \Delta t\right) \neq 0.$$

Apply (2.3.4) with

$$\alpha = \frac{|f(t)|^p}{\left(\int_a^b |f(s)|^p\, \Delta s\right)}, \quad \text{and } \beta = \frac{|g(t)|^q}{\int_a^b |g(s)|^q\, \Delta s},$$

and integrate the obtained inequality between a and b (this is possible since all functions are rd-continuous), we find that

$$\int_a^b \frac{|f(t)|}{\left(\int_a^b |f(s)|^p \Delta s\right)^{1/p}} \frac{|g(t)|}{\left(\int_a^b |g(s)|^q \Delta s\right)^{1/q}} \Delta t = \int_a^b \alpha^{1/p}(t)\beta^{1/q}(t)\Delta t$$

$$\leq \int_a^b \left(\frac{\alpha(t)}{p} + \frac{\beta(t)}{q}\right)\Delta t = \int_a^b \left[\frac{|f(t)|^p}{p\left(\int_a^b |f(s)|^p \Delta s\right)} + \frac{|g(t)|^q}{q\int_a^b |g(s)|^q \Delta s}\right]\Delta t$$

$$= \frac{\int_a^b |f(t)|^p \Delta t}{p\left(\int_a^b |f(s)|^p \Delta s\right)} + \frac{\int_a^b |g(t)|^q \Delta t}{q\int_a^b |g(s)|^q \Delta s} = \frac{1}{p} + \frac{1}{q} = 1,$$

which is the desired inequality (2.3.3). The proof is complete. ∎

As a special case when $p = q = 2$, we have the following Schwarz's inequality.

Theorem 2.3.2 *Let* $a, b \in \mathbb{T}$. *For* $f, g \in C_{rd}(\mathbb{I}, \mathbb{R})$, *we have*

$$\int_a^b |f(t)g(t)|\, \Delta t \leq \left[\int_a^b |f(t)|^2\, \Delta t\right]^{\frac{1}{2}} \left[\int_a^b |g(t)|^2\, \Delta t\right]^{\frac{1}{2}}. \qquad (2.3.5)$$

Setting

$$\alpha = \frac{|h(t)|^{1/p}\,|f(t)|}{\left(\int_a^b |h(s)|\,|f(s)|^p \Delta s\right)^{1/p}}, \text{ and } \beta = \frac{|h(t)|^{1/q}\,|f(t)|}{\left(\int_a^b |h(s)|\,|g(s)|^q \Delta s\right)^{1/q}},$$

in the proof of Theorem 2.3.1 and applying the Young inequality, we have the following inequality.

Theorem 2.3.3 *Let* $h, f, g \in C_r([a,b]_\mathbb{T}, [0, \infty))$. *If* $1/p + 1/q = 1$, *with* $p > 1$, *then*

$$\int_a^b h(t)f(t)g(t)\Delta t \leq \left(\int_a^b h(t)f^p(t)\Delta t\right)^{1/p}\left(\int_a^b h(t)g^q(t)\Delta t\right)^{1/q}. \qquad (2.3.6)$$

Now we give the nabla Hölder type inequality on time scales.

Theorem 2.3.4 *Let* $a, b \in \mathbb{T}$. *For* $f, g, h \in C_{ld}([a,b]_\mathbb{T}, \mathbb{R})$, *we have*

$$\int_a^b |h(t)|\,|f(t)g(t)|\,\nabla t \leq \left[\int_a^b |h(t)|\,|f(t)|^p\,\nabla t\right]^{\frac{1}{p}}\left[\int_a^b |h(t)|\,|g(t)|^q\,\nabla t\right]^{\frac{1}{q}},$$

$$(2.3.7)$$

where $p > 1$ *and* $\frac{1}{p} + \frac{1}{q} = 1$.

Proof. Setting

$$A = \frac{|h(t)|^{1/p}\,|f(t)|}{\left(\int_a^b |h(s)|\,|f(s)|^p\,\nabla s\right)^{1/p}}, \text{ and } B = \frac{|h(t)|^{1/q}\,|f(t)|}{\left(\int_a^b |h(s)|\,|g(s)|^q\,\nabla s\right)^{1/q}},$$

and applying the Young inequality $AB \le \frac{A^p}{p} + \frac{B^q}{q}$, where A, B are nonnegative, $p > 1$ and $\frac{1}{p} + \frac{1}{q} = 1$, we see that

$$
\begin{aligned}
\int_a^b A(t)B(t)\nabla t &\le \int_a^b \left(\frac{A^p}{p} + \frac{B^q}{q}\right)\nabla t \\
&= \int_a^b \left[\frac{|h(t)|\,|f(t)|^p}{p\left(\int_a^b |h(s)|\,|f(s)|^p\,\nabla s\right)} + \frac{|h(t)|\,|g(t)^q|}{q\int_a^b |h(s)|\,|g(s)|^q\,\Delta s}\right]\nabla t \\
&= \frac{\int_a^b |h(t)|\,|f(t)|^p\,\nabla t}{p\left(\left(\int_a^b |h(s)|\,|f(s)|^p\,\nabla s\right)\right)} + \frac{\int_a^b |h(t)|\,|g(t)^q|\,\nabla t}{q\int_a^b |h(s)|\,|g(s)|^q\,\nabla s} \\
&= \frac{1}{p} + \frac{1}{q} = 1,
\end{aligned}
$$

which is the desired inequality (2.3.7). The proof is complete. ∎

As a special case of Theorem 2.3.4 when $p = q = 2$, we have the following result.

Theorem 2.3.5 *Let* $a, b \in \mathbb{T}$. *For* $f, g, h \in C_{ld}([a,b]_\mathbb{T}, \mathbb{R})$, *we have*

$$\int_a^b |h(t)|\,|f(t)g(t)|\,\nabla t \le \left[\int_a^b |h(t)|\,|f(t)|^2\,\nabla t\right]^{\frac{1}{2}}\left[\int_a^b |h(t)|\,|g(t)|^2\,\nabla t\right]^{\frac{1}{2}}.$$

$$(2.3.8)$$

Theorem 2.3.6 *Let* $a, b \in \mathbb{T}$. *For* $f, g, h \in C_{ld}([a,b]_\mathbb{T}, \mathbb{R})$, *we have*

$$\int_a^b |h(t)|\,|f(t)g(t)|\,\nabla t \ge \left[\int_a^b |h(t)|\,|f(t)|^p\,\nabla t\right]^{\frac{1}{p}}\left[\int_a^b |h(t)|\,|g(t)|^q\,\nabla t\right]^{\frac{1}{q}},$$

$$(2.3.9)$$

where $p < 0$ *or* $q < 0$ *and* $\frac{1}{p} + \frac{1}{q} = 1$.

Proof. Without loss of generality, we assume that $p < 0$. Set $P = -p/q$ and $Q = 1/q$. Then $1/P + 1/Q = 1$ with $P > 1$ and $Q > 1$. From (2.3.7) we have

$$\int_a^b |h(t)|\,|F(t)G(t)|\,\nabla t \le \left[\int_a^b |h(t)|\,|F(t)|^P\,\nabla t\right]^{\frac{1}{P}}\left[\int_a^b |h(t)|\,|G(t)|^Q\,\nabla t\right]^{\frac{1}{Q}}.$$

Letting $F(t) = f^{-q}(t)$ and $G(t) = f^q(t)g^q(t)$ in the last inequality, we get the desired inequality (2.3.9). The proof is complete. ∎

As an application of Hölder inequality (2.3.3), we have the following theorem.

Theorem 2.3.7 *Let a, $b \in \mathbb{T}$ with $a < b$ and f and g be two positive functions defined on the interval $[a, b]_\mathbb{T}$ such that $0 < m \leq f/g \leq M < \infty$. Then for $p > 1$ and $q > 1$ with $1/p + 1/q = 1$, we have*

$$\int_a^b f^{1/p}(t)g^{1/q}(t)\Delta t \leq \frac{M^{1/p^2}}{m^{1/q^2}} \int_a^b f^{1/q}(t)g^{1/p}(t)\Delta t, \qquad (2.3.10)$$

and then

$$\int_a^b f^{1/p}(t)g^{1/q}(t)\Delta t \leq \frac{M^{1/p^2}}{m^{1/q^2}} \left(\int_a^b f(t)\Delta t \right)^{1/q} \left(\int_a^b g(t)\Delta t \right)^{1/p}.$$

Proof. From inequality (2.3.3), we obtain

$$\int_a^b f^{1/p}(t)g^{1/q}(t)\Delta t \leq \left(\int_a^b f(t)\Delta t \right)^{1/p} \left(\int_a^b g(t)\Delta t \right)^{1/q},$$

that is

$$\int_a^b f^{1/p}(t)g^{1/q}(t)\Delta t \leq \left(\int_a^b f^{1/p}(t)f^{1/q}(t)\Delta t \right)^{1/p} \left(\int_a^b g^{1/q}(t)g^{1/p}(t)\Delta t \right)^{1/q}.$$

Since $f^{1/p}(t) \leq M^{1/p}g^{1/p}(t)$ and $g^{1/q}(t) \leq m^{-1/q}f^{1/q}(t)$, then from the above inequality it follows that

$$\int_a^b f^{1/p}(t)g^{1/q}(t)\Delta t \leq M^{1/p^2}m^{-1/q^2} \left(\int_a^b f^{1/q}(t)g^{1/p}(t)\Delta t \right)^{1/p}$$
$$\times \left(\int_a^b f^{1/q}(t)g^{1/p}(t)\Delta t \right)^{1/q},$$

that is

$$\int_a^b f^{1/p}(t)g^{1/q}(t)\Delta t \leq M^{1/p^2}m^{-1/q^2} \int_a^b f^{1/q}(t)g^{1/p}(t)\Delta t. \qquad (2.3.11)$$

Hence, the inequality (2.3.10) is proved. The proof is complete. ∎

The following theorems give the reverse Hölder type inequality on time scales.

Theorem 2.3.8 *Let a, $b \in \mathbb{T}$ with $a < b$ and f and g be two positive functions defined on the interval $[a,b]_\mathbb{T}$ such that $0 < m \le f^p/g^q \le M < \infty$. Then for $p > 1$ and $q > 1$ with $1/p + 1/q = 1$, we have*

$$\left(\int_a^b f^p(t)\Delta t \right)^{1/p} \left(\int_a^b g^q(t)\Delta t \right)^{1/q} \le \left(\frac{M}{m} \right)^{\frac{1}{pq}} \int_a^b f(t)g(t)\Delta t. \quad (2.3.12)$$

Proof. Since $f^p/g^q \le M$, then we have $g \ge M^{-1/q} f^{p/q}$. Therefore

$$fg \ge M^{\frac{-1}{q}} f^{\frac{p}{q}+1} = M^{\frac{-1}{q}} f^{\frac{p+q}{q}} = M^{\frac{-1}{q}} f^p,$$

and so

$$\left(\int_a^b f^p(t)\Delta t \right)^{\frac{1}{p}} \le M^{\frac{1}{pq}} \left(\int_a^b f(t)g(t)\Delta t \right)^{\frac{1}{p}}. \quad (2.3.13)$$

Also since $m \le f^p/g^q$, then we have $f \ge m^{1/p} g^{q/p}$. Then

$$\int_a^b f(t)g(t)\Delta t \ge m^{1/p} \int_a^b g^{1+q/p}(t)\Delta t = m^{1/p} \int_a^b g^q(t)\Delta t,$$

and so

$$\left(\int_a^b f(t)g(t)\Delta t \right)^{1/q} \ge m^{\frac{1}{pq}} \left(\int_a^b g^q(t)\Delta t \right)^{\frac{1}{q}}. \quad (2.3.14)$$

Combining (2.3.13) and (2.3.14), we have the desired inequality (2.3.12). The proof is complete. ∎

In Theorem 2.3.8, if we replace f^p and g^q by f and g, we obtain the reverse Hölder type inequality

$$\left(\int_a^b f(t)\Delta t \right)^{1/p} \left(\int_a^b g(t)\Delta t \right)^{1/q} \le \left(\frac{M}{m} \right)^{\frac{1}{pq}} \int_a^b f^{1/p}(t)g^{1/q}(t)\Delta t. \quad (2.3.15)$$

Theorem 2.3.9 *Let a, $b \in \mathbb{T}$ with $a < b$ and f and g be two positive functions defined on the interval $[a,b]_\mathbb{T}$ such that $0 < m \le f^p \le M < \infty$. Then for $p > 1$ and $q > 1$ with $1/p + 1/q = 1$, we have*

$$\left(\int_a^b f^{1/p}(t)\Delta t \right)^p \ge (b-a)^{\frac{p+1}{q}} \left(\frac{m}{M} \right)^{\frac{p+1}{pq}} \left(\int_a^b f^p(t)\Delta t \right)^{1/p}. \quad (2.3.16)$$

Proof. Putting $g = 1$ in Theorem 2.3.8, we obtain

$$\left(\int_a^b f^p(t)\Delta t \right)^{1/p} (b-a)^{1/q} \le \left(\frac{m}{M} \right)^{\frac{-1}{pq}} \int_a^b f(t)\Delta t.$$

Therefore, we get

$$\left(\int_a^b f^p(t)\Delta t\right)^{1/p} \leq \left(\frac{m}{M}\right)^{\frac{-1}{pq}} (b-a)^{-1/q} \int_a^b f(t)\Delta t. \qquad (2.3.17)$$

Substituting g in (2.3.15) leads to

$$\left(\int_a^b f(t)\Delta t\right)^{1/p} \leq \left(\frac{m}{M}\right)^{\frac{-1}{pq}} (b-a)^{-1/q} \int_a^b f^{1/p}(t)\Delta t,$$

and so

$$\int_a^b f(t)\Delta t \leq \left(\frac{m}{M}\right)^{\frac{-1}{q}} (b-a)^{-p/q} \left(\int_a^b f^{1/p}(t)\Delta t\right)^p. \qquad (2.3.18)$$

Combining (2.3.17) with (2.3.18), we obtain

$$\left(\int_a^b f^{1/p}(t)\Delta t\right)^p \geq \left(\frac{m}{M}\right)^{\frac{p+1}{pq}} (b-a)^{(p+1)/q} \left(\int_a^b f^p(t)\Delta t\right)^{1/p},$$

which is the desired inequality (2.3.16). The proof is complete. ∎
Next we prove a Hölder type inequality in two dimensionals on time scales.

Theorem 2.3.10 Let $a, \ b \ \in \ \mathbb{T}$ with $a \ < \ b$ and f and g be two rd-continuous functions defined on the interval $[a,b]_{\mathbb{T}} \times [a,b]_{\mathbb{T}}$. Then

$$\int_a^b \int_a^b |f(x,y)g(x,y)|\, \Delta x \Delta y \qquad (2.3.19)$$

$$\leq \left(\int_a^b \int_a^b |f(x,y)|^p\, \Delta x \Delta y\right)^{1/p} \left(\int_a^b \int_a^b |g(x,y)|^q\, \Delta x \Delta y\right)^{1/q},$$

where $p > 1$ and $q = p/((p-1))$.

Proof. Suppose without loss of generality that

$$\left(\int_a^b \int_a^b |f(x,y)|^p\, \Delta x \Delta y\right) \int_a^b \int_a^b |g(x,y)|^q\, \Delta x \Delta y \neq 0.$$

Apply the Young inequality $\alpha^{1/p}\beta^{1/q} \leq \frac{\alpha}{p} + \frac{\beta}{q}$ (2.3.4) with

$$\alpha(x,y) = \frac{|f(x,y)|^p}{\int_a^b \int_a^b |f(\tau_1,\tau_2)|^p\, \Delta\tau_1\Delta\tau_2},$$

$$\beta(x,y) = \frac{|g(x,y)|^q}{\int_a^b \int_a^b |g(\tau_1,\tau_2)|^q\, \Delta\tau_1\Delta\tau_2},$$

and integrate the obtained inequality between a and b to get

$$\int_a^b \int_a^b \alpha^{1/p}(x,y)\beta^{1/q}(x,y)\Delta x \Delta y$$

$$\leq \int_a^b \int_a^b \left(\frac{\alpha(x,y)}{p} + \frac{\beta(x,y)}{q} \right) \Delta x \Delta y$$

$$= \frac{\int_a^b \int_a^b |f(x,y)|^p \, \Delta x \Delta y}{p \int_a^b \int_a^b |f(\tau_1,\tau_2)|^p \, \Delta \tau_1 \Delta \tau_2} + \frac{\int_a^b \int_a^b |g(x,y)|^q \, \Delta x \Delta y}{q \int_a^b \int_a^b |g(\tau_1,\tau_2)|^q \, \Delta \tau_1 \Delta \tau_2}$$

$$= \frac{1}{p} + \frac{1}{q} = 1.$$

The proof is complete. ∎

Now, we give the diamond α-Hölder inequalities on time scales by applying the diamond α-Jensen inequalities on time scales. As an application of the diamond α-Jensen inequality proved in Theorem 2.2.6 by taking $F(t) = t^p$ for $p > 1$ and g and $|h|$ be replaced by $ug^{-p/q}$ and hg^q, we have the following Hölder inequality.

Theorem 2.3.11 Let h, u, $g \in C([a,b]_{\mathbb{T}}, \mathbb{R})$ with $\int_a^b h(t)g^q(t)\Diamond_\alpha t > 0$. If $1/p + 1/q = 1$, with $p > 1$, then

$$\int_a^b |h(t)| \, |u(t)g(t)| \, \Diamond_\alpha t \leq \left(\int_a^b |h(t)| \, |u(t)|^p \, \Diamond_\alpha t \right)^{1/p} \left(\int_a^b |h(t)| \, |g(t)|^q \, \Diamond_\alpha t \right)^{1/q}.$$

$$(2.3.20)$$

In the particular case $h = 1$, Theorem 2.3.11 gives the diamond-α version of the classical Hölder inequality:

$$\int_a^b |u(t)g(t)| \, \Diamond_\alpha t \leq \left(\int_a^b |u(t)|^p \, \Diamond_\alpha t \right)^{1/p} \left(\int_a^b |g(t)|^q \, \Diamond_\alpha t \right)^{1/q}, \quad (2.3.21)$$

where $p > 1$ and $q = p/(p-1)$. In the special case $p = q = 2$, the inequality (2.3.21) reduces to the following diamond-α Cauchy–Schwarz integral inequality on time scales

$$\int_a^b |u(t)g(t)| \, \Diamond_\alpha t \leq \sqrt{\left(\int_a^b |u(t)|^2 \, \Diamond_\alpha t \right) \left(\int_a^b |g(t)|^2 \, \Diamond_\alpha t \right)}. \quad (2.3.22)$$

Theorem 2.3.12 Let h, u, $g \in C([a,b]_\mathbb{T}, \mathbb{R})$ with $\int_a^b h(t) g^q(t) \Diamond_\alpha t > 0$. If $1/p + 1/q = 1$, with $p < 0$ or $q < 0$, then

$$\int_a^b |h(t)| |u(t) g(t)| \Diamond_\alpha t \geq \left(\int_a^b |h(t)| |u(t)|^p \Diamond_\alpha t \right)^{1/p}$$

$$\times \left(\int_a^b |h(t)| |g(t)|^q \Diamond_\alpha t \right)^{1/q}.$$

Theorem 2.3.13 Let a, $b \in \mathbb{T}$ with $a < b$ and f and g be two positive functions defined on the interval $[a,b]_\mathbb{T}$ such that $0 < m \leq f^p/g^q \leq M < \infty$. Then for $p > 1$ with $1/p + 1/q = 1$, we have

$$\left(\int_a^b f^p(t) \Diamond_\alpha t \right)^{1/p} \left(\int_a^b g^q(t) \Diamond_\alpha t \right)^{1/q} \leq \left(\frac{M}{m} \right)^{\frac{1}{pq}} \int_a^b f(t) g(t) \Diamond_\alpha t. \quad (2.3.23)$$

Proof. As in the proof of Theorem 2.3.8, we get that

$$\left(\int_a^b f^p(t) \Diamond_\alpha t \right)^{\frac{1}{p}} \leq M^{\frac{1}{pq}} \left(\int_a^b f(t) g(t) \Diamond_\alpha t \right)^{\frac{1}{p}},$$

and

$$\left(\int_a^b f(t) g(t) \Diamond_\alpha t \right)^{1/q} \geq (m)^{\frac{1}{pq}} \left(\int_a^b g^q(t) \Diamond_\alpha t \right)^{\frac{1}{q}}.$$

Combining these two inequalities, we have the desired inequality (2.3.23). The proof is complete. ∎

Now, we give the diamond α-Hölder type inequality in two dimensions on time scales. In this case, we assume that the double integral is defined as an iterated integral. Let $\mathbb{T}$ be a time scale with $a, b \in \mathbb{T}$, $a < b$, and f be a real-valued function on $\mathbb{T} \times \mathbb{T}$. Because we need notation for partial derivatives with respect to time scale variables x and y we denote the time scale partial derivative of $f(x,y)$ with respect to x by $f^{\Diamond_\alpha^1}(x,y)$ and let $f^{\Diamond_\alpha^2}(x,y)$ denote the time scale partial derivative with respect to y. Fix an arbitrary $y \in \mathbb{T}$. Then the diamond-α derivative of the function

$$\mathbb{T} \to \mathbb{R}, \qquad x \to f(x,y)$$

is denoted by $f^{\Diamond_\alpha^1}$. Let now $x \in \mathbb{T}$. The diamond-α derivative of the function

$$\mathbb{T} \to \mathbb{R}, \qquad y \to f(x,y)$$

is denoted by $f^{\Diamond_\alpha^{21}}$. If the function f has a $\Diamond_\alpha^1$ antiderivative A, i.e., $A^{\Diamond_\alpha^1} = f$, and A has a $\Diamond_\alpha^2$ antiderivative B, i.e., $B^{\Diamond_\alpha^2} = A$, then

$$\int_a^b \int_a^b f(x,y) \Diamond_\alpha x \Diamond_\alpha y = \int_a^b (A(b,y) - A(a,y)) \Diamond_\alpha y$$

$$= B(b,b) - B(b,a) - B(a,b) + B(a,a).$$

Note that $\left(B^{\Diamond_\alpha^2} \right)^{\Diamond_\alpha^1} = (A)^{\Diamond_\alpha^1} = f$.

Now we are ready to state and prove the diamond α-Hölder inequality in two dimensions on time scales.

Theorem 2.3.14 *Let $\mathbb{T}$ be a time scale, $a, b \in \mathbb{T}$, with $a < b$, f, g, h : $[a, b]_{\mathbb{T}} \times [a, b]_{\mathbb{T}} \to \mathbb{R}$, be $\Diamond_\alpha$ integrable functions, and $1/p + 1/q = 1$ with $p > 1$. Then,*

$$\int_a^b \int_a^b |h(x, y)f(x, y)g(x, y)| \, \Diamond_\alpha x \Diamond_\alpha y \qquad (2.3.24)$$

$$\leq \left(\int_a^b \int_a^b |h(x, y)f(x, y)|^p \, \Diamond_\alpha x \Diamond_\alpha y \right)^{1/p}$$

$$\left(\int_a^b \int_a^b |h(x, y)g(x, y)|^q \, \Diamond_\alpha x \Diamond_\alpha y \right)^{1/q}.$$

Proof. Inequality (2.3.24) is trivially true in the case when f, or g, or h is identically zero. Suppose that

$$\left(\int_a^b \int_a^b |h(x, y)f(x, y)|^{1/p} \, \Diamond_\alpha x \Diamond_\alpha y \right) \left(\int_a^b \int_a^b |h(x, y)g(x, y)|^{1/q} \, \Diamond_\alpha x \Diamond_\alpha y \right) \neq 0,$$

and let

$$A(x, y) = \frac{|h(x, y)|^{1/p} |f(x, y)|}{\left(\int_a^b \int_a^b |h(x, y)| \, |f(x, y)|^p \, \Diamond_\alpha x \Diamond_\alpha y \right)^{1/p}},$$

$$B(x, y) = \frac{|h(x, y)|^{\frac{1}{q}} |g(x, y)|}{\left(\int_a^b \int_a^b |h(x, y)| \, |g(x, y)|^q \, \Diamond_\alpha x \Diamond_\alpha y \right)^{1/q}}.$$

Applying the Young inequality $AB \leq \frac{A^p}{p} + \frac{B^q}{q}$, we have that

$$\int_a^b \int_a^b A(x, y)B(x, y)\Diamond_\alpha x \Diamond_\alpha y \leq \frac{1}{p} \frac{\int_a^b \int_a^b |h(x, y)| \, |f(x, y)|^p \, \Diamond_\alpha x \Diamond_\alpha y}{\left(\int_a^b \int_a^b |h(x, y)| \, |f(x, y)|^p \, \Diamond_\alpha x \Diamond_\alpha y \right)}$$

$$+ \frac{1}{q} \frac{\int_a^b \int_a^b |h(x, y)| \, |g(x, y)|^q \, \Diamond_\alpha x \Diamond_\alpha y}{\left(\int_a^b \int_a^b |h(x, y)| \, |g(x, y)|^q \, \Diamond_\alpha x \Diamond_\alpha y \right)}$$

$$= \frac{1}{p} + \frac{1}{q} = 1,$$

and the desired inequality follows. The proof is complete. $\blacksquare$

As a special case of Theorem 2.3.14, when $p = q = 2$, we get the two dimensional diamond-α Cauchy Schwartz's inequality.

Corollary 2.3.1 *Let* $\mathbb{T}$ *be a time scale,* $a, b \in \mathbb{T}$, *with* $a < b$, f, g, h : $[a, b]_{\mathbb{T}} \times [a, b]_{\mathbb{T}} \to \mathbb{R}$, *be* $\Diamond_{\alpha}$ *integrable functions, and* $1/p + 1/q = 1$ *with* $p > 1$. *Then,*

$$\int_a^b \int_a^b |h(x, y) f(x, y) g(x, y)| \, \Diamond_{\alpha} x \Diamond_{\alpha} y$$
$$\leq \left(\int_a^b \int_a^b |h(x, y) f(x, y)|^2 \, \Diamond_{\alpha} x \Diamond_{\alpha} y \right)^{1/2} \left(\int_a^b \int_a^b |h(x, y) g(x, y)|^2 \, \Diamond_{\alpha} x \Diamond_{\alpha} y \right)^{1/2}.$$

Now, we apply the theory of isotonic linear functional which was presented in Sect. 2.2 to derive a Hölder type inequality on time scales. The results are adapted from [30]. We need the following theorem to prove the main results [117].

Theorem 2.3.15 *Let* E, L, *and* A *be such that* (L_1), (L_2), (A_1) *and* (A_2) *in Definition 2.2.1 are satisfied. For* $p \neq 1$, *define* $q = p/(p-1)$. *Assume* $|\omega| |f|^p$, $|\omega| |g|^q$, $|\omega f g| \in L$. *If* $p > 1$, *then*

$$A(|\omega f g|) \leq A^{1/p}(|\omega| |f|^p) A^{1/q}(|\omega| |g|^q).$$

This inequality is reversed if $0 < p < 1$ *and* $A(|\omega| |g|^q) > 0$ *and also it is reversed if* $p < 0$ *and* $A(|\omega| |f|^p) > 0$.

Now, the application of Theorems 2.2.12 and 2.3.15 gives us the following Hölder's inequality.

Theorem 2.3.16 *For* $p > 1$, *define* $q = p/(p-1)$. *Let* $E \subset \mathbb{R}^n$ *be as in Theorem 2.2.12. Assume that* $|\omega| |f|^p$, $|\omega| |g|^q$, $|\omega f g|$ *are* Δ-*integrable on* E. *If* $p > 1$, *then*

$$\int_E |\omega(t) f(t) g(t)| \, \Delta t \leq \left(\int_E |\omega(t)| |f(t)|^p \, \Delta t \right)^{1/p} \left(\int_E |\omega(t)| |g(t)|^q \, \Delta t \right)^{1/q}.$$

This inequality is reversed if $0 < p < 1$ *and* $\int_E |\omega(t)| |g(t)|^q \, \Delta t > 0$ *and also it is reversed if* $p < 0$ *and* $\int_E |\omega(t)| |f(t)|^p \, \Delta t > 0$.

2.4 Minkowski Inequalities

The well-known Minkowski integral inequality is given in [3, 72, 110]. Let f and g be real-valued functions defined on $[a, b]$ such that the functions $|f(x)|^p$ and $|g(x)|^p$ for $p > 1$ are integrable on $[a, b]$. Then

$$\left(\int_a^b |f(x) + g(x)|^p \, dx \right)^{1/p} \leq \left(\int_a^b |f(x)|^p \, dx \right)^{1/p} + \left(\int_a^b |g(x)|^p \, dx \right)^{1/p}.$$

Equality holds if and only if $f(x) = 0$ almost everywhere or $g(x) = \lambda f(x)$ almost everywhere with a constant $\lambda \geq 0$. The discrete version of Minkowski inequality is given by

$$\left(\sum_{i=1}^{n} |f(i) + g(i)|^p \right)^{1/p} \leq \left(\sum_{i=1}^{n} |f(i)|^p \right)^{1/p} + \left(\sum_{i=1}^{n} |g(i)|^p \right)^{1/p},$$

where $f(n)$ and $g(n)$ are two positive-tuples and $p > 1$. Equality holds if and only f and g are proportional.

In this section we establish the Minkowski integral inequality and its extensions on time scales. The results in this section are adapted from [23, 30, 39, 45, 115, 150, 155].

Theorem 2.4.1 *Let f, g, $h \in C_{rd}([a,b]_\mathbb{T}, \mathbb{R})$ and $p > 1$. Then*

$$\left(\int_a^b |h(x)| \, |f(x) + g(x)|^p \, \Delta x \right)^{1/p} \leq \left(\int_a^b |h(x)| \, |f(x)|^p \, \Delta x \right)^{1/p}$$

$$+ \left(\int_a^b |h(x)| \, |g(x)|^p \, \Delta x \right)^{1/p}. \quad (2.4.1)$$

Proof. Note

$$\int_a^b |h(x)| \, |f(x) + g(x)|^p \, \Delta x = \int_a^b |h(x)| \, |f(x) + g(x)|^{p-1} \, |f(x) + g(x)| \, \Delta x$$

$$\leq \int_a^b |h(x)| \, |f(x) + g(x)|^{p-1} \, |f(x)| \, \Delta x$$

$$+ \int_a^b |h(x)| \, |f(x) + g(x)|^{p-1} \, |g(x)| \, \Delta x.$$

Applying the Hölder inequality (2.3.6), we get that

$$\int_a^b |h(x)| \, |f(x) + g(x)|^p \, \Delta x$$

$$\leq \left(\int_a^b |h(x)| \left(|f(x) + g(x)|^{p-1} \right)^q \Delta x \right)^{1/q} \left(\int_a^b |h(x)| \, |f(x)|^p \, \Delta x \right)^{1/p}$$

$$+ \left(\int_a^b |h(x)| \left(|f(x) + g(x)|^{p-1} \right)^q \Delta x \right)^{1/q} \left(\int_a^b |h(x)| \, |g(x)|^p \, \Delta x \right)^{1/p}$$

$$= \left(\int_a^b |h(x)| \, |f(x) + g(x)|^p \, \Delta x \right)^{1/q}$$

$$\times \left[\left(\int_a^b |h(x)| \, |f(x)|^p \, \Delta x \right)^{1/p} + \left(\int_a^b |h(x)| \, |g(x)|^p \, \Delta x \right)^{1/p} \right].$$

Therefore

$$\left(\int_a^b |h(x)|\, |f(x) + g(x)|^p\, \Delta x \right)^{1/p}$$

$$= \left(\int_a^b |h(x)|\, |f(x) + g(x)|^p\, \Delta x \right)^{1-1/q}$$

$$= \left[\left(\int_a^b |h(x)|\, |f(x)|^p\, \Delta x \right)^{1/p} + \left(\int_a^b |h(x)|\, |g(x)|^p\, \Delta x \right)^{1/p} \right],$$

which is the desired inequality (2.4.1). The proof is complete. ∎

As a special case when $h(x) = 1$, we obtain the time scale classical Minkowski inequality

$$\left(\int_a^b |f(x) + g(x)|^p\, \Delta x \right)^{1/p} \leq \left(\int_a^b |f(x)|^p\, dx \right)^{1/p} + \left(\int_a^b |g(x)|^p\, dx \right)^{1/p}.$$

As in the proof of Theorem 2.4.1 (using (2.3.7)) we obtain the following nabla Minkowski inequality.

Theorem 2.4.2 Let f, g, $h \in C_{ld}([a,b]_\mathbb{T}, \mathbb{R})$ and $p > 1$. Then

$$\left(\int_a^b |h(x)|\, |f(x) + g(x)|^p\, \nabla x \right)^{1/p}$$

$$\leq \left(\int_a^b |h(x)|\, |f(x)|^p\, \nabla x \right)^{1/p} + \left(\int_a^b |h(x)|\, |g(x)|^p\, \nabla x \right)^{1/p}.$$

Applying the diamond-α Hölder inequality (2.3.20) we have the following diamond-α Minkowski's inequality.

Theorem 2.4.3 Let f, g, $h \in C([a,b]_\mathbb{T}, \mathbb{R})$ and $p > 1$. Then

$$\left(\int_a^b |h(x)|\, |f(x) + g(x)|^p\, \Diamond_\alpha x \right)^{1/p}$$

$$\leq \left(\int_a^b |h(x)|\, |f(x)|^p\, \Diamond_\alpha x \right)^{1/p} + \left(\int_a^b |h(x)|\, |g(x)|^p\, \Diamond_\alpha x \right)^{1/p}.$$

Theorem 2.4.4 Let f, $g : [a,b]_\mathbb{T} \to \mathbb{R}$, are positive rd-continuous functions and satisfying $0 < m \leq f/g \leq M < \infty$ on $[a,b]_\mathbb{T}$ and for $p > 1$ define $q = p/(p-1)$. Then

$$\left(\int_a^b f^p(x)\Delta x \right)^{1/p} + \left(\int_a^b g^p(x)\Delta x \right)^{1/p} \leq c \left(\int_a^b (f(x) + g(x))^p\, \Delta x \right)^{\frac{1}{p}},$$

$$(2.4.2)$$

where $c = \left(\frac{m}{M} \right)^{\frac{1}{pq}}$.

Proof. To prove the inequality (2.4.2), we apply Theorem 2.3.8. The inner term in the right-hand side can be rewritten as

$$\int_a^b (f(x) + g(x))^p \, \Delta x$$

$$= \int_a^b (f(x) + g(x))^{p-1} f(x) \Delta x$$

$$+ \int_a^b (f(x) + g(x))^{p-1} g(x) \Delta x$$

$$\geq \left(\frac{M}{m}\right)^{\frac{1}{pq}} \left(\int_a^b f^p(x) \Delta x\right)^{\frac{1}{p}} \left(\int_a^b (f(x) + g(x))^{q(p-1)} \Delta x\right)^{\frac{1}{q}}$$

$$+ \left(\frac{M}{m}\right)^{\frac{1}{pq}} \left(\int_a^b g^p(x) \Delta x\right)^{\frac{1}{p}} \left(\int_a^b (f(x) + g(x))^{q(p-1)} \Delta x\right)^{\frac{1}{q}}$$

$$= \left(\frac{M}{m}\right)^{\frac{1}{pq}} \left(\int_a^b (f(x) + g(x))^p \, \Delta x\right)^{\frac{1}{q}}$$

$$\times \left[\left(\int_a^b f^p(x) \Delta x\right)^{\frac{1}{p}} + \left(\int_a^b g^p(x) \Delta x\right)^{\frac{1}{p}}\right].$$

Therefore, we obtain

$$\left(\int_a^b f^p(x) \Delta x\right)^{\frac{1}{p}} + \left(\int_a^b g^p(x) \Delta x\right)^{\frac{1}{p}} \leq \left(\frac{m}{M}\right)^{\frac{1}{pq}} \left(\int_a^b (f(x) + g(x))^p \, \Delta x\right)^{1-\frac{1}{q}}$$

$$= \left(\frac{m}{M}\right)^{\frac{1}{pq}} \left(\int_a^b (f(x) + g(x))^p \, \Delta x\right)^{\frac{1}{p}},$$

which is the desired inequality (2.4.2). The proof is complete. ∎

Now, we apply the theory of isotonic linear functional that was presented in Sect. 2.2 to derive a Minkowski inequality on time scales. To do this we need the following theorem as given in [117].

Theorem 2.4.5 *Let E, L, and A be such that (L_1), (L_2), (A_1) and (A_2), as in Definition 2.2.1, are satisfied. For $p \in \mathbb{R}$, assume $|\omega| |f|^p$, $|\omega| |g|^p$, $|\omega| |f + g|^p \in L$. If $p > 1$, then*

$$A^{1/p}(|\omega| |f + g|^p) \leq A^{1/p}(|\omega| |f|^p) + A^{1/p}(|\omega| |g|^p).$$

This inequality is reversed if $0 < p < 1$ or $p < 0$ provided that $A(|\omega| |g|^p) > 0$ and $A(|\omega| |f|^p) > 0$ hold.

Now, the application of Theorems 2.2.12 and 2.4.5 gives us the following Minkowski inequality.

Theorem 2.4.6 *Let $E \subset \mathbb{R}^n$ be as in Theorem 2.2.12. For $p \in \mathbb{R}$, assume $|\omega| |f|^p$, $|\omega| |g|^p$, $|\omega| |f + g|^p$ are Δ-integrable on E. If $p > 1$, then*

$$\left(\int_E |\omega(t)| |f(t) + g(t)|^p \Delta t \right)^{1/p} \leq \left(\int_E |\omega(t)| |f(t)|^p \Delta t \right)^{1/p}$$
$$+ \left(\int_E |\omega(t)| |g(t)|^p \Delta t \right)^{1/p}. \quad (2.4.3)$$

This inequality is reversed if $0 < p < 1$ or $p < 0$ provided that $\int_E |\omega(t)| |g(t)|^q \Delta t > 0$ and $\int_E |\omega(t)| |f(t)|^p \Delta t > 0$.

In the following we obtain generalizations of Minkowski inequalities on time scales. The inequalities will be proved for several variables and based on the definitions of the multiple Riemann and Lebesgue Δ-integration on time scales given in [53].

Let $n \in \mathbb{N}$ be fixed. For $i \in \{1, 2, \ldots, n\}$, let $\mathbb{T}_i$ denote a time scale and

$$\Lambda^n = \mathbb{T}_1 \times \mathbb{T}_2 \times \ldots \times \mathbb{T}_n = \{t = (t_1, t_2, \ldots, t_n) : t_i \in \mathbb{T}_i, \ 1 \leq i \leq n\},$$

as the n-dimensional time scale. Let μ_Δ be the σ-additive Lebesuge Δ-measure on Λ^n and $\mathcal{F}$ be the family of Δ-measurable subsets of Λ^n. Let $E \subset \mathcal{F}$ and $(E, \mathcal{F}, \mu_\Delta)$ be a time scale measure space. Then for a Δ-measurable function $f : E \to \mathbb{R}$, the corresponding Δ-integral of f over E will be denoted by

$$\int_E f(t_1, t_2, \ldots, t_n) \Delta_1 t_1 \Delta_2 t_2 \ldots \Delta_n t_n, \text{ or } \int_E f(t) \Delta t,$$

$$\text{or } \int_E f d\mu_\Delta, \text{ or } \int_E f(t) d\mu_\Delta(t).$$

Here, we state the Fubini theorem for integrals. It is used in the proofs of our main results.

Theorem 2.4.7 *Let (X, M, μ_Δ) and $(Y, \mathcal{L}, \nu_\Delta)$ be two finite-dimensional time scale measure space. If $f : X \times Y \to \mathbb{R}$ is a Δ-integrable function. Setting*

$$\varphi(y) = \int_X f(x, y) d\mu_\Delta(x), \text{ for } y \in Y,$$

and

$$\psi(x) = \int_Y f(x, y) d\nu_\Delta(y), \text{ for } x \in X,$$

then φ is Δ-integrable on Y and ψ is Δ-integrable on X and

$$\int_X d\mu_\Delta(x) \int_Y f(x, y) d\nu_\Delta(y) = \int_Y d\nu_\Delta(y) \int_X f(x, y) d\mu_\Delta(x). \quad (2.4.4)$$

We mention here that all theorems in Lebesgue integration theory, including the Lebesgue dominated convergence theorem, hold also for Lebesgue Δ-integral on Λ^n. This means that all the classical inequalities including Jensen's inequalities, Hölder inequalities, Minkowski inequalities, and their converses for multiple integration on time scales hold for both Riemann and Lebesuge integrals on time scales.

Theorem 2.4.8 *Let $(E, \mathcal{F}, \mu_\Delta)$ be a time scale measure space. For $p \in \mathbb{R}$, assume w, f, g are nonnegative functions such that wf^p, wg^p, $w(f+g)^p$ are Δ-integrable on E. If $p > 1$, then*

$$\left(\int_E \omega(t)\,(f(t) + g(t))^p \, d\mu_\Delta t \right)^{1/p} \leq \left(\int_E \omega(t) f^p(t) d\mu_\Delta t \right)^{1/p}$$
$$+ \left(\int_E \omega(t) g^p(t) d\mu_\Delta t \right)^{1/p}.$$

Note that Theorem 2.4.8 also holds if we have a finite number of functions. The next theorem gives an inequality of Minkowski type for infinitely many functions. We assume that all integrals are finite.

Theorem 2.4.9 *Let (X, L, μ_Δ) and (Y, λ, ν_Δ) be two finite-dimensional time scale measure space and let u, v f be Δ-integrable functions on X, Y and $X \times Y$, respectively. If $p > 1$, then*

$$\left[\int_X \left(\int_Y f(x,y) v(y) d\nu_\Delta y \right)^p u(x) d\mu_\Delta x \right]^{1/p}$$
$$\leq \int_Y \left(\int_X f^p(x,y) u(x) d\mu_\Delta x \right)^{1/p} v(y) d\nu_\Delta y, \qquad (2.4.5)$$

holds provided all integrals in (2.4.5) exists. If $0 < p < 1$ and

$$\int_X \left(\int_Y fv d\nu_\Delta \right)^p u d\mu_\Delta > 0 \quad and \quad \int_Y fv d\nu_\Delta > 0, \qquad (2.4.6)$$

holds, then (2.4.5) is reversed. If $p < 0$ and (2.4.6) and

$$\int_X f^p(x,y) u(x) d\mu_\Delta x > 0, \qquad (2.4.7)$$

hold, then (2.4.5) is reversed as well.

Proof. Let $p > 1$. Put

$$H(x) = \int_Y f(x,y) v(y) d\nu_\Delta y.$$

Now, by using Fubini's Theorem 2.4.7 and Hölder inequality in Theorem 2.3.16 on time scales, we have

$$
\begin{aligned}
\int_X H^p(x)u d\mu_\Delta &= \int_X H^{p-1}(x)H(x)u(x)d\mu_\Delta x \\
&= \int_X \left(\int_Y f(x,y)v(y)d\nu_\Delta y \right) H^{p-1}(x)u(x)d\mu_\Delta x \\
&= \int_Y \left(\int_X f(x,y)H^{p-1}(x)u(x)d\mu_\Delta x \right) v(y)d\nu_\Delta y \\
&\leq \int_Y \left(\int_X f^p(x,y)u(x)d\mu_\Delta x \right)^{1/p} \\
&\quad \times \left(\int_X H^p(x)u(x)d\mu_\Delta x \right)^{\frac{p-1}{p}} v(y)d\nu_\Delta y \\
&= \int_Y \left(\int_X f^p(x,y)u(x)d\mu_\Delta x \right)^{1/p} v(y)d\nu_\Delta y \\
&\quad \times \left(\int_X H^p(x)u(x)d\mu_\Delta x \right)^{\frac{p-1}{p}},
\end{aligned}
$$

and hence

$$
\left(\int_X H^p(x)u(x)d\mu_\Delta x \right)^{1/p} \leq \int_Y \left(\int_X f^p(x,y)u(x)d\mu_\Delta x \right)^{1/p} v(y)d\nu_\Delta y,
$$

which is the desired inequality (2.4.5). For $p < 0$ and $0 < p < 1$, the corresponding result can be obtained similarly. The proof is complete. ∎

2.5   Steffensen Inequalities

In 1918 Steffensen [142] proved the following inequality. Let a and b be real numbers such that $a < b$, f, and g are integrable functions from $[a, b]$ into $\mathbb{R}$ such that f is decreasing and for every $t \in [a, b]$, $0 \leq g(t) \leq 1$. Then

$$
\int_a^{a-\lambda} f(t)dt \leq \int_a^b f(t)g(t)dt \leq \int_a^{a+\lambda} f(t)dt, \tag{2.5.1}
$$

where $\lambda = \int_a^b g(t)dt$. The discrete analogue of Steffensen's inequality is given by

$$
\sum_{i=n-k_2+1}^{n} x_i \leq \sum_{i=1}^{n} x_i y_i \leq \sum_{i=1}^{k_1} x_i,
$$

where $(x_i)_{i=1}^n$ is a nonincreasing finite sequence of nonnegative real numbers and $(y_i)_{i=1}^n$ is a finite sequence of real numbers such that for every i, $0 \leq y_i \leq 1$ and $k_2 \leq \sum_{i=1}^n y_i \leq k_1$ for $k_1, k_2 \in \{1, 2, \dots, n\}$.

In this section, we prove some Steffensen inequalities on time scales The results in this section are adapted from [26, 114].

Theorem 2.5.1 *Let* $a, b \in \mathbb{T}_k^k$ *with* $a < b$ *and* $f, g : [a, b]_\mathbb{T} \to \mathbb{R}$ *be* Δ-*integrable functions such that* f *of one sign and decreasing and* $0 \le g(t) \le 1$ *for every* $t \in [a, b]_\mathbb{T}$. *Suppose that also* $l, \gamma \in [a, b]_\mathbb{T}$ *such that*

$$b - l \le \int_a^b g(t)\Delta t \le \gamma - a, \quad \text{if } f > 0 \text{ for all } t \in [a, b]_\mathbb{T},$$

$$\gamma - a \le \int_a^b g(t)\Delta t \le b - l, \quad \text{if } f < 0 \text{ for all } t \in [a, b]_\mathbb{T},$$

then

$$\int_l^b f(t)\Delta t \le \int_a^b f(t)g(t)\Delta t \le \int_a^\gamma f(t)\Delta t. \tag{2.5.2}$$

Proof. We consider the case when $f > 0$ and prove the left inequality. Now

$$\int_a^b f(t)g(t)\Delta t - \int_l^b f(t)\Delta t$$

$$= \int_a^l f(t)g(t)\Delta t + \int_l^b f(t)g(t)\Delta t - \int_l^b f(t)\Delta t$$

$$= \int_a^l f(t)g(t)\Delta t - \int_l^b f(t)[1 - g(t)]\Delta t$$

$$\ge \int_a^l f(t)g(t)\Delta t - f(l)\int_l^b [1 - g(t)]\Delta t$$

$$= \int_a^l f(t)g(t)\Delta t - f(l)(b - l) + f(l)\int_l^b g(t)\Delta t$$

$$\ge \int_a^l f(t)g(t)\Delta t - f(l)\int_a^b g(t)\Delta t + f(l)\int_l^b g(t)\Delta t$$

$$= \int_a^l f(t)g(t)\Delta t - f(l)[\int_a^b g(t)\Delta t - \int_l^b g(t)\Delta t]$$

$$= \int_a^l f(t)g(t)\Delta t - f(l)\int_a^l g(t)\Delta t = \int_a^l [f(t) - f(l)]\int_a^l g(t)\Delta t \ge 0,$$

since f is decreasing and g is nonnegative. The proof of the right inequality is similar. The proof is complete. ∎

Note that in Theorem 2.5.1 above we could easily replace the delta integral with the nabla integral under the same hypotheses.

Theorem 2.5.2 *Let* $a, b \in \mathbb{T}_k^k$ *with* $a < b$ *and* $f, g : [a,b]_\mathbb{T} \to \mathbb{R}$ *be* ∇*-integrable functions such that* f *is of one sign and decreasing and* $0 \le g(t) \le 1$ *on* $[a,b]_\mathbb{T}$. *Suppose that also* $l, \gamma \in [a,b]_\mathbb{T}$ *such that*

$$b - l \quad \le \quad \int_a^b g(t)\nabla t \le \gamma - a, \quad \text{if } f > 0 \text{ for all } t \in [a,b]_\mathbb{T},$$

$$\gamma - a \quad \le \quad \int_a^b g(t)\nabla t \le b - l, \quad \text{if } f < 0 \text{ for all } t \in [a,b]_\mathbb{T}.$$

Then

$$\int_l^b f(t)\nabla t \le \int_a^b f(t)g(t)\nabla t \le \int_a^\gamma f(t)\nabla t. \tag{2.5.3}$$

The following theorems more closely resemble the theorem in the continuous case (the proofs are identical to that above and omitted).

Theorem 2.5.3 *Let* $a, b \in \mathbb{T}_k^k$ *with* $a < b$ *and* $f, g : [a,b]_\mathbb{T} \to \mathbb{R}$ *be* Δ*-integrable functions such that* f *is of one sign and decreasing and* $0 \le g \le 1$ *for every* $t \in [a,b]_\mathbb{T}$. *Assume that* $\lambda = \int_a^b g(t)\Delta t$ *such that* $b - \lambda$, $a + \lambda \in \mathbb{T}$. *Then*

$$\int_{b-\lambda}^b f(t)\Delta t \le \int_a^b f(t)g(t)\Delta t \le \int_a^{a+\lambda} f(t)\Delta t.$$

Theorem 2.5.4 *Let* $a, b \in \mathbb{T}_k^k$ *with* $a < b$ *and* $f, g : [a,b]_\mathbb{T} \to \mathbb{R}$ *be* ∇*-integrable functions such that* f *is of one sign and decreasing and* $0 \le g \le 1$ *for every* $t \in [a,b]_\mathbb{T}$. *Assume that* $\lambda = \int_a^b g(t)\nabla t$ *such that* $b - \lambda$, $a + \lambda \in \mathbb{T}$. *Then*

$$\int_{b-\lambda}^b f(t)\nabla t \le \int_a^b f(t)g(t)\nabla t \le \int_a^{a+\lambda} f(t)\nabla t.$$

In the following, we prove the diamond-α Steffensen inequality using the diamond-α derivative on time scales. We begin with the following lemma that will be needed later.

Lemma 2.5.1 *Let* $a, b \in \mathbb{T}_k^k$ *with* $a < b$ *and* $f, g, h : [a,b]_\mathbb{T} \to \mathbb{R}$ *be* $\Diamond_\alpha$*-integrable functions. Suppose that also* $l, \gamma \in [a,b]_\mathbb{T}$ *such that*

$$\int_a^\gamma h(t)\Diamond_\alpha t = \int_a^b g(t)\Diamond_\alpha t = \int_l^b h(t)\Diamond_\alpha t. \tag{2.5.4}$$

Then

$$\int_a^b f(t)g(t)\Diamond_\alpha t \quad = \quad \int_\gamma^b [f(t) - f(\gamma)]g(t)\Diamond_\alpha t \tag{2.5.5}$$

$$+ \int_a^\gamma \{f(t)h(t) - [f(t) - f(\gamma)][h(t) - g(t)]\}\,\Diamond_\alpha t,$$

$$\int_a^b f(t)g(t)\Diamond_\alpha t = \int_a^l [f(t) - f(l)]g(t)\Diamond_\alpha t \tag{2.5.6}$$

$$+ \int_l^b \{f(t)h(t) - [f(t) - f(l)][h(t) - g(t)]\} \Diamond_\alpha t.$$

Proof. We prove (2.5.5). By direct computation, we have

$$\int_a^\gamma \{f(t)h(t) - [f(t) - f(\gamma)][h(t) - g(t)]\} \Diamond_\alpha t - \int_a^b f(t)g(t)\Diamond_\alpha t$$

$$= \int_a^\gamma \{f(t)h(t) - f(t)g(t) - [f(t) - f(\gamma)][h(t) - g(t)]\} \Diamond_\alpha t$$

$$+ \int_a^\gamma f(t)g(t)\Diamond_\alpha t - \int_a^b f(t)g(t)\Diamond_\alpha t$$

$$= \int_a^\gamma f(\gamma)[h(t) - g(t)]\Diamond_\alpha t - \int_\gamma^b f(t)g(t)\Diamond_\alpha t$$

$$= f(\gamma) \int_a^\gamma h(t)\Diamond_\alpha t - f(\gamma) \int_a^\gamma g(t)\Diamond_\alpha t - \int_\gamma^b f(t)g(t)\Diamond_\alpha t.$$

Applying the assumption $\int_a^\gamma h(t)\Diamond_\alpha t = \int_a^b g(t)\Diamond_\alpha t$, we see that

$$\int_a^\gamma \{f(t)h(t) - [f(t) - f(\gamma)][h(t) - g(t)]\} \Diamond_\alpha t - \int_a^b f(t)g(t)\Diamond_\alpha t$$

$$= f(\gamma) \int_a^b g(t)\Diamond_\alpha t - f(\gamma) \int_a^\gamma g(t)\Diamond_\alpha t - \int_\gamma^b f(t)g(t)\Diamond_\alpha t$$

$$= f(\gamma) \left(\int_a^b g(t)\Diamond_\alpha t - \int_a^\gamma g(t)\Diamond_\alpha t \right) - \int_\gamma^b f(t)g(t)\Diamond_\alpha t$$

$$= f(\gamma) \int_\gamma^b g(t)\Diamond_\alpha t - \int_\gamma^b f(t)g(t)\Diamond_\alpha t = \int_\gamma^b [f(\gamma) - f(t)]g(t)\Diamond_\alpha t,$$

which is the desired inequality (2.5.5). The proof of (2.5.6) is similar and thus is omitted. The proof is complete. ∎

Theorem 2.5.5 *Let a, $b \in \mathbb{T}_k^k$ with $a < b$ and f, g, $h : [a, b]_\mathbb{T} \to \mathbb{R}$ be $\Diamond_\alpha$-integrable functions such that f is of one sign and decreasing and $0 \leq g(t) \leq h(t)$ for every $t \in [a, b]_\mathbb{T}$. Assume l, $\gamma \in [a, b]_\mathbb{T}$ such that*

$$\begin{cases} \int_l^\gamma h(t)\Diamond_\alpha t \leq \int_a^b g(t)\Diamond_\alpha t \leq \int_a^\gamma h(t)\Diamond_\alpha t, & \text{if } f \geq 0, \ t \in [a, b]_\mathbb{T}, \\ \int_a^\gamma h(t)\Diamond_\alpha t \leq \int_a^b g(t)\Diamond_\alpha t \leq \int_l^\gamma h(t)\Diamond_\alpha t, & \text{if } f \leq 0, \ t \in [a, b]_\mathbb{T}. \end{cases} \tag{2.5.7}$$

Then

$$\int_l^b f(t)h(t)\Diamond_\alpha t \leq \int_a^b f(t)g(t)\Diamond_\alpha t \leq \int_a^\gamma f(t)h(t)\Diamond_\alpha t. \tag{2.5.8}$$

Proof. We prove the left inequality in (2.5.8), in the case $f \geq 0$. The proofs of the other cases are similar. Since f is decreasing and g is nonnegative, we see that

$$\int_a^b f(t)g(t)\Diamond_\alpha t - \int_l^b f(t)h(t)\Diamond_\alpha t$$

$$= \int_a^l f(t)g(t)\Diamond_\alpha t + \int_l^b f(t)g(t)\Diamond_\alpha t - \int_l^b f(t)h(t)\Diamond_\alpha t$$

$$= \int_a^l f(t)g(t)\Diamond_\alpha t - \int_l^b f(t)\left[h(t) - g(t)\right]\Diamond_\alpha t$$

$$\geq \int_a^l f(t)g(t)\Diamond_\alpha t - f(l)\int_l^b \left[h(t) - g(t)\right]\Diamond_\alpha t$$

$$= \int_a^l f(t)g(t)\Diamond_\alpha t - f(l)\int_l^b h(t)\Diamond_\alpha t + f(l)\int_l^b g(t)\Diamond_\alpha t$$

$$\geq \int_a^l f(t)g(t)\Diamond_\alpha t - f(l)\int_a^b g(t)\Diamond_\alpha t + f(l)\int_l^b g(t)\Diamond_\alpha t$$

$$= \int_a^l f(t)g(t)\Diamond_\alpha t - f(l)\left[\int_a^b g(t)\Diamond_\alpha t - \int_l^b g(t)\Diamond_\alpha t\right]$$

$$= \int_a^l f(t)g(t)\Diamond_\alpha t - f(l)\int_a^l g(t)\Diamond_\alpha t$$

$$= \int_a^l \left[f(t) - f(l)\right]g(t)\Diamond_\alpha t \geq 0.$$

∎

As a special case of Theorem 2.5.5 when $\alpha = 1$ and $\alpha = 0$, we have the following results.

Corollary 2.5.1 *Let a, $b \in \mathbb{T}^k$ with $a < b$ and f, g, $h : [a,b]_\mathbb{T} \to \mathbb{R}$ be Δ-integrable functions such that f is of one sign and decreasing and $0 \leq g(t) \leq h(t)$ for every $t \in [a,b]_\mathbb{T}$. Assume $l, \gamma \in [a,b]_\mathbb{T}$ such that*

$$\begin{cases} \int_l^\gamma h(t)\Delta t \leq \int_a^b g(t)\Delta t \leq \int_a^\gamma h(t)\Delta t, & \text{if } f \geq 0, \ t \in [a,b]_\mathbb{T}, \\ \int_a^\gamma h(t)\Delta t \leq \int_a^b g(t)\Delta t \leq \int_l^\gamma h(t)\Delta t, & \text{if } f \leq 0, \ t \in [a,b]_\mathbb{T}. \end{cases} \quad (2.5.9)$$

Then

$$\int_l^b f(t)h(t)\Delta t \leq \int_a^b f(t)g(t)\Delta t \leq \int_a^\gamma f(t)h(t)\Delta t. \quad (2.5.10)$$

Corollary 2.5.2 *Let a, $b \in \mathbb{T}^k$ with $a < b$ and f, g, $h : [a,b]_\mathbb{T} \to \mathbb{R}$ be ∇-integrable functions such that f is of one sign and decreasing and $0 \leq g(t) \leq h(t)$ for every $t \in [a,b]_\mathbb{T}$. Assume $l, \gamma \in [a,b]_\mathbb{T}$ such that*

$$\begin{cases} \int_l^\gamma h(t)\nabla t \le \int_a^b g(t)\nabla t \le \int_a^\gamma h(t)\nabla t, \ if \ f \ge 0, \ t \in [a,b]_\mathbb{T}, \\ \int_a^\gamma h(t)\nabla t \le \int_a^b g(t)\nabla t \le \int_l^\gamma h(t)\nabla t, \ if \ f \le 0, \ t \in [a,b]_\mathbb{T}. \end{cases} \tag{2.5.11}$$

Then

$$\int_l^b f(t)h(t)\nabla t \le \int_a^b f(t)g(t)\nabla t \le \int_a^\gamma f(t)h(t)\nabla t. \tag{2.5.12}$$

Theorem 2.5.6 *Let* $a, b \in \mathbb{T}_k^k$ *with* $a < b$ *and* $f, g, h : [a,b]_\mathbb{T} \to \mathbb{R}$ *be* $\Diamond_\alpha$-*integrable functions such that* f *is of one sign and decreasing and* $0 \le g(t) \le h(t)$ *for every* $t \in [a,b]_\mathbb{T}$. *Assume* $l, \gamma \in [a,b]_\mathbb{T}$ *such that*

$$\int_a^\gamma h(t)\Diamond_\alpha t = \int_a^b g(t)\Diamond_\alpha t = \int_l^b h(t)\Diamond_\alpha t. \tag{2.5.13}$$

Then

$$\begin{aligned} \int_l^b f(t)h(t)\Diamond_\alpha t &\le \int_l^b (f(t)h(t) - [f(t) - f(l)][h(t) - g(t)])\Diamond_\alpha t \\ &\le \int_a^b f(t)g(t)\Diamond_\alpha t \qquad\qquad (2.5.14) \\ &\le \int_a^\gamma (f(t)h(t) - [f(t) - f(\gamma)][h(t) - g(t)])\Diamond_\alpha t \\ &\le \int_a^\gamma f(t)h(t)\Diamond_\alpha t. \end{aligned}$$

Proof. In view of the assumption that the function f is decreasing and that $0 \le g(t) \le h(t)$ on $[a,b]_\mathbb{T}$, we see that

$$\int_a^l [f(t) - f(l)]g(t)\Diamond_\alpha t \ge 0, \quad \int_l^b [f(l) - f(t)][h(t) - g(t)]\Diamond_\alpha t \ge 0. \tag{2.5.15}$$

Using the integral identity (2.5.6) together with the integrals in (2.5.15), we have

$$\begin{aligned} \int_l^b f(t)h(t)\Diamond_\alpha t &\le \int_l^b (f(t)h(t) - [f(t) - f(l)][h(t) - g(t)])\Diamond_\alpha t \qquad (2.5.16) \\ &\le \int_a^b f(t)g(t)\Diamond_\alpha t. \end{aligned}$$

In the same way as above, we obtain that

$$\begin{aligned} \int_a^b f(t)g(t)\Diamond_\alpha t &\le \int_a^\gamma (f(t)h(t) - [f(t) - f(\gamma)][h(t) - g(t)])\Diamond_\alpha t \\ &\qquad\qquad\qquad\qquad\qquad\qquad\qquad\qquad (2.5.17) \\ &\le \int_a^\gamma f(t)h(t)\Diamond_\alpha t. \end{aligned}$$

The proof of (2.5.14) is completed by combining (2.5.16) and (2.5.17). The proof is complete. ∎

As a special case of Theorem 2.5.6, when $\alpha = 1$ and $\alpha = 0$, we have the following results.

Corollary 2.5.3 *Let a, $b \in \mathbb{T}^k$ with $a < b$ and f, g, $h : [a,b]_{\mathbb{T}} \to \mathbb{R}$ be Δ-integrable functions such that f is of one sign and decreasing and $0 \le g(t) \le h(t)$ for every $t \in [a,b]_{\mathbb{T}}$. Assume l, $\gamma \in [a,b]_{\mathbb{T}}$ such that*

$$\int_a^\gamma h(t)\Delta t = \int_a^b g(t)\Delta t = \int_l^b h(t)\Delta t. \qquad (2.5.18)$$

Then

$$\int_l^b f(t)h(t)\Delta t$$
$$\le \int_l^b (f(t)h(t) - [f(t) - f(l)][h(t) - g(t)])\Delta t \le \int_a^b f(t)g(t)\Delta t$$
$$\le \int_a^\gamma (f(t)h(t) - [f(t) - f(\gamma)][h(t) - g(t)])\Delta t \le \int_a^\gamma f(t)h(t)\Delta t.$$

Corollary 2.5.4 *Let a, $b \in \mathbb{T}_k$ with $a < b$ and f, g, $h : [a,b]_{\mathbb{T}} \to \mathbb{R}$ be ∇-integrable functions such that f is of one sign and decreasing and $0 \le g(t) \le h(t)$ for every $t \in [a,b]_{\mathbb{T}}$. Assume l, $\gamma \in [a,b]_{\mathbb{T}}$ such that*

$$\int_a^\gamma h(t)\nabla t = \int_a^b g(t)\nabla t = \int_l^b h(t)\nabla t. \qquad (2.5.19)$$

Then

$$\int_l^b f(t)h(t)\nabla t$$
$$\le \int_l^b (f(t)h(t) - [f(t) - f(l)][h(t) - g(t)])\nabla t \le \int_a^b f(t)g(t)\nabla t$$
$$\le \int_a^\gamma (f(t)h(t) - [f(t) - f(\gamma)][h(t) - g(t)])\nabla t \le \int_a^\gamma f(t)h(t)\nabla t.$$

Theorem 2.5.7 *Let a, $b \in \mathbb{T}_k^k$ with $a < b$ and f, g, h and $\varphi : [a,b]_{\mathbb{T}} \to \mathbb{R}$ be $\Diamond_\alpha$-integrable functions such that f is of one sign and decreasing and $0 \le \varphi(t) \le g(t) \le h(t) - \varphi(t)$ for every $t \in [a,b]_{\mathbb{T}}$. Assume l, $\gamma \in [a,b]_{\mathbb{T}}$ such that*

$$\int_a^\gamma h(t)\Diamond_\alpha t = \int_a^b g(t)\Diamond_\alpha t = \int_l^b h(t)\Diamond_\alpha t. \qquad (2.5.20)$$

Then

$$\int_l^b f(t)h(t)\Diamond_\alpha t + \int_a^b |[f(t) - f(l)]\varphi(t)|\,\Diamond_\alpha t$$
$$\le \int_a^b f(t)g(t)\Diamond_\alpha t \le \int_a^\gamma f(t)h(t) - \int_a^\gamma |[f(t) - f(\gamma)]\varphi(t)|\,\Diamond_\alpha t. \qquad (2.5.21)$$

Proof. From the assumption that the function f is decreasing and that

$$0 \leq \varphi(t) \leq g(t) \leq h(t) - \varphi(t) \text{ on } [a, b]_{\mathbb{T}},$$

it follows that

$$\int_a^\gamma [f(t) - f(\gamma)][h(t) - g(t)] \Diamond_\alpha t + \int_\gamma^b [f(\gamma) - f(t)]g(t) \Diamond_\alpha t$$

$$= \int_a^\gamma |f(t) - f(\gamma)| \, [h(t) - g(t)] \Diamond_\alpha t + \int_\gamma^b |f(\gamma) - f(t)| \, g(t) \Diamond_\alpha t$$

$$\geq \int_a^\gamma |f(t) - f(\gamma)| \, \varphi(t) \Diamond_\alpha t + \int_\gamma^b |f(\gamma) - f(t)| \, \varphi(t) \Diamond_\alpha t$$

$$= \int_a^b |f(t) - f(\gamma)| \, \varphi(t) \Diamond_\alpha t. \tag{2.5.22}$$

Similarly, we find that

$$\int_a^l [f(t) - f(l)]g(t) \Diamond_\alpha t + \int_l^b [f(l) - f(t)][h(t) - g(t)] \Diamond_\alpha t$$

$$\geq \int_a^b |f(t) - f(l)| \, \varphi(t) \Diamond_\alpha t. \tag{2.5.23}$$

By combining the integrals in (2.5.5) and (2.5.6) and the inequalities (2.5.22) and (2.5.23), we have the inequality (2.5.21). The proof is complete. ∎

2.6 Hermite–Hadamard Inequalities

The Hermite–Hadamard inequality was published in [70]. For the convex function $f : [a, b] \to \mathbb{R}$, the integral of f can be estimated by the inequality

$$f\left(\frac{a+b}{2}\right) \leq \frac{1}{b-a} \int_a^b f(x) dx \leq \frac{f(a) + f(b)}{2}.$$

We note that the left-hand side of the Hermite–Hadamard inequality is a special case of the Jensen inequality.

The results in this section are adapted from [26, 63, 64]. First, we begin with an inequality containing the delta derivative on time scales.

Theorem 2.6.1 Let $f : [a, b]_{\mathbb{T}} \to \mathbb{R}$ be delta differentiable function such that $m \leq f^\Delta(t) \leq M$ for every $t \in [a, b]_{\mathbb{T}}$ for some numbers $m < M$. If there exist $l, \gamma \in [a, b]_{\mathbb{T}}$ such that

$$\gamma - a \leq \frac{[f(b) - f(a) - m(b - a)]}{M - m} \leq b - l,$$

then

$$mh_2(a,b) + (M-m)h_2(a,\gamma) \quad \leq \quad (b-a)f(b) - \int_a^b f(t)\Delta t$$

$$\leq \quad Mh_2(a,b) + (m-M)h_2(a,l), \quad (2.6.1)$$

where $h(t,s)$ is defined as in (1.4.5).

Proof. Let

$$k(t) := \frac{[f(t) - m(t-b)]}{M-m}, \quad F(t) := h_1(a,\sigma(t)),$$

and

$$G(t) := k^\Delta(t) = \frac{[f^\Delta(t) - m]}{M-m} \in [0,1].$$

Clearly F is decreasing and nonpositive, and

$$\int_a^b G(t)\Delta t = \frac{[f(b) - f(a) - m(b-a)]}{M-m} \in [\gamma - a, b - l].$$

Note

$$\int_l^b F(t)\Delta t = \int_l^b h_1(a,\sigma(t))\Delta t = -\, h_2(a,t)|_l^b = -h_2(a,b) + h_2(a,l),$$

and

$$\int_a^\gamma F(t)\Delta t = -\, h_2(a,t)|_a^\gamma = -h_2(a,\gamma).$$

Moreover, using the formula for integration by parts for delta integrals, we see that

$$\int_a^b F(t)G(t)\Delta t \quad = \quad \int_a^b F(t)k^\Delta(t)\Delta t = h_1(a,t)k(t)|_a^b - \int_a^b h_1^\Delta(a,t)k(t)\Delta t$$

$$= \quad \frac{1}{M-m}\left[-(b-a)f(b) + \int_a^b f(t)\Delta t + mh_2(a,b)\right].$$

Using Steffensen's inequality for delta integrals, we obtain that

$$-h_2(a,b) + h_2(a,l) \quad \leq \quad \frac{1}{M-m}\left[-(b-a)f(b) + \int_a^b f(t)\Delta t + mh_2(a,b)\right]$$

$$\leq \quad -h_2(a,\gamma),$$

which yields the desired inequality (2.6.1). The proof is complete. ∎

Suppose that f is $(n+1)$ times nabla differentiable on $\mathbb{T}_{\kappa^{n+1}}$. Using Taylor's Theorem 1.4.4, we define the remainder function by

$$\check{R}_{-1,f}(.,s) = f(s),$$

and for $n > -1$,

$$\check{R}_{n,f}(t,s) = f(s) - \sum_{k=0}^{n} \hat{h}_k(t,s) f^{\nabla^k}(s) = \int_t^s \hat{h}_n(s,\rho(\xi)) f^{\nabla^{n+1}}(\xi) \nabla \xi.$$

The proof of the next result is by induction (and we omit the proof).

Lemma 2.6.1 *Suppose f is $(n+1)$ times nabla differentiable on $\mathbb{T}_{\kappa^{n+1}}$. Then*

$$\int_a^b \hat{h}_{n+1}(t,\rho(s)) f^{\nabla^{n+1}}(s) \nabla s = \int_a^t \check{R}_{n,f}(a,s) \nabla s + \int_t^b \check{R}_{n,f}(b,s) \nabla s.$$

Corollary 2.6.1 *Suppose f is $(n+1)$ times nabla differentiable on $\mathbb{T}_{\kappa^{n+1}}$. Then*

$$\int_a^b \hat{h}_{n+1}(a,\rho(s)) f^{\nabla^{n+1}}(s) \nabla s = \int_a^b \check{R}_{n,f}(b,s) \nabla s,$$

$$\int_a^b \hat{h}_{n+1}(b,\rho(s)) f^{\nabla^{n+1}}(s) \nabla s = \int_a^b \check{R}_{n,f}(a,s) \nabla s.$$

Our next result follows by induction (we leave the details to the reader).

Lemma 2.6.2 *Suppose f is $(n+1)$ times delta differentiable on $\mathbb{T}^{\kappa^{n+1}}$. Then*

$$\int_a^b h_{n+1}(t,\sigma(s)) f^{\Delta^{n+1}}(s) \Delta s = \int_a^t R_{n,f}(a,s) \Delta s + \int_t^b R_{n,f}(b,s) \Delta s,$$

where

$$R_{n,f}(t,s) = f(s) - \sum_{j=0}^{n} h_j(s,t) f^{\Delta^j}(t).$$

Theorem 2.6.2 *Let f be an $(n+1)$ times nabla differentiable function such that $f^{\nabla^{n+1}}(s)$ is increasing and f^{∇^n} is monotonic (either increasing or decreasing) on $[a,b]_\mathbb{T}$. Assume $l, \gamma \in [a,b]_\mathbb{T}$ such that*

$$b - l \;\leq\; \frac{\hat{h}_{n+2}(b,a))}{\hat{h}_{n+1}(b,\rho(a))} \leq \gamma - a, \text{ if } f^{\nabla^n} \text{ is decreasing,}$$

$$\gamma - a \;\leq\; \frac{\hat{h}_{n+2}(b,a))}{\hat{h}_{n+1}(b,\rho(a))} \leq b - l, \text{ if } f^{\nabla^n} \text{ is increasing.}$$

Then

$$f^{\nabla^n}(\gamma) - f^{\nabla^n}(a) \leq \frac{\int_a^b \check{R}_{n,f}(a,s) \nabla s}{\hat{h}_{n+1}(b,\rho(a))} \leq f^{\nabla^n}(b) - f^{\nabla^n}(l). \tag{2.6.2}$$

Proof. Assume that f^{∇^n} is decreasing (the case where f^{∇^n} is increasing is similar and is omitted). Let $F = -f^{\nabla^{n+1}}$. Now, since f^{∇^n} is decreasing, we have $F \geq 0$ and decreasing on $[a,b]_\mathbb{T}$. Define

$$g(t) = \frac{\hat{h}_{n+1}(b,\rho(t))}{\hat{h}_{n+1}(b,\rho(a))} \in [0,1], \quad \text{for } t \in [a,b]_\mathbb{T} \text{ and } n \geq -1.$$

We will apply Steffensen's inequality (see Theorem 2.5.2). Using the fact that

$$\hat{h}^{\nabla}_{k+1}(t,s) = -\hat{h}_k(t,\rho(s)), \tag{2.6.3}$$

we see that

$$\int_a^b g(t)\nabla t = \frac{1}{\hat{h}_{n+1}(b,\rho(a))} \int_a^b \hat{h}_{n+1}(b,\rho(t))\nabla t = \frac{\hat{h}_{n+2}(b,a))}{\hat{h}_{n+1}(b,\rho(a))}.$$

That is

$$b - l \leq \frac{\hat{h}_{n+2}(b,a))}{\hat{h}_{n+1}(b,\rho(a))} \leq \gamma - a,$$

then

$$\int_l^b F(t)\nabla t \leq \int_a^b g(t)F(t)\nabla t \leq \int_a^\gamma F(t)\nabla t.$$

By Corollary 2.6.1 this simplifies to

$$f^{\nabla^n}(t)\Big|^\gamma_{t=a} \leq \frac{1}{\hat{h}_{n+1}(b,\rho(a))} \int_a^b \check{R}_{n,f}(a,s)\nabla s \leq f^{\nabla^n}(t)\Big|^\gamma_{t=l},$$

which gives the desired inequality (2.6.2). The proof is complete. ∎

It is evident that an analogous result can be found for the delta integral case using the delta results in Corollary 2.5.1 by putting $h(t) = 1$. As usual a twice nabla differentiable function $f : [a,b]_\mathbb{T} \to \mathbb{R}$ is convex on $[a,b]_\mathbb{T}$ if and only if $f^{\nabla^2} \geq 0$ on $[a,b]_\mathbb{T}$.

Corollary 2.6.2 Let $f : [a,b]_\mathbb{T} \to \mathbb{R}$ be convex and monotonic. Assume l, $\gamma \in [a,b]_\mathbb{T}$ such that

$$l \geq b - \frac{\hat{h}_2(b,a)}{b - \rho(a)}, \quad \gamma \geq \frac{\hat{h}_2(b,a)}{b - \rho(a)} + a, \text{ if } f \text{ is decreasing,}$$

$$l \leq b - \frac{\hat{h}_2(b,a)}{b - \rho(a)}, \quad \gamma \leq \frac{\hat{h}_2(b,a)}{b - \rho(a)} + a, \text{ if } f \text{ is increasing.}$$

Then

$$f(\gamma) + \frac{\rho(a) - a}{b - \rho(a)} f(a) \leq \frac{1}{b - \rho(a)} \int_a^b f(t)\nabla t \leq f(b) + \frac{b - a}{b - \rho(a)} f(a) - f(l).$$

Another slightly different form of the Hermite–Hadamard inequality is the following inequality which is given by applying the Steffensen inequality proved in Theorem 2.5.2.

Theorem 2.6.3 *Let* $f : [a,b]_\mathbb{T} \to \mathbb{R}$ *be convex and monotonic. Assume* l, $\gamma \in [a,b]_\mathbb{T}$ *such that*

$$l \geq a + \frac{\hat{h}_2(b,a)}{b-a}, \quad \gamma \geq b - \frac{\hat{h}_2(b,a)}{b-a}, \text{ if } f \text{ is decreasing,}$$

$$l \leq a + \frac{\hat{h}_2(b,a)}{b-a}, \quad \gamma \leq b - \frac{\hat{h}_2(b,a)}{b-a}, \text{ if } f \text{ is increasing.}$$

Then

$$f(\gamma) \leq \frac{1}{b-a} \int_a^b f^\rho(t) \nabla t \leq f(b) + f(a) - f(l). \tag{2.6.4}$$

Proof. Assume that f is decreasing and convex. Then $f^{\nabla^2} \geq 0$ and $f^\nabla \leq 0$. Then $F = -f^\nabla$ is decreasing and satisfies $F \geq 0$. For $G(t) = \frac{b-t}{b-a}$, we see for every $t \in [a,b]$ that $0 \leq G(t) \leq 1$ and F and G satisfy the hypotheses in Theorem 2.5.2. Now, the inequality

$$b - l \leq \int_a^b G(t) \nabla t \leq \gamma - a,$$

can be rewritten in the form

$$b - l \leq \frac{1}{b-a} \int_a^b (b-t) \nabla t \leq \gamma - a.$$

We consider the left hand inequality which takes the form

$$l \geq b - \frac{1}{b-a} \int_a^b (b-t) \nabla t = b - \frac{1}{b-a} \int_a^b (b - a + t - a) \nabla t,$$

which simplifies to

$$l \geq a + \frac{\hat{h}_2(b,a)}{b-a}.$$

Similarly

$$\gamma \geq b - \frac{\hat{h}_2(b,a)}{b-a}.$$

Furthermore, note that $\int_r^s F(t) \nabla t = f(r) - f(s)$, and integrating by parts yields that

$$\int_a^b F(t) G(t) \nabla t = \int_a^b \frac{(t-b)}{b-a} f^\nabla(t) \nabla t = f(a) - \frac{1}{b-a} \int_a^b f^\rho(t) \nabla t.$$

It follows that Steffensen's inequality takes the form

$$f(l) - f(b) \leq f(a) - \frac{1}{b-a} \int_a^b f^\rho(t) \nabla t \leq f(a) - f(\gamma),$$

which can be arranged to match the desired inequality (2.6.4). The case where f is increasing is similar and is omitted. The proof is complete. ∎

Theorem 2.6.4 *Let* $f : [a,b]_\mathbb{T} \to \mathbb{R}$ *be an* $n+1$ *times nabla differentiable function such that* $m \leq f^{\nabla^{n+1}}(t) \leq M$ *for every* $t \in [a,b]_\mathbb{T}$ *for some numbers* $m < M$. *If there exist* $l, \gamma \in [a,b]_\mathbb{T}$ *such that*

$$b - l \leq \frac{\left[f^{\nabla^n}(b) - f^{\nabla^n}(a) - m(b-a) \right]}{M - m} \leq \gamma - a,$$

then

$$m\hat{h}_{n+2}(b,a) + (M-m)\hat{h}_{n+2}(b,l) \leq \int_a^b \check{R}_{n,f}(a,t)\nabla t$$

$$\leq \quad M\hat{h}_{n+2}(b,a) + (m-M)\hat{h}_{n+2}(b,\gamma). \tag{2.6.5}$$

where $\hat{h}_n(t,s)$ *is defined as in (1.4.7).*

Proof. Let

$$k(t) = \frac{1}{M-m}\left[f(t) - m\hat{h}_{n+1}(t,a) \right], \quad F(t) = \hat{h}_{n+1}(b,\rho(t)),$$

and

$$G(t) = k^{\nabla^{n+1}}(t) = \frac{1}{M-m}\left[f^{\nabla^{n+1}}(t) - m \right] \in [0,1].$$

Observe that F is nonnegative and decreasing, and

$$\int_a^b G(t)\nabla t = \frac{1}{M-m}\left[f^{\nabla^n}(b) - f^{\nabla^n}(a) - m(b-a) \right].$$

Now by (2.6.3), we get that

$$\int_l^b F(t)\nabla t = \int_l^b \hat{h}_{n+1}(b,\rho(t))\nabla t = \hat{h}_{n+2}(b,l),$$

and

$$\int_a^\gamma F(t)\nabla t = \hat{h}_{n+2}(b,a) - \hat{h}_{n+2}(b,\gamma).$$

Moreover, using Corollary 2.6.1, we have

$$
\begin{aligned}
\int_a^b G(t)F(t)\nabla t &\leq \frac{1}{M-m}\int_a^b \hat{h}_{n+1}(b,\rho(t))\left(f^{\nabla^n}(t)-m\right)\nabla t \\
&= \frac{1}{M-m}\int_a^b \check{R}_{n,f}(a,t)\nabla t + \frac{m}{M-m}\left.\hat{h}_{n+2}(b,t)\right|_a^b \\
&= \frac{1}{M-m}\int_a^b \check{R}_{n,f}(a,t)\nabla t - \frac{m}{M-m}\hat{h}_{n+2}(b,a).
\end{aligned}
$$

Using Steffensen's inequality (2.5.3), we have

$$
\begin{aligned}
\hat{h}_{n+2}(b,l) &\leq \frac{1}{M-m}\left[\int_a^b \check{R}_{n,f}(a,t)\nabla t - m\hat{h}_{n+2}(b,a)\right] \\
&\leq \hat{h}_{n+2}(b,a) - \hat{h}_{n+2}(b,\gamma),
\end{aligned}
$$

which yields the desired inequality (2.6.5). The proof is complete. ∎

The following inequality is an inequality of Hermite–Hadamard type for nabla derivative and is derived from Theorem 2.6.4 with $n = 0$.

Theorem 2.6.5 *Let $f : [a,b]_\mathbb{T} \to \mathbb{R}$ be nabla differentiable function such that $m \leq f^\nabla \leq M$ for every $t \in [a,b]_\mathbb{T}$ for some numbers $m < M$. If there exist $l, \gamma \in [a,b]_\mathbb{T}$ such that*

$$
b - l \leq \frac{[f(b) - f(a) - m(b-a)]}{M-m} \leq \gamma - a,
$$

then

$$
\begin{aligned}
m\hat{h}_2(b,a) + (M-m)\hat{h}_2(b,l) &\leq \int_a^b f(t)\nabla t - (b-a)f(a) \\
&\leq M\hat{h}_2(b,a) + (m-M)\hat{h}_2(b,\gamma),
\end{aligned}
$$

where $\hat{h}_n(t,s)$ is defined as in (1.4.7).

Next we present some inequalities of Hermite–Hadamard type for diamond-α derivative on time scales. We start with a few technical lemmas. The first lemma gives the relation between the integrals of delta, nabla, and classical integrals on $\mathbb{R}$ and we present it without proof.

Lemma 2.6.3 *Let $f : \mathbb{T} \to \mathbb{R}$ be a continuous function and $a, b \in \mathbb{T}$.*

(i) If f is nondecreasing on $\mathbb{T}$, then

$$
(b-a)f(a) \leq \int_a^b f(t)\Delta t \leq \int_a^b \tilde{f}(t)dt \leq \int_a^b f(t)\nabla t \leq (b-a)f(b),
$$

where $\tilde{f} : \mathbb{R} \to \mathbb{R}$ is a continuous nondecreasing function such that $f(t) = \tilde{f}(t)$ for all $t \in \mathbb{T}$.

(ii) If f is nonincreasing on $\mathbb{T}$, *then*

$$(b-a)f(a) \geq \int_a^b f(t)\Delta t \geq \int_a^b \tilde{f}(t)dt \geq \int_a^b f(t)\nabla t \geq (b-a)f(b),$$

where $\tilde{f} : \mathbb{R} \to \mathbb{R}$ *is a continuous nonincreasing function such that* $f(t) = \tilde{f}(t)$ *for all* $t \in \mathbb{T}$.

In both cases, there exists an

$$\alpha_T = \frac{\int_a^b \tilde{f}(t)dt - \int_a^b f(t)\nabla t}{\int_a^b f(t)\Delta t - \int_a^b f(t)\nabla t} \in [0,1],$$

such that

$$\int_a^b f(t)\Diamond_{\alpha_T}t = \int_a^b \tilde{f}(t)dt.$$

Remark 2.6.1   *(i). If f is nondecreasing on* $\mathbb{T}$, *then for* $\alpha \leq \alpha_T$, *we have*

$$\int_a^b f(t)\Diamond_\alpha t \geq \int_a^b \tilde{f}(t)dt,$$

while if $\alpha \geq \alpha_T$, *we have*

$$\int_a^b f(t)\Diamond_\alpha t \leq \int_a^b \tilde{f}(t)dt.$$

(ii). If f is nonincreasing on $\mathbb{T}$, *then for* $\alpha \leq \alpha_T$, *we have*

$$\int_a^b f(t)\Diamond_\alpha t \leq \int_a^b \tilde{f}(t)dt,$$

while if $\alpha \geq \alpha_T$, *we have*

$$\int_a^b f(t)\Diamond_\alpha t \geq \int_a^b \tilde{f}(t)dt.$$

(iii) If $\mathbb{T} = [a,b]$ *or f is a constant, then* α_T *can be any real number from* $[0,1]$. *Otherwise* $\alpha_T \in (0,1)$.

Next we present a lemma which gives a relation between the existence of the delta integral of a linear function and its corresponding nabla integral.

Lemma 2.6.4 *Let* $f : \mathbb{T} \to \mathbb{R}$ *be linear function and let* $\tilde{f} : [a,b] \to \mathbb{R}$ *be the corresponding linear function. If* $\int_a^b f(t)\Delta t = \int_a^b \tilde{f}(t)dt - C$, *with* $C \in \mathbb{R}$, *then* $\int_a^b f(t)\nabla t = \int_a^b \tilde{f}(t)dt + C$.

Let

$$x_\alpha = \frac{1}{b-a} \int_a^b t \lozenge_\alpha t,$$

and call it the α-center of the time scale interval $[a,b]_{\mathbb{T}}$. Now, we are in a position to state and prove diamond-α Hermite–Hadamard type inequalities on time scales.

Theorem 2.6.6 *Let $\mathbb{T}$ be a time scale and $a, b \in \mathbb{T}$. Let $f : [a,b]_{\mathbb{T}} \to \mathbb{R}$ be a continuous convex function. Then*

$$f(x_\alpha) \le \frac{1}{b-a} \int_a^b f(t) \lozenge_\alpha t \le \frac{b - x_\alpha}{b - a} f(a) + \frac{x_\alpha - a}{b - a} f(b). \qquad (2.6.6)$$

Proof. For every convex function, we have

$$f(t) \le f(a) + \frac{f(b) - f(a)}{b - a}(t - a). \qquad (2.6.7)$$

By taking the diamond-α integral we get

$$\int_a^b f(t) \lozenge_\alpha t \;\le\; \int_a^b f(a) \lozenge_\alpha t + \int_a^b \frac{f(b) - f(a)}{b - a}(t - a) \lozenge_\alpha t$$

$$= (b-a)f(a) + \frac{f(b) - f(a)}{b - a} \left(\int_a^b t \lozenge_\alpha t - a(b - a) \right),$$

that is

$$\frac{1}{b-a} \int_a^b f(t) \lozenge_\alpha t \le \frac{b - x_\alpha}{b - a} f(a) + \frac{x_\alpha - a}{b - a} f(b),$$

which is the right-hand side of (2.6.6). For the left-hand side, we use Theorem 2.2.5, by taking $g(s) = s$ and $F = f$ to get that

$$f \left(\frac{\int_a^b s \lozenge_\alpha s}{b - a} \right) \le \frac{\int_a^b f(s) \lozenge_\alpha s}{b - a}.$$

Hence, we have

$$f(x_\alpha) \le \frac{1}{b-a} \int_a^b f(s) \lozenge_\alpha s,$$

which is the right-hand side of (2.6.6). The proof is complete. ∎

Remark 2.6.2 *The right-hand side of the Hermite–Hadamard inequality (2.6.6) remains true for all $0 \le \alpha \le \lambda$, including the nabla integral, if $f(b) \le f(a)$ and for all $\lambda \le \alpha \le 1$, including the delta derivative, if $f(b) \ge f(a)$, where x_λ is the λ-center of the time scale interval $[a,b]_{\mathbb{T}}$.*

Let us suppose that $f(b) \geq f(a)$. Then by taking the diamond-α integral of the inequality (2.6.7), we get that

$$
\begin{aligned}
\int_a^b f(t)\Diamond_\alpha t &\leq (b-a)f(a) + \frac{f(b)-f(a)}{b-a}\left(\int_a^b t\Diamond_\alpha t - a(b-a)\right) \\
&\leq (b-a)f(a) + (f(b)-f(a))(x_\lambda - a) \\
&\leq (b-x_\lambda)f(a) + f(b)(x_\lambda - a).
\end{aligned}
$$

According to Lemma 2.6.3, the last inequality is true for $\int_a^b t\Diamond_\alpha t \leq \int_a^b t\Diamond_\lambda t$, that is for $\alpha \geq \lambda$. The same arguments work for $\lambda \geq \alpha$.

Remark 2.6.3 *The left-hand side of the Hermite–Hadamard inequality (2.6.6) remains true for all $0 \leq \alpha \leq \lambda$, including the nabla integral, if f is nonincreasing for all $\lambda \leq \alpha \leq 1$, including the delta derivative, if f is nondecreasing*

Let us suppose that f is nonincreasing. Then using Theorem 2.2.5, by taking $g(s) = s$ and $F = f$, we have

$$
f\left(\frac{\int_a^b s\Diamond_\alpha s}{b-a}\right) \leq \frac{\int_a^b f(s)\Diamond_\alpha s}{b-a}.
$$

For $\alpha \geq \lambda$, we have $\int_a^b t\Diamond_\alpha t \leq \int_a^b t\Diamond_\lambda t$ and so

$$
f\left(\frac{\int_a^b s\Diamond_\lambda s}{b-a}\right) \leq f\left(\frac{\int_a^b s\Diamond_\alpha s}{b-a}\right) \leq \frac{\int_a^b f(s)\Diamond_\alpha s}{b-a},
$$

that is

$$
f(x_\lambda) \leq \frac{1}{b-a}\int_a^b f(s)\Diamond_\alpha s.
$$

The same arguments are used to prove the case when f is nondecreasing.

Theorem 2.6.7 *Let $\mathbb{T}$ be a time scale, α, $\lambda \in [0,1]$ and a, $b \in \mathbb{T}$. Let $f : [a,b]_\mathbb{T} \to \mathbb{R}$ be a continuous convex function. Then*

(i). if f is nondecreasing on $[a,b]_\mathbb{T}$, then for all $\alpha \in [0,\lambda]$ one has

$$
f(x_\lambda) \leq \frac{1}{b-a}\int_a^b f(t)\Diamond_\alpha t, \tag{2.6.8}
$$

and for all $\alpha \in [\lambda, 1]$, one has

$$
\frac{1}{b-a}\int_a^b f(t)\Diamond_\alpha t \leq \frac{b-x_\lambda}{b-a}f(a) + \frac{x_\lambda - a}{b-a}f(b). \tag{2.6.9}
$$

(ii). *if* f *is nonincreasing on* $[a, b]_\mathbb{T}$, *then for all* $\alpha \in [0, \lambda]$ *one has the inequality (2.6.9), and for all* $\alpha \in [\lambda, 1]$, *one has the inequality (2.6.8).*

Now we prove an inequality of Hermite–Hadamard type with a weight function.

Theorem 2.6.8 *Let* $\mathbb{T}$ *be a time scale and* $a, b \in \mathbb{T}$. *Let* $f : [a, b]_\mathbb{T} \to \mathbb{R}$ *be a continuous convex function and let* $w : [a, b]_\mathbb{T} \to \mathbb{R}$ *be a continuous function such that* $w(t) \geq t$ *for all* $t \in \mathbb{T}$ *and* $\int_a^b w(t) \Diamond_\alpha t > 0$. *Then*

$$f(x_{w,\alpha}) \leq \frac{1}{\int_a^b w(t) \Diamond_\alpha t} \int_a^b f(t) w(t) \Diamond_\alpha t$$

$$\leq \frac{b - x_{w,\alpha}}{b - a} f(a) + \frac{x_{w,\alpha} - a}{b - a} f(b), \qquad (2.6.10)$$

where $x_{w,\alpha} = \int_a^b t w(t) \Diamond_\alpha t / \int_a^b w(t) \Diamond_\alpha t$.

Proof. For the convex function $f(t)$, we have

$$f(t) \leq f(a) + \frac{f(b) - f(a)}{b - a} (t - a).$$

Multiplying this inequality by $w(t)$ which is nonnegative, we get after integration that

$$\int_a^b w(t) f(t) \Diamond_\alpha t \leq f(a) \int_a^b w(t) \Diamond_\alpha t$$

$$+ \frac{f(b) - f(a)}{b - a} \left[\int_a^b t w(t) \Diamond_\alpha t - a \int_a^b w(t) \Diamond_\alpha t \right],$$

that is

$$\frac{1}{\int_a^b w(t) \Diamond_\alpha t} \int_a^b f(t) \Diamond_\alpha t \leq \frac{b - x_{w,\alpha}}{b - a} f(a) + \frac{x_{w,\alpha} - a}{b - a} f(b),$$

which is the right-hand side of (2.6.10). For the left-hand side, we use Theorem 2.2.6, by taking $g(s) = s$ and $h(t) = w(t)$ and $F = f$ to get that

$$f \left(\frac{\int_a^b w(s) s \Diamond_\alpha s}{\int_a^b w(s) \Diamond_\alpha s} \right) \leq \frac{\int_a^b f(s) w(s) \Diamond_\alpha s}{\int_a^b w(s) s \Diamond_\alpha s}.$$

Hence, we have

$$f(x_{w,\alpha}) \leq \frac{1}{\int_a^b w(t) \Diamond_\alpha t} \int_a^b w(t) f(t) \Diamond_\alpha t,$$

which is the left-hand side of (2.6.10). The proof is complete. ∎

Remark 2.6.4 *If we consider concave functions instead of the convex functions, the inequalities (2.6.6), (2.6.8)–(2.6.10) are reversed.*

2.7   Čebyšev Inequalities

The Čebyšev inequality (see [110]) is given by

$$\int_a^b p(x)dx \int_a^b p(x)f(x)g(x)dx \geq \int_a^b p(x)f(x)dx \int_a^b p(x)g(x)dx, \quad (2.7.1)$$

where f, g : $[a,b] \to \mathbb{R}$ are integrable functions both increasing or both decreasing and $p : [a,b] \to \mathbb{R}^+$ is an integrable function. If one of the functions f or g is nonincreasing and the other nondecreasing then the inequality in (2.7.1) is reversed. The special case of (2.7.1), when $p = 1$ is given by

$$\int_a^b f(x)g(x)dx \geq \frac{1}{b-a} \int_a^b f(x)dx \int_a^b g(x)dx. \quad (2.7.2)$$

For each of the above inequalities there exists a corresponding discrete analogue. The discrete version of (2.7.1) is given by

$$\sum_{i=1}^n p(i) \sum_{i=1}^n p(i)a(i)b(i) \geq \sum_{i=1}^n p(i)a(i) \sum_{i=1}^n p(i)g(i), \quad (2.7.3)$$

where $a = (a(1),\, a(2),\ldots,a(n))$, $b = (b(1),\, b(2),\ldots,b(n))$ are two nondecreasing (or nonincreasing) sequences and $p = (p(1),\, p(2),\ldots,p(n))$ is a nonnegative sequence with equality if and only if at least one of the sequences a or b is constant. The discrete version of (2.7.2) is given by

$$\sum_{i=1}^n p(i)a(i)b(i) \geq \frac{1}{n} \sum_{i=1}^n a(i) \sum_{i=1}^n g(i), \quad (2.7.4)$$

and is also called the discrete Čebyšev's inequality.

In this section we obtain Čebyšev's type inequalities on time scales which as special cases contain the above continuous and discrete inequalities. The results are adapted from [26, 156].

Theorem 2.7.1 *Suppose that $p \in C_{rd}([a,b]_\mathbb{T}, [0,\infty))$. Let f_1, f_2, k_1, $k_2 \in C_{rd}([a,b]_\mathbb{T}, \mathbb{R})$ satisfy the following two conditions:*

(C_1). $f_2(x)k_2(x) > 0$ *on* $[a,b]_\mathbb{T}$,

(C_2). $\frac{f_1(x)}{f_2(x)}$ *and* $\frac{k_1(x)}{k_2(x)}$ *are similarly ordered (or oppositely ordered), that is, for all x, $y \in [a,b]_\mathbb{T}$*

$$\left(\frac{f_1(x)}{f_2(x)} - \frac{f_1(y)}{f_2(y)} \right) \left(\frac{k_1(x)}{k_2(x)} - \frac{k_1(y)}{k_2(y)} \right) \geq 0 \ (or \ \leq 0).$$

Then

$$\frac{1}{2} \int_a^b \int_a^b p(x)p(y) \begin{vmatrix} f_1(x) & f_1(y) \\ f_2(x) & f_2(y) \end{vmatrix} \begin{vmatrix} k_1(x) & k_1(y) \\ k_2(x) & k_2(y) \end{vmatrix} \Delta x \Delta y$$

$$= \begin{vmatrix} \int_a^b p(x)f_1(x)k_1(x)\Delta x & \int_a^b p(x)f_1(x)k_2(x)\Delta x \\ \int_a^b p(x)f_2(x)k_1(x)\Delta x & \int_a^b p(x)f_2(x)k_2(x)\Delta x \end{vmatrix} \geq 0 \ (\leq 0).$$

$$(2.7.5)$$

Proof. Let x, $y \in [a, b]_\mathbb{T}$. Then it follows from (C_1), (C_2) and the identity

$$p(x)p(y) \begin{vmatrix} f_1(x) & f_1(y) \\ f_2(x) & f_2(y) \end{vmatrix} \begin{vmatrix} k_1(x) & k_1(y) \\ k_2(x) & k_2(y) \end{vmatrix}$$

$$= p(x)p(y)f_2(x)f_2(y)k_2(x)k_2(y) \left(\frac{f_1(x)}{f_2(x)} - \frac{f_1(y)}{f_2(y)} \right) \left(\frac{k_1(x)}{k_2(x)} - \frac{k_1(y)}{k_2(y)} \right),$$

that (2.7.5) holds. The proof is complete. ∎

Putting $f_1(x) = f(x)$, $k_1(x) = g(x)$ and $f_2(x) = k_2(x) = 1$ in Theorem 2.7.1, we have the following delta Čebyšev's type inequality on time scales.

Corollary 2.7.1 *Suppose that p, f, $g \in C_{rd}([a, b]_\mathbb{T}, \mathbb{R})$ with $p(x) > 0$ on $[a, b]_\mathbb{T}$. Let $f(x)$ and $g(x)$ be similarly ordered (or oppositely ordered). Then*

$$\int_a^b p(x)\Delta x \int_a^b p(x)f(x)g(x)\Delta x \geq (\leq) \int_a^b p(x)f(x)\Delta x \int_a^b p(x)g(x)\Delta x.$$

$$(2.7.6)$$

Remark 2.7.1 *Let p, $\gamma \in C_{rd}([a, b]_\mathbb{T}, [0, \infty))$. If $f(x)$ and $g(x)$ are similarly ordered (or oppositely ordered), then it follows from (2.7.6) that*

$$\int_a^b p(x)\Delta x \int_a^b p(x)f(\gamma(x))g(\gamma(x))\Delta x$$

$$\geq (\leq) \int_a^b p(x)f(\gamma(x))\Delta x \int_a^b p(x)g(\gamma(x))\Delta x.$$

Remark 2.7.2 *Let p, $f_i \in C_{rd}([a, b]_\mathbb{T}, \mathbb{R})$ for $i = 1, 2, \ldots, n$ with $p(x) > 0$ on $[a, b]_\mathbb{T}$. Suppose that $f_1(x)$, $f_2(x), \ldots, f_n(x)$ are similarly ordered. Then we have from (2.7.6) that*

$$\left(\int_a^b p(x)\Delta x\right)^{n-1} \int_a^b p(x)(f_1(x)f_2(x)\ldots .f_n(x))\Delta x$$

$$= \left(\int_a^b p(x)\Delta x\right)^{n-2} \left(\int_a^b p(x)\Delta x\right) \left(\int_a^b p(x)(f_1(x)f_2(x)\ldots .f_n(x))\Delta x\right)$$

$$\geq \left(\int_a^b p(x)\Delta x\right)^{n-2} \left(\int_a^b p(x)f_1(x)\Delta x\right) \left(\int_a^b p(x)(f_2(x)\ldots .f_n(x))\Delta x\right)$$

$$\geq \left(\int_a^b p(x)\Delta x\right)^{n-3} \left(\int_a^b p(x)f_1(x)\Delta x\right) \left(\int_a^b p(x)f_2(x)\Delta x\right)$$

$$\times \left(\int_a^b p(x)(f_3(x)\ldots .f_n(x))\Delta x\right)$$

$$\geq \ldots \geq \left(\int_a^b p(x)f_1(x)\Delta x\right) \left(\int_a^b p(x)f_2(x)\Delta x\right) \ldots \left(\int_a^b p(x)f_n(x)\Delta x\right).$$

This gives us that

$$\left(\int_a^b p(x)\Delta x\right)^{n-1} \int_a^b p(x)(f_1(x)f_2(x)\ldots .f_n(x))\Delta x \geq \left(\int_a^b p(x)f_1(x)\Delta x\right)$$

$$\times \left(\int_a^b p(x)f_2(x)\Delta x\right) \ldots \left(\int_a^b p(x)f_n(x)\Delta x\right). \tag{2.7.7}$$

In particular, if $f_1 = f_2 = \ldots = f_n$, then

$$\left(\int_a^b p(x)\Delta x\right)^{n-1} \int_a^b p(x)(f_n(x))^n \Delta x \geq \left(\int_a^b p(x)f(x)\Delta x\right)^n.$$

Putting $f(x) = \frac{f_1(x)}{f_2(x)}$, $g(x) = \frac{g_1(x)}{g_2(x)}$ and $p(x) = f_2(x)g_2(x)$ in (2.7.6), we have the following delta Čebyšev's type inequality on time scales.

Corollary 2.7.2 *Suppose that f_1, f_2, g_1, $g_2 \in C_{rd}([a,b]_\mathbb{T}, \mathbb{R})$ with $f_2(x)g_2(x) > 0$ on $[a,b]_\mathbb{T}$. If $\frac{f_1(x)}{f_2(x)}$ and $\frac{g_1(x)}{g_2(x)}$ are both increasing or both decreasing, then*

$$\int_a^b f_1(x)g_1(x)\Delta x \int_a^b f_2(x)g_2(x)\Delta x \geq \int_a^b f_1(x)g_2(x)\Delta x \int_a^b f_2(x)g_1(x)\Delta x. \tag{2.7.8}$$

If one of $\frac{f_1(x)}{f_2(x)}$ or $\frac{g_1(x)}{g_2(x)}$ is nonincreasing and the other nondecreasing then the inequality in (2.7.8) is reversed.

We notice that if $f_1(x) = f(x)f_2(x)$, $g_1(x) = g(x)g_2(x)$ and $p(x) = f_2(x)g_2(x)$, then the inequality (2.7.8) reduces to the inequality (2.7.6).

Theorem 2.7.2 *Let* $f \in C_{rd}([a,b]_\mathbb{T}, [0, \infty))$ *be decreasing (or increasing) with* $\int_a^b xp(x)f(x)\Delta x > 0$ *and* $\int_a^b p(x)f(x)\Delta x > 0$. *Then*

$$\frac{\int_a^b xp(x)f^2(x)\Delta x}{\int_a^b xp(x)f(x)\Delta x} \geq (\leq) \frac{\int_a^b p(x)f^2(x)\Delta x}{\int_a^b p(x)f(x)\Delta x}. \qquad (2.7.9)$$

Proof. Clearly, for any x, $y \in [a,b]_\mathbb{T}$,

$$\int_a^b \int_a^b f(x)f(y)p(x)p(y)(y-x)(f(x)-f(y))\Delta x \Delta y \geq (\leq) 0,$$

which implies inequality (2.7.9). The proof is complete. ∎

Remark 2.7.3 *Let* $f \in C_{rd}([a,b]_\mathbb{T}, [0, \infty))$ *and* n *be a positive integer. If* p *and* g *are replaced by* p/f *and* f^n *respectively, then the Čebyšev inequality (2.7.6) is reduced to the inequality*

$$\int_a^b p(x)(f(x))^n \Delta x \int_a^b \frac{p(x)}{f(x)}\Delta x \geq \int_a^b p(x)\Delta x \int_a^b p(x)(f(x))^{n-1}\Delta x,$$

which implies that

$$\int_a^b p(x)(f(x))^n \Delta x \left(\int_a^b \frac{p(x)}{f(x)}\Delta x\right)^2$$

$$\geq \int_a^b p(x)\Delta x \int_a^b p(x)(f(x))^{n-1}\Delta x \int_a^b \frac{p(x)}{f(x)}\Delta x$$

$$\geq \left(\int_a^b p(x)\Delta x\right)^2 \int_a^b p(x)(f(x))^{n-2}\Delta x,$$

provided f *and* f^n *are similarly ordered. Proceeding we get*

$$\int_a^b p(x)(f(x))^n \Delta x \left(\int_a^b \frac{p(x)}{f(x)}\Delta x\right)^n \geq \left(\int_a^b p(x)\Delta x\right)^{n+1}.$$

Theorem 2.7.3 *If* $p, f \in C_{rd}([a,b]_\mathbb{T}, [0, \infty))$ *with* $f(x) > 0$ *on* $[a,b]_\mathbb{T}$ *and* n *a positive integer, then*

$$\left(\int_a^b \frac{p(x)}{f(x)}\Delta x\right)^n \left(\int_a^b p(x)f^n(x)\Delta x\right) \geq \left(\int_a^b p(x)\Delta x\right)^n. \qquad (2.7.10)$$

Proof. It follows from $f(x) > 0$ on $[a,b]_\mathbb{T}$ that $f^n(x)$ and $1/f(x)$ are oppositely ordered on $[a,b]_\mathbb{T}$. Hence by (2.7.6) we have

$$\int_a^b p(x)\,(f(x))^n\,\Delta x \left(\int_a^b \frac{p(x)}{f(x)}\Delta x\right)^n$$

$$\geq \int_a^b p(x)\Delta x \left(\int_a^b \frac{p(x)}{f(x)}\Delta x\right)^{n-1} \int_a^b p(x)\,(f(x))^{n-1}\,\Delta x$$

$$\geq \left(\int_a^b p(x)\Delta x\right)^2 \left(\int_a^b \frac{p(x)}{f(x)}\Delta x\right)^{n-2} \int_a^b p(x)\,(f(x))^{n-2}\,\Delta x$$

$$\geq \ldots \geq \left(\int_a^b p(x)\Delta x\right)^n,$$

which is the desired inequality (2.7.10). The proof is complete. ∎

Theorem 2.7.4 *Let* $g_1, g_2, \ldots, g_n \in C_{rd}([a,b]_\mathbb{T}, \mathbb{R})$ *and* $p, h_1, h_2, \ldots, h_{n-1} \in C_{rd}([a,b]_\mathbb{T}, [0,\infty))$ *with* $g_n(x) > 0$ *on* $[a,b]_\mathbb{T}$*. If*

$$\frac{g_1(x)g_2(x)\ldots g_{n-1}(x)}{h_1(x)h_2(x)\ldots h_{n-1}(x)} \quad \text{and} \quad \frac{h_{n-1}(x)}{g_n(x)},$$

are similarly ordered (or oppositely ordered), then

$$\int_a^b p(x)g_n(x)\Delta x \int_a^b \frac{p(x)g_1(x)g_2(x)\ldots g_{n-1}(x)}{h_1(x)h_2(x)\ldots h_{n-1}(x)}\Delta x$$

$$\geq (\leq)\int_a^b p(x)h_{n-1}(x)\Delta x \int_a^b \frac{p(x)g_1(x)g_2(x)\ldots g_n(x)}{h_1(x)h_2(x)\ldots h_{n-1}(x)}\Delta x. \tag{2.7.11}$$

Proof. Taking

$$f_1(x) = \frac{g_1(x)g_2(x)\ldots g_{n-1}(x)}{h_1(x)h_2(x)\ldots h_{n-1}(x)}, \ k_1(x) = h_{n-1}(x), \ f_2(x) = 1, \text{ and } k_2(x) = g_n(x),$$

in Theorem 2.7.1, we get the desired inequality (2.7.11). The proof is complete. ∎

Theorem 2.7.5 *Let* $p, f_1, f_2, \ldots, f_n \in C_{rd}([a,b]_\mathbb{T}, [0,\infty))$ *and* $g_1, g_2, \ldots,$ $g_n \in C_{rd}([a,b]_\mathbb{T}, [0,\infty))$*. If the functions* $f_1, \frac{f_2}{g_1}, \ldots, \frac{f_n}{g_{n-1}}$ *are similarly ordered and for each pair* $\frac{f_k}{g_{k-1}}$*,* g_{k-1} *is oppositely ordered for* $k = 2, 3, \ldots, n$*, then*

$$\int_a^b p(x)f_1(x)\frac{f_2(x)f_3(x)\ldots f_n(x)}{g_1(x)g_2(x)\ldots g_{n-1}(x)}\Delta x$$

$$\geq \frac{\int_a^b p(x)f_1(x)\Delta x \int_a^b p(x)f_2(x)\Delta x \ldots \int_a^b p(x)f_n(x)\Delta x}{\int_a^b p(x)g_1(x)\Delta x \int_a^b p(x)g_2(x)\Delta x \ldots \int_a^b p(x)g_n(x)\Delta x}. \tag{2.7.12}$$

Proof. Let $f_1, f_2, \ldots, f_n$ be replaced by $f_1, \frac{f_2}{g_1}, \ldots, \frac{f_n}{g_{n-1}}$ in (2.7.7), and we obtain

$$\left(\int_a^b p(x)\Delta x\right)^{n-1} \int_a^b p(x)f_1(x)\frac{f_2(x)f_3(x)\ldots f_n(x)}{g_1(x)g_2(x)\ldots g_{n-1}(x)}\Delta x$$

$$\geq \left(\int_a^b p(x)f_1(x)\Delta x\right) \prod_{k=2}^n \int_a^b p(x)\frac{f_k(x)}{g_{k-1}(x)}\Delta x. \tag{2.7.13}$$

Also, since $\frac{f_k}{g_{k-1}}$, g_{k-1} is oppositely ordered for $k = 2, 3, \ldots, n$, it follows from (2.7.6), that

$$\int_a^b p(x)\Delta x \left(\int_a^b p(x)f_k(x)\Delta x\right) \leq \left(\int_a^b p(x)g_{k-1}(x)\Delta x\right)\int_a^b p(x)\frac{f_k(x)}{g_{k-1}(x)}\Delta x.$$

Thus

$$\int_a^b p(x)\frac{f_k(x)}{g_{k-1}(x)}\Delta x \geq \frac{\int_a^b p(x)\Delta x \left(\int_a^b p(x)f_k(x)\Delta x\right)}{\int_a^b p(x)g_{k-1}(x)\Delta x}.$$

This and (2.7.13) imply (2.7.12). The proof is complete. ■

Theorem 2.7.6 Let $p, f_1, f_2, \ldots, f_n \in C_{rd}([a,b]_\mathbb{T}, [0,\infty))$ and $k_1, k_2, \ldots,$ $k_{n-1} \in C_{rd}([a,b]_\mathbb{T}, \mathbb{R})$. If

$$\frac{f_1(x)f_2(x)\ldots f_{i-1}(x)}{k_1(x)k_2(x)\ldots k_{i-1}(x)} \quad \text{and} \quad \frac{k_{i-1}(x)}{f_i(x)},$$

are similarly ordered (or oppositely ordered) for $i = 2, 3, \ldots, n$, then

$$\left(\int_a^b p(x)f_1(x)\Delta x\right)\left(\int_a^b p(x)f_2(x)\Delta x\right)\ldots\left(\int_a^b p(x)f_n(x)\Delta x\right)$$

$$\geq (\leq) \left(\int_a^b p(x)k_1(x)\Delta x\right)\left(\int_a^b p(x)k_2(x)\Delta x\right)\ldots\left(\int_a^b p(x)k_{n-1}(x)\Delta x\right)$$

$$\times \int_a^b p(x)\frac{f_1(x)f_2(x)\ldots f_n(x)}{k_1(x)k_2(x)\ldots k_{n-1}(x)}\Delta x. \tag{2.7.14}$$

Proof. If $f_1(x), k_1(x), f_2(x)$ and $k_2(x)$ are replaced by $f_1(x), 1, k_1(x),$ $\frac{f_2(x)}{k_1(x)}$ in Theorem 2.7.1, then we obtain

$$\int_a^b p(x)f_1(x)\Delta x \int_a^b p(x)f_2(x)\Delta x \geq (\leq) \int_a^b p(x)k_1(x)\Delta x \int_a^b p(x)\frac{f_1(x)f_2(x)}{k_1(x)}\Delta x.$$

Thus the theorem holds for $n = 2$. Suppose that the theorem holds for $n - 1$, that is

$$\left(\int_a^b p(x)f_1(x)\Delta x \right) \left(\int_a^b p(x)f_2(x)\Delta x \right) \cdots \left(\int_a^b p(x)f_{n-1}(x)\Delta x \right)$$

$$\geq \; (\leq) \left(\int_a^b p(x)k_1(x)\Delta x \right) \left(\int_a^b p(x)k_2(x)\Delta x \right) \cdots \left(\int_a^b p(x)k_{n-2}(x)\Delta x \right)$$

$$\times \int_a^b p(x)\frac{f_1(x)f_2(x)\cdots f_{n-1}(x)}{k_1(x)k_2(x)\cdots k_{n-2}(x)}\Delta x, \tag{2.7.15}$$

if

$$\frac{f_1(x)f_2(x)\cdots f_{i-1}(x)}{k_1(x)k_2(x)\cdots k_{i-1}(x)} \quad \text{and} \quad \frac{k_{i-1}(x)}{f_i(x)},$$

are similarly ordered (or oppositely ordered) for $i = 2, 3, .., n-1$. Multiplying both sides of (2.7.15) by $\int_a^b p(x)f_n(x)\Delta x$, we get that

$$\int_a^b p(x)f_1(x)\Delta x \int_a^b p(x)f_2(x)\Delta x \cdots \int_a^b p(x)f_{n-1}(x)\Delta x \int_a^b p(x)f_n(x)\Delta x$$

$$\geq \; (\leq) \left(\int_a^b p(x)k_1(x)\Delta x \right) \left(\int_a^b p(x)k_2(x)\Delta x \right) \cdots \left(\int_a^b p(x)k_{n-2}(x)\Delta x \right)$$

$$\times \int_a^b p(x)\frac{f_1(x)f_2(x)\cdots f_{n-1}(x)}{k_1(x)k_2(x)\cdots k_{n-2}(x)}\Delta x \int_a^b p(x)f_n(x)\Delta x. \tag{2.7.16}$$

It follows from Theorem 2.7.5 that

$$\int_a^b p(x)\frac{f_1(x)f_2(x)\cdots f_{n-1}(x)}{k_1(x)k_2(x)\cdots k_{n-2}(x)}\Delta x \int_a^b p(x)f_n(x)\Delta x$$

$$\geq \; (\leq) \int_a^b p(x)\frac{f_1(x)f_2(x)\cdots f_n(x)}{k_1(x)k_2(x)\cdots k_{n-1}(x)}\Delta x \int_a^b p(x)k_{n-1}(x)\Delta x.$$

This and (2.7.16) imply

$$\int_a^b p(x)f_1(x)\Delta x \int_a^b p(x)f_2(x)\Delta x \cdots \int_a^b p(x)f_{n-1}(x)\Delta x \int_a^b p(x)f_n(x)\Delta x$$

$$\geq \; (\leq) \left(\int_a^b p(x)k_1(x)\Delta x \right) \left(\int_a^b p(x)k_2(x)\Delta x \right) \cdots \left(\int_a^b p(x)k_{n-1}(x)\Delta x \right)$$

$$\times \int_a^b p(x)\frac{f_1(x)f_2(x)\cdots f_n(x)}{k_1(x)k_2(x)\cdots k_{n-1}(x)}\Delta x.$$

Then, by induction we have the desired inequality (2.7.14). The proof is complete. ∎

Remark 2.7.4 Let $k_n \in C_{rd}([a,b]_\mathbb{T}, \mathbb{R})$. If $f_1(x), f_2(x), \ldots, f_n(x)$ and $k_1(x), k_2(x)., \ldots, k_{n-1}(x)$ are replaced by $f_1(x)f_2(x) \ldots f_n(x), k_1(x)k_2(x) \ldots k_n(x), f_1(x)k_2(x) \ldots k_n(x), k_1(x)f_2(x)k_3(x) \ldots k_n(x), \ldots, k_1(x)k_2(x) \ldots k_{n-2}(x)f_{n-1}(x)k_n(x)$ in Theorem 2.7.6, respectively, then

$$\int_a^b p(x)f_1(x)f_2(x) \ldots f_n(x)\Delta x \left(\int_a^b p(x)k_1(x)k_2(x) \ldots k_n(x)\Delta x \right)^{-1}$$

$$\geq \left(\int_a^b p(x)f_1(x)k_2(x) \ldots k_n(x)\Delta x \right) \left(\int_a^b p(x)k_1(x)f_2(x)k_3(x) \ldots k_n(x)\Delta x \right)$$

$$\ldots \int_a^b p(x)k_1(x)k_2(x) \ldots k_{n-1}(x)f_n(x)\Delta x, \tag{2.7.17}$$

if $\frac{f_i(x)}{k_i(x)} > 0$ for $i = 1, 2, \ldots, n$ and $k_1(x)k_2(x) \ldots k_{n-1}(x) > 0$ on $[a,b]_\mathbb{T}$.

Remark 2.7.5 Letting $f_1(x) = f_2(x) = \ldots = f_n(x) = f(x)$ and $k_1(x) = k_2(x) = \ldots = k_n(x) = k^{\frac{1}{n-1}}(x)$ in (2.7.17) with $k(x) > 0$ on $[a,b]_\mathbb{T}$, we obtain a Hölder type inequality on time scales

$$\left(\int_a^b p(x)f(x)k(x)\Delta x \right)^n \leq \int_a^b p(x)(f(x))^n \Delta x \left(\int_a^b p(x)k^{\frac{n}{n-1}}(x)\Delta x \right)^{n-1}.$$

Remark 2.7.6 Let $p, f, g \in C_{rd}([a,b]_\mathbb{T}, [0, \infty))$. Putting $f_1(x) = (f(x))^n g(x)$, $f_2(x) = f_3(x) = \ldots = f_n(x) = g(x)$, and $k_1(x) = k_2(x) = \ldots = k_{n-1}(x) = f(x)g(x)$ in (2.7.14), we see that

$$\left(\int_a^b p(x)f(x)g(x)\Delta x \right)^n \leq \int_a^b p(x)(f(x))^n g(x)\Delta x \left(\int_a^b p(x)g(x)\Delta x \right)^{n-1}.$$

Remark 2.7.7 Taking $k_1(x) = k_2(x) = \ldots = k_{n-1}(x) = (f_1(x)f_2(x) \ldots f_n(x))^{\frac{1}{n}}$ in (2.7.14), we obtain

$$\left(\int_a^b p(x)f_1(x)\Delta x \right) \left(\int_a^b p(x)f_2(x)\Delta x \right) \ldots \left(\int_a^b p(x)f_n(x)\Delta x \right)$$

$$\geq \left(\int_a^b p(x)(f_1(x)f_2(x) \ldots f_n(x))^{\frac{1}{n}} \Delta x \right)^n,$$

if $f_i > 0$ on $[a,b]_\mathbb{T}$ and $\frac{1}{f_i(x)}(f_1(x)f_2(x) \ldots f_n(x))^{\frac{1}{n}}$ $(i = 1, 2, \ldots, n)$ are similarly ordered.

Remark 2.7.8 *Taking $k_1(x) = k_2(x) = \ldots = k_{n-1}(x) = 1$ in (2.7.14), we get the Čebyšev type inequality*

$$\left(\int_a^b p(x)f_1(x)\Delta x\right)\left(\int_a^b p(x)f_2(x)\Delta x\right)\ldots\left(\int_a^b p(x)f_n(x)\Delta x\right)$$

$$\leq \left(\int_a^b p(x)\Delta x\right)^{n-1}\int_a^b p(x)f_1(x)f_2(x)\ldots f_n(x)\Delta x,$$

if $f_i > 0$ on $[a,b]_{\mathbb{T}}$ and $f_i(x)$ $(i = 1, 2, \ldots, n)$ are similarly ordered.

We end this section by considering the Čebyšev inequality in the case of nabla integrals; see [26].

Theorem 2.7.7 *Let f and g be both increasing or both decreasing in $[a,b]_{\mathbb{T}}$. Then*

$$\int_a^b f(t)g(t)\nabla t \geq \frac{1}{b-a}\int_a^b f(t)\nabla t \int_a^b g(t)\nabla t. \qquad (2.7.18)$$

If one of the functions is increasing and the other is decreasing, then the inequality is reversed.

Now, we give some applications of Theorem 2.7.7.

Theorem 2.7.8 *Assume that $f^{\nabla^{n+1}}$ is monotonic on $[a,b]_{\mathbb{T}}$ and let*

$$\check{R}_{n,f}(t,s) = f(s) - \sum_{k=0}^n \hat{h}_k(t,s)f^{\nabla^k}(s) = \int_t^s \hat{h}_n(s,\rho(\xi))f^{\nabla^{n+1}}(\xi)\nabla\xi.$$

(i). If $f^{\nabla^{n+1}}$ is increasing, then

$$\int_a^b \check{R}_{n,f}(a,t)\nabla t - \left[\frac{f^{\nabla^n}(b) - f^{\nabla^n}(a)}{b-a}\right]\hat{h}_{n+2}(b,a)$$

$$\geq \left[f^{\nabla^{n+1}}(a) - f^{\nabla^{n+1}}(b)\right]\hat{h}_{n+2}(b,a). \qquad (2.7.19)$$

(ii). If $f^{\nabla^{n+1}}$ is decreasing, then

$$\int_a^b \check{R}_{n,f}(a,t)\nabla t - \left[\frac{f^{\nabla^n}(b) - f^{\nabla^n}(a)}{b-a}\right]\hat{h}_{n+2}(b,a)$$

$$\leq \left[f^{\nabla^{n+1}}(a) - f^{\nabla^{n+1}}(b)\right]\hat{h}_{n+2}(b,a).$$

Proof. The proof of (ii) is analogous to that of (i) so we will just consider (i). Let $F(t) = f^{\nabla^{n+1}}(t)$ and $G(t) = \hat{h}_n(b,\rho(t))$. Then F is increasing and G is decreasing by assumption. From inequality (2.7.18), we see that

$$\int_a^b F(t)G(t)\nabla t \leq \frac{1}{b-a}\int_a^b F(t)\nabla t \int_a^b G(t)\nabla t. \qquad (2.7.20)$$

By Corollary 2.6.1, we see that

$$\int_a^b F(t)G(t)\nabla t = \int_a^b \hat{h}_{n+1}(b,\rho(t))f^{\nabla^{n+1}}(t)\nabla t = \int_a^b \check{R}_{n,f}(a,t)\nabla t.$$

We also have

$$\int_a^b F(t)\nabla t = f^{\nabla^n}(b) - f^{\nabla^n}(a), \text{ and } \int_a^b G(t)\nabla t = \int_a^b \hat{h}_{n+1}(b,\rho(t))\nabla t = \hat{h}_{n+2}(b,a).$$

Thus the inequality (2.7.20) implies that

$$\int_a^b \check{R}_{n,f}(a,t)\nabla t \le \frac{1}{b-a}\left(f^{\nabla^n}(b) - f^{\nabla^n}(a)\right)\hat{h}_{n+2}(b,a).$$

Since $f^{\nabla^{n+1}}$ is increasing on $[a,b]_\mathbb{T}$,

$$f^{\nabla^{n+1}}(a)\hat{h}_{n+2}(b,a) \le \frac{1}{b-a}\left(f^{\nabla^n}(b) - f^{\nabla^n}(a)\right)\hat{h}_{n+2}(b,a)$$
$$\le f^{\nabla^{n+1}}(b)\hat{h}_{n+2}(b,a),$$

and, we have

$$\int_a^b \check{R}_{n,f}(a,t)\nabla t - \frac{1}{b-a}\left(f^{\nabla^n}(b) - f^{\nabla^n}(a)\right)\hat{h}_{n+2}(b,a)$$
$$\ge \int_a^b \check{R}_{n,f}(a,t)\nabla t - f^{\nabla^{n+1}}(b)\hat{h}_{n+2}(b,a).$$

Now Corollary 2.6.1 and $f^{\nabla^{n+1}}$ is increasing imply that

$$f^{\nabla^{n+1}}(b)\int_a^b \hat{h}_{n+1}(b,\rho(t))\nabla t \ge \int_a^b \check{R}_{n,f}(a,t)\nabla t \ge f^{\nabla^{n+1}}(a)\int_a^b \hat{h}_{n+1}(b,\rho(t))\nabla t,$$

which simplifies to

$$f^{\nabla^{n+1}}(b)\hat{h}_{n+2}(b,a) \ge \int_a^b \check{R}_{n,f}(a,t)\nabla t \ge f^{\nabla^{n+1}}(a)\int_a^b \hat{h}_{n+2}(b,a)\nabla t.$$

We now have inequality (2.7.19). The proof is complete. ∎

Theorem 2.7.9 *Assume that $f^{\nabla^{n+1}}$ is monotonic on $[a,b]_\mathbb{T}$.*

(i) If $f^{\Delta^{n+1}}$ is increasing, then

$$0 \le (-1)^{n+1}\int_a^b R_{n,f}(b,t)\Delta t - \left[\frac{f^{\Delta^n}(b) - f^{\Delta^n}(a)}{b-a}\right]g_{n+2}(b,a)$$
$$\le \left[f^{\Delta^{n+1}}(b) - f^{\Delta^{n+1}}(a)\right]g_{n+2}(b,a). \tag{2.7.21}$$

(ii). If $f^{\Delta^{n+1}}$ is decreasing, then

$$0 \geq (-1)^{n+1} \int_a^b R_{n,f}(b,t)\nabla t - \left[\frac{f^{\Delta^n}(b) - f^{\Delta^n}(a)}{b-a} \right] g_{n+2}(b,a)$$

$$\geq \left[f^{\Delta^{n+1}}(b) - f^{\Delta^{n+1}}(a) \right] g_{n+2}(b,a).$$

Proof. The proof of (ii) is analogous to that of (i) so we only consider (i). Let $F(t) = f^{\Delta^{n+1}}(t)$ and $G(t) = (-1)^{n+1}h_{n+1}(a,\sigma(t))$. Then F and G are increasing. Inequality (2.7.6) with $p = 1$, $f = F$ and $g = G$, gives

$$\int_a^b F(t)G(t)\Delta t \geq \frac{1}{b-a} \int_a^b F(t)\Delta t \int_a^b G(t)\Delta t. \qquad (2.7.22)$$

By Lemma 2.6.2 with $t = a$,

$$\int_a^b F(t)G(t)\Delta t = (-1)^{n+1} \int_a^b h_{n+1}(a,\sigma(t))f^{\Delta^{n+1}}(t)\Delta t$$

$$= (-1)^{n+1} \int_a^b R_{n,f}(b,t)\Delta t.$$

We also have $\int_a^b F(t)\Delta t = f^{\Delta^n}(b) - f^{\Delta^n}(b)$ and

$$\int_a^b G(t)\Delta t = (-1)^{n+1} \int_a^b h_{n+1}(a,\sigma(t))\Delta t = g_{n+2}(b,a).$$

Thus by (2.7.22), we have

$$0 \leq (-1)^{n+1} \int_a^b R_{n,f}(b,t)\Delta t - \frac{1}{b-a} \left[f^{\Delta^n}(b) - f^{\Delta^n}(b) \right] g_{n+2}(b,a).$$

Since $f^{\Delta^{n+1}}$ is increasing on $[a,b]_{\mathbb{T}}$,

$$f^{\Delta^{n+1}}(a)g_{n+2}(b,a) \leq \frac{1}{b-a} \left[f^{\Delta^n}(b) - f^{\Delta^n}(b) \right] g_{n+2}(b,a) \leq f^{\Delta^{n+1}}(b)g_{n+2}(b,a),$$

and we have

$$(-1)^{n+1} \int_a^b R_{n,f}(b,t)\Delta t - f^{\Delta^{n+1}}(a)g_{n+2}(b,a)$$

$$\geq (-1)^{n+1} \int_a^b R_{n,f}(b,t)\Delta t - \frac{\left[f^{\Delta^n}(b) - f^{\Delta^n}(b) \right]}{b-a} g_{n+2}(b,a).$$

Now, from Definition 1.4.1, since

$$g_n(t,s) = (-1)^n h_n(s,t),$$

we have by Lemma 2.6.2 with $t = a$ that

$$(-1)^{n+1} \int_a^b R_{n,f}(b,t)\Delta t = \int_a^b g_{n+1}(\sigma(t),a) f^{\Delta^{n+1}}(t)\Delta t.$$

Since $f^{\Delta^{n+1}}$ is increasing, we get that

$$f^{\Delta^{n+1}}(b) \int_a^b g_{n+1}(\sigma(t),a)\Delta t \geq (-1)^{n+1} \int_a^b R_{n,f}(b,t)\Delta t$$

$$\geq f^{\Delta^{n+1}}(a) \int_a^b g_{n+1}(\sigma(t),a)\Delta t$$

which simplifies to

$$f^{\Delta^{n+1}}(b) g_{n+1}(b,a) \geq (-1)^{n+1} \int_a^b R_{n,f}(b,t)\Delta t \geq f^{\Delta^{n+1}}(a) g_{n+2}(b,a).$$

We now have (2.7.21). The proof is complete. ∎

Remark 2.7.9 *In Theorem 2.7.8 (i), if $n = 0$, we obtain*

$$\int_a^b f(t)\nabla t \leq (b-a)f(a) + \frac{\hat{h}_2(b,a)}{b-a}(f(b) - f(a)). \tag{2.7.23}$$

Theorem 2.7.10 *Assume that f is nabla convex on $[a,b]_{\mathbb{T}}$, that is, $f^{\nabla^2} \geq 0$ on $[a,b]_{\mathbb{T}}$. Then*

$$\int_a^b f^\rho(t)(t-a)\nabla t \leq (b-a)f(b) - \frac{\hat{h}_2(b,a)}{b-a}(f(b) - f(a)). \tag{2.7.24}$$

Proof. If $F = f^\nabla$ and $G = t - a = \hat{h}_1(t,a)$, then both F and G are increasing functions. By Čebyšev's inequality we see that

$$\int_a^b f^\rho(t)(t-a)\nabla t \geq \frac{1}{b-a} \int_a^b f^\nabla(t)\nabla t \int_a^b \hat{h}_1(t,a)\nabla t.$$

Using nabla integration by parts on the left-hand side we get the desired inequality (2.7.24). The proof is complete. ∎

The following result is a Hermite–Hadamard type inequality for time scales and is obtained by a combination of (2.7.23) and (2.7.24).

Corollary 2.7.3 *Let f be nabla convex on $[a,b]_{\mathbb{T}}$. Then*

$$\frac{1}{b-a} \int_a^b \frac{f^\rho(t) + f(t)}{2}\nabla t \leq \frac{f(a) + f(b)}{2}.$$

Chapter 3

Opial Inequalities

All human knowledge begins with intuitions proceeds thence to concepts and ends with ideas.

Kant (1724–1804).

In 1960 Opial proved that if x is absolutely continuous on $[a, b]$ with $x(a) = x(b) = 0$, then

$$\int_a^b |x(t)| \left|x'(t)\right| dt \le \frac{(b-a)}{4} \int_a^b \left|x'(t)\right|^2 dt. \qquad (3.0.1)$$

We refer the reader to [9] for results on Opial type inequalities. We also note if x is absolutely continuous on $(0, b)$ with $x(0) = 0$, then

$$\int_0^b |x(t)| \left|x'(t)\right| dt \le \frac{b}{2} \int_0^b \left|x'(t)\right|^2 dt. \qquad (3.0.2)$$

The discrete version of (3.0.1) was proved by Lasota [101] and is given by

$$\sum_{i=1}^{h-1} |x_i \Delta x_i| \le \frac{1}{2} \left[\frac{h+1}{2}\right] \sum_{i=1}^{h-1} |\Delta x_i|^2, \qquad (3.0.3)$$

where $\{x_i\}_{0 \le i \le h}$ is a sequence of real numbers with $x_0 = x_h = 0$ and $[x]$ is the greatest integer function. For a real sequence $\{x_i\}_{0 \le i \le h}$ with $x_0 = 0$, we have

$$\sum_{i=1}^{h-1} |x_i \Delta x_i| \le \frac{h-1}{2} \sum_{i=0}^{h-1} |\Delta x_i|^2. \qquad (3.0.4)$$

© Springer International Publishing Switzerland 2014
R. Agarwal et al., *Dynamic Inequalities On Time Scales*,
DOI 10.1007/978-3-319-11002-8_3

The chapter is organized as follows. In Sect. 3.1 we establish some first order
Opial type inequalities and in Sect. 3.2 we establish some generalizations.
Section 3.3 discusses inequalities with two different weight functions and
in Sect. 3.4 we present some Opial type inequalities involving higher order
derivatives. In Sect. 3.5 we obtain a sequence of Opial type inequalities for
first and higher order diamond alpha derivatives on time scales.

Throughout this chapter (usually without mentioning) the integrals in the
statements of the theorems are assumed to exist.

3.1 Opial Type Inequalities I

In this section, we will present some inequalities of Opial's type on time
scales with first order derivatives. The results in this section are adapted
from [49, 90, 123, 138, 139, 158].

Theorem 3.1.1 *Let $\mathbb{T}$ be a time scale with 0, $h \in \mathbb{T}$. For a delta differentiable $x : [0, h]_{\mathbb{T}} \to \mathbb{R}$ with $x(0) = 0$, then*

$$\int_0^h |x(t) + x^\sigma(t)| \left|x^\Delta(t)\right| \Delta t \leq h \int_0^h \left|x^\Delta(t)\right|^2 \Delta t, \qquad (3.1.1)$$

with equality when $x(t) = ct$.

Proof. Consider $y(t) = \int_0^t |x^\Delta(s)| \, \Delta s$. Then $y^\Delta(t) = |x^\Delta(t)|$ and $|x| \leq y$.
By the Cauchy–Schwarz inequality, we have that

$$\int_0^h |x(t) + x^\sigma(t)| \left|x^\Delta(t)\right| \Delta t \leq \int_0^h (|x(t)| + |x^\sigma(t)|) \, x^\Delta(t) \Delta t$$

$$\leq \int_0^h (y(t) + y^\sigma(t)) \, y^\Delta(t) \Delta t = \int_0^h \left(y^2(t)\right)^\Delta \Delta t = y^2(h)$$

$$= \left(\int_0^h |x^\Delta(t)| \, \Delta t\right)^2 \leq h \int_0^h \left|x^\Delta(t)\right|^2 \Delta t,$$

which is the desired inequality (3.1.1). Now, let $x(t) = ct$ for some $c \in \mathbb{R}$.
Then $x^\Delta(t) = c$ and it is easy to check that equality holds in (3.1.1). The
proof is complete. ∎

Example 3.1.1 *Consider the initial value problem*

$$y^\Delta(t) = 1 - t + \frac{1}{t} y^2(t), \ \ 0 \leq t \leq 1, \ \ y(0) = 0. \qquad (3.1.2)$$

Let $y(t)$ be a solution of (3.1.2) and let $R(t) = 1 - t + \int_0^t \left|y^\Delta(s)\right|^2 \Delta s$. Let $t \in [0,1]_{\mathbb{T}}$. Then using (3.1.1), we have

$$
\begin{aligned}
\left|y^\Delta(t)\right| &= \left|1 - t + \frac{1}{t}y^2(t)\right| \le |1 - t| + \frac{1}{t}y^2(t) \\
&= 1 - t + \frac{1}{t}\left|\int_0^t \left(y^2(s)\right)^\Delta \Delta s\right| \\
&\le 1 - t + \frac{1}{t}\int_0^t \left|\left(y^2(s)\right)^\Delta\right| \Delta s \\
&= 1 - t + \frac{1}{t}\int_0^t \left|(y(s) + y^\sigma(s))\, y^\Delta(t)\right| \Delta s \\
&\le 1 - t + \int_0^t \left(y^\Delta(t)\right)^2 \Delta s = R(t).
\end{aligned}
$$

Hence $R^\Delta(t) = -1 + \left|y^\Delta(t)\right| = -1 + \left|y^\Delta(t)\right|^2 \le R^2(t) - 1$ and $R(0) = 1$. Let w be the unique solution of

$$
w^\Delta(t) = (1 + R(t))w(t), \quad w(0) = 1.
$$

Now, because $w(t) > 0$ and $(R - 1)^\Delta = R^\Delta \le R^2(t) - 1$, we have

$$
\begin{aligned}
\left(\frac{R-1}{w}\right)^\Delta &= \frac{w(R-1)^\Delta - (R-1)w^\Delta}{ww^\sigma} = \frac{wR^\Delta - (1 + R^2)w}{ww^\sigma} \\
&\le \frac{wR^2 - (1 + R^2)w}{ww^\sigma} \le 0.
\end{aligned}
$$

Thus

$$
\frac{R(t) - 1}{w(t)} = \frac{R(0) - 1}{w(0)}\int_0^t \left(\frac{R-1}{w}\right)^\Delta \Delta t \le 0,
$$

and hence $R(t) \le 1$. Therefore $y^\Delta(t) \le \left|y^\Delta(t)\right| \le R(t) \le 1$ and hence $y(t) \le t$.

The following theorem gives the nabla Opial inequality on time scales.

Theorem 3.1.2 Let $\mathbb{T}$ be a time scale with 0, $h \in \mathbb{T}$. For a nabla differentiable $x : [0, h]_{\mathbb{T}} \to \mathbb{R}$ with $x^\nabla(t)$ ld-continuous and $x(0) = 0$, then

$$
\int_0^h \left|x(t) + x^\rho(t)\right| \left|x^\nabla(t)\right| \Delta t \le h \int_0^h \left|x^\nabla(t)\right|^2 \nabla t, \tag{3.1.3}
$$

with equality when $x(t) = ct$.

Proof. Consider $y(t) = \int_0^t \left| x^\nabla(s) \right| \nabla s$. Then $y^\nabla(t) = \left| x^\nabla(t) \right|$ and $|x| \le y$. By the Cauchy–Schwarz inequality, we have that

$$
\int_0^h \left| x(t) + x^\rho(t) \right| \left| x^\nabla(t) \right| \nabla t \le \int_0^h \left(|x(t)| + |x^\rho(t)| \right) x^\nabla(t) \nabla t
$$
$$
\le \int_0^h \left(y(t) + y^\rho(t) \right) y^\nabla(t) \nabla t = \int_0^h \left(y^2(t) \right)^\nabla \nabla t = y^2(h) - y(0)
$$
$$
= \left(\int_0^h \left| x^\nabla(t) \right| \nabla t \right)^2 \le h \int_0^h \left| x^\nabla(t) \right|^2 \nabla t,
$$

which is the desired inequality (3.1.3). Now, let $x(t) = ct$ for some $c \in \mathbb{R}$. Then $x^\nabla(t) = c$ and it is easy to check that equality holds in (3.1.3). The proof is complete. ■

We next present a generalization of Theorem 3.1.1 when $x(0)$ need not be equal to 0.

Theorem 3.1.3 Let $\mathbb{T}$ be a time scale with $0, \ h \in \mathbb{T}$. For a delta differentiable $x : [0, h]_\mathbb{T} \to \mathbb{R}, \ \ then$

$$
\int_0^h |x(t) + x^\sigma(t)| \left| x^\Delta(t) \right| \Delta t \le \alpha \int_0^h \left| x^\Delta(t) \right|^2 \Delta t + 2\beta \int_0^h \left| x^\Delta(t) \right| \Delta t, \quad (3.1.4)
$$

where

$$
\alpha \in \mathbb{T} \ with \ dist\left(\frac{h}{2}, \alpha \right) = dist\left(\frac{h}{2}, \mathbb{T} \right), \quad (3.1.5)
$$

and $\beta = \max\{x(0), \ x(h)\}$.

Proof. We consider

$$
y(t) = \int_0^t \left| x^\Delta(t) \right| \Delta t, \quad and \quad z(t) = \int_t^h \left| x^\Delta(t) \right| \Delta t.
$$

Then $y^\Delta(t) = \left| x^\Delta(t) \right|$ and $z^\Delta(t) = - \left| x^\Delta(t) \right|,$

$$
|x(t)| \le |x(t) - x(0)| + |x(0)| = \left| \int_0^t x^\Delta(t) \Delta t \right| + |x(0)|
$$
$$
\le \int_0^t \left| x^\Delta(t) \right| \Delta t + |x(0)| = y(t) + |x(0)|,
$$

and similarly $|x(t)| \leq z(t) + |x(h)|$. Let $u \in [0, h]_{\mathbb{T}}$. Then applying the Cauchy–Schwarz inequality, we get that

$$\int_0^u |x(t) + x^\sigma(t)| \, |x^\Delta(t)| \, \Delta t$$

$$\leq \int_0^u [y(t) + y^\sigma(t) + 2 |x(0)|] \, y^\Delta(t) \Delta t$$

$$= \int_0^u (y(t) + y^\sigma(t)) \, \Delta t + 2 |x(0)| \int_0^u y^\Delta(t) \Delta t = y^2(u) + 2 |x(0)| \, y(u)$$

$$\leq u \int_0^u |x^\Delta(t)|^2 \, \Delta t + 2 |x(0)| \int_0^u |x^\Delta(t)| \, \Delta t. \tag{3.1.6}$$

Similarly, we obtain that

$$\int_u^h |x(t) + x^\sigma(t)| \, |x^\Delta(t)| \, \Delta t \leq z^2(u) + 2 |x(h)| \, z(u)$$

$$\leq (h - u) \int_u^h |x^\Delta(t)|^2 \, \Delta t + 2 |x(h)| \int_u^h |x^\Delta(t)| \, \Delta t. \tag{3.1.7}$$

By putting $v(u) = \max\{u, h - u\}$ and adding (3.1.6) and (3.1.7), we have

$$\int_0^h |x(t) + x^\sigma(t)| \, |x^\Delta(t)| \, \Delta t \leq v(u) \int_0^h |x^\Delta(t)|^2 \, \Delta t + 2\beta \int_u^h |x^\Delta(t)| \, \Delta t.$$

This is true for any $u \in [0, h]_{\mathbb{T}}$, so it is also true if $v(u)$ is replaced by $\min_{u \in [0, h]_{\mathbb{T}}} v(u)$. However, this last quantity is easily seen to be equal to α. The proof is complete. ∎

The proof of the following theorem follows from the proof of Theorem 3.1.3 when $\beta = 0$.

Theorem 3.1.4 *Let $\mathbb{T}$ be a time scale with 0, $h \in \mathbb{T}$. For a delta differentiable $x : [0, h]_{\mathbb{T}} \to \mathbb{R}$, with $x(0) = x(h) = 0$, then*

$$\int_0^h |x(t) + x^\sigma(t)| \, |x^\Delta(t)| \, \Delta t \leq \alpha \int_0^h |x^\Delta(t)|^2 \, \Delta t, \tag{3.1.8}$$

where α is given as in (3.1.5).

In the following, we give some generalizations of the above inequalities which lead to Opial type inequalities with weight functions.

Theorem 3.1.5 *Let $\mathbb{T}$ be a time scale with 0, $h \in \mathbb{T}$ and $w(t)$ be a positive and rd-continuous function on $[0, h]_{\mathbb{T}}$ such that $\int_0^h w^{1-q}(t) \Delta t < \infty$, $q > 1$. For a delta differentiable $x : [0, h]_{\mathbb{T}} \to \mathbb{R}$ with $x(0) = 0$, then*

$$\int_0^h |x(t) + x^\sigma(t)| \, |x^\Delta(t)| \, \Delta t \leq \left(\int_0^h w^{1-q}(t) \Delta t \right)^{\frac{2}{q}} \left(\int_0^h w(t) \, |x^\Delta(t)|^p \, \Delta t \right)^{\frac{2}{p}},$$

$$\tag{3.1.9}$$

where $p > 1$ and $1/p + 1/q = 1$ and with equality when $x(t) = c \int_0^t w^{1-q}(s) \Delta s$.

Proof. Consider $y(t) = \int_0^t |x^\Delta(t)|\, \Delta t$. Then $y^\Delta(t) = |x^\Delta(t)|$ and $|x| \le y$. By the Cauchy–Schwarz inequality, we have that

$$
\int_0^h |x(t) + x^\sigma(t)| \, |x^\Delta(t)| \, \Delta t \le \int_0^h (|x(t)| + |x^\sigma(t)|) \, x^\Delta(t) \Delta t
$$

$$
\le \int_0^h (y(t) + y^\sigma(t)) \, y^\Delta(t) \Delta t = \int_0^h (y^2(t))^\Delta \Delta t = y^2(h)
$$

$$
= \left(\int_0^h |x^\Delta(t)| \, \Delta t \right)^2 = \left(\int_0^h w^{\frac{-1}{p}}(t) w^{\frac{1}{p}}(t) \, |x^\Delta(t)| \, \Delta t \right)^2
$$

$$
\le \left(\int_0^h \left(w^{\frac{-1}{p}}(t) \right)^q \right)^{\frac{2}{q}} \left(\int_0^h w \, |x^\Delta(t)|^p \, \Delta t \right)^{\frac{2}{p}}.
$$

The proof is complete. ∎

Theorem 3.1.6 *Let $\mathbb{T}$ be a time scale with $0,\ h \in \mathbb{T}$ and $w(t)$ be a positive and rd-continuous function on $[0, h]_\mathbb{T}$ such that $\int_0^h w^{1-q}(t)\Delta t < \infty$, $q > 1$. For a delta differentiable $x : [0, h]_\mathbb{T} \to \mathbb{R}$, then*

$$
\int_0^h |x(t) + x^\sigma(t)| \, \left| x^\Delta(t) \right| \Delta t \le v^{2/q} \left(\int_0^h w \left| x^\Delta(t) \right|^p \Delta t \right)^{2/p} + 2\beta \int_0^h \left| x^\Delta(t) \right| \Delta t,
$$

where $p > 1$ and $1/p + 1/q = 1$, $\beta = \max\{x(0),\ x(h)\}$, and

$$
v = \max\{ \int_0^\alpha w^{1-q}(t)\Delta t, \int_\alpha^h w^{1-q}(t)\Delta t\},
$$

and

$$
\alpha \in \mathbb{T} \text{ with } dist(\frac{h}{2}, \alpha) = dist(\frac{h}{2}, \mathbb{T}).
$$

Proof. The proof is a combination of Theorems 3.1.3 and 3.1.5 and hence is omitted. ∎

Corollary 3.1.1 *Let $\mathbb{T}$ be a time scale with $0,\ h \in \mathbb{T}$ and $w(t)$ be a positive and rd-continuous function on $[0, h]_\mathbb{T}$ such that $\int_0^h w^{1-q}(t)\Delta t < \infty$, $q > 1$. For a delta differentiable $x : [0, h]_\mathbb{T} \to \mathbb{R}$ with $x(0) = x(h) = 0$, then*

$$
\int_0^h |x(t) + x^\sigma(t)| \, |x^\Delta(t)| \, \Delta t \le v^{2/q} \left(\int_0^h w \, |x^\Delta(t)|^p \, \Delta t \right)^{2/p}, \qquad (3.1.10)
$$

where $p > 1$ and $1/p + 1/q = 1$ and

$$
v = \max\{ \int_0^\alpha w^{1-q}(t)\Delta t, \int_\alpha^h w^{1-q}(t)\Delta t\},
$$

and

$$\alpha \in \mathbb{T} \text{ with } dist(\frac{h}{2}, \alpha) = dist(\frac{h}{2}, \mathbb{T}). \tag{3.1.11}$$

Theorem 3.1.7 *Let* $a, b \in \mathbb{T}$ *and* $q, f \in C^1_{rd}([a, b]_\mathbb{T}, \mathbb{R})$ *with* $f(a) = 0$. *Then*

$$\int_a^b q(\eta) \left| [f(\eta) + f^\sigma(\eta)] f^\Delta(\eta) \right| \Delta\eta \le K_q(b, a) \int_a^b \left(f^\Delta(\eta) \right)^2 \Delta\eta, \tag{3.1.12}$$

where

$$K_q(b, a) := \left(2 \int_a^b q^2(t) \left[\sigma(t) - a \right] \Delta t \right)^{1/2}, \text{ for } t \in [a, b]_\mathbb{T}. \tag{3.1.13}$$

Proof. We have

$$f(t) = \int_a^t f^\Delta(\eta) \Delta\eta \le \left((t - a)g(t) \right)^{1/2}, \tag{3.1.14}$$

where

$$g(t) := \int_a^t \left| f^\Delta(\eta) \right|^2 \Delta\eta, \text{ for } t \in [a, b]_\mathbb{T}. \tag{3.1.15}$$

Hence, we get

$$\left[g^\Delta(t) \right]^{1/2} = \left| f^\Delta(t) \right|, \text{ for all } t \in [a, b]_\mathbb{T}. \tag{3.1.16}$$

Then using

$$\alpha^{1/2} + \beta^{1/2} \le 2(\alpha + \beta)^{1/2} \quad \text{for all} \quad \alpha, \beta \in \mathbb{R}^+$$

and Hölder's inequality we have

$$\int_a^b q(\eta) \left| [f(\eta) + f^\sigma(\eta)] f^\Delta(\eta) \right| \Delta\eta$$

$$\le \int_a^b |q(\eta)| \left(|f(\eta)| + |f^\sigma(\eta)| \right) \left| f^\Delta(\eta) \right| \Delta\eta$$

$$\le \int_a^b |q(\eta)| \left(\sigma(\eta) - a \right)^{1/2} \left((g(\eta))^{1/2} + (g^\sigma(\eta))^{1/2} \right) \left(g^\Delta(\eta) \right)^{1/2} \Delta\eta$$

$$\le 2 \int_a^b |q(\eta)| \left(\sigma(\eta) - a \right)^{1/2} \left([g(\eta) + g^\sigma(\eta)] g^\Delta(\eta) \right)^{1/2} \Delta\eta$$

$$= \int_a^b |q(\eta)| (\sigma(\eta) - a)^{1/2} \left(\left([g(\eta)]^2 \right)^\Delta \right)^{1/2} \Delta\eta$$

$$\le K_q(b, a) \left(\int_a^b \left[(g(\eta))^2 \right]^\Delta \Delta\eta \right)^{1/2}$$

$$= K_q(b, a)g(b). \tag{3.1.17}$$

Thus, substituting (3.1.15) into (3.1.17), we see that (3.1.12) is true. The proof is complete. ∎

The following result is complementary to Theorem 3.1.7.

Theorem 3.1.8 *Let* $a, b \in \mathbb{T}$ *and* $q, f \in C_{rd}^1([a, b]_{\mathbb{T}}, \mathbb{R})$ *with* $f(b) = 0$. *Then*

$$\int_a^b q(\eta) \left| (f(\eta) + f^\sigma(\eta)) f^\Delta(\eta) \right| \Delta\eta \leq L_q(b, a) \int_a^b \left(f^\Delta(\eta) \right)^2 \Delta\eta, \quad (3.1.18)$$

where

$$L_q(b, a) := \left(2 \int_a^b q^2(t) [b - t] \Delta t \right)^{1/2}, \, for \quad t \in [a, b]_{\mathbb{T}}. \quad (3.1.19)$$

Proof. Setting

$$h(t) := \int_t^b \left| f^\Delta(\eta) \right|^2 \Delta\eta, \quad for \quad t \in [a, b]_{\mathbb{T}},$$

we have

$$f(t) = \int_t^b \left[-f^\Delta(\eta) \right] \Delta\eta \leq \left((b - t) h(t) \right)^{1/2}$$

for all $t \in [a, b]_{\mathbb{T}}$. Following the steps in the proof of Theorem 3.1.7 we obtain the required result. ∎

The next result combines Theorem 3.1.7 and Theorem 3.1.8 on the segments $[a, c]_{\mathbb{T}}$ and $[c, b]_{\mathbb{T}}$, respectively.

Theorem 3.1.9 *Let* $a, b \in \mathbb{T}$ *and* $q, f \in C_{rd}^1([a, b]_{\mathbb{T}}, \mathbb{R})$ *with* $f(a) = f(b) = 0$. *Then*

$$\int_a^b q(\eta) \left| [f(\eta) + f^\sigma(\eta)] f^\Delta(\eta) \right| \Delta\eta \leq \max \{ K_q(b, c), L_q(c, a) \} \int_a^b \left(f^\Delta(\eta) \right)^2 \Delta\eta,$$

holds for any $c \in [a, b]_{\mathbb{T}}$, *where* K_q, L_q *are as defined in* (3.1.13) *and* (3.1.19), *respectively.*

Corollary 3.1.2 *Let* $a, b \in \mathbb{T}$ *and* $q, f \in C_{rd}^1([a, b]_{\mathbb{T}}, \mathbb{R})$ *with* $f(a) = f(b) = 0$. *Then we have*

$$\int_a^b q(\eta) \left| [f(\eta) + f^\sigma(\eta)] f^\Delta(\eta) \right| \Delta\eta$$

$$\leq \min_{c \in [a, b]_{\mathbb{T}}} \left\{ \max \{ K_q(b, c), L_q(c, a) \} \right\} \int_a^b \left(f^\Delta(\eta) \right)^2 \Delta\eta,$$

where K, L *are as defined in* (3.1.13) *and* (3.1.19), *respectively.*

In the following, we establish some Opial dynamic inequalities with two different weight functions.

Theorem 3.1.10 *If r and q are positive rd-continuous functions on $[0,h]_{\mathbb{T}}$, $\int_0^h (\Delta t / r(t)) < \infty$, q nonincreasing and $x : [0,h]_{\mathbb{T}} \to \mathbb{R}$ is delta differentiable with $x(0) = 0$, then*

$$\int_0^h q^\sigma(t) \left| (x(t) + x^\sigma(t)) \, x^\Delta(t) \right| \Delta t \le \int_0^h \frac{\Delta t}{r(t)} \int_0^h r(t) q(t) \left| x^\Delta(t) \right|^2 \Delta t.$$
$$(3.1.20)$$

Proof. We consider

$$y(t) = \int_0^t \sqrt{q^\sigma(s)} \left| x^\Delta(s) \right| \Delta s.$$

Then $y^\Delta(t) = \sqrt{q^\sigma(t)} \left| x^\Delta(t) \right|$ and since for $0 \le s < t$ we have that $\sigma(s) \le t$. This implies that $q^\sigma(s) \ge q(t)$, and then we get

$$|x(t)| \le \int_0^t \left| x^\Delta(s) \right| \Delta s \le \int_0^t \sqrt{\frac{q^\sigma(s)}{q(t)}} \left| x^\Delta(s) \right| \Delta s = \frac{y(t)}{\sqrt{q(t)}}.$$

Apply the Cauchy–Schwarz inequality and we have

$$\int_0^h q^\sigma(t) \left| (x(t) + x^\sigma(t)) \, x^\Delta(t) \right| \Delta t$$

$$\le \int_0^h q^\sigma(t) \left(\frac{y(t)}{\sqrt{q(t)}} + \frac{y^\sigma(t)}{\sqrt{q^\sigma(t)}} \right) \frac{y^\Delta(t)}{\sqrt{q^\sigma(t)}} \Delta t$$

$$\le \int_0^h \left(y(t) + y^\sigma(t) \right) y^\Delta(t) \Delta t$$

$$= y^2(h) = \left[\int_0^t \frac{1}{\sqrt{r(s)}} \sqrt{r(s) q^\sigma(s)} \left| x^\Delta(s) \right| \Delta s \right]^2$$

$$\le \left(\int_0^t \frac{\Delta s}{r(s)} \right) \left(\int_0^t r(s) q^\sigma(s) \left| x^\Delta(s) \right|^2 \Delta s \right).$$

The proof is complete. ∎

Remark 3.1.1 *Note that in the case when $\mathbb{T} = \mathbb{R}$, the inequality (3.1.20) reduces to the Yang [152] inequality*

$$\int_0^h q(t) \left| x(t) \right| \left| x'(t) \right| dt \le \frac{1}{2} \int_0^h \frac{1}{r(t)} dt \int_0^h r(t) q(t) \left| x'(t) \right|^2 dt, \qquad (3.1.21)$$

where $r(t)$ is a positive and continuous function with $\int_0^t ds / r(s) < \infty$ and $q(t)$ is a positive, bounded, and nonincreasing function on $[0,h]_{\mathbb{R}}$. When $q(t) = 1$, we get the Beesack [41] inequality

$$\int_0^h \left| x(t) \right| \left| x'(t) \right| dt \le \frac{1}{2} \int_0^h \frac{1}{r(t)} dt \int_0^h r(t) \left| x'(t) \right|^2 dt. \qquad (3.1.22)$$

Theorem 3.1.11   *Let* $\mathbb{T}$ *be a time scale with* $a,\ \tau \in \mathbb{T}$. *Assume that* $s \in C_{rd}([a,\tau]_{\mathbb{T}}, \mathbb{R})$ *and* r *be a positive rd-continuous function on* $(a,\tau)_{\mathbb{T}}$ *such that* $\int_a^\tau r^{-1}(t)\Delta t < \infty$. *If* $y : [a,\tau]_{\mathbb{T}} \to \mathbb{R}$ *is delta differentiable with* $y(a) = 0$ *(and* y^Δ *does not change sign in* $(a,\tau)_{\mathbb{T}}$) *then we have*

$$\int_a^\tau s(x) \left| y(x) + y^\sigma(x) \right| \left| y^\Delta(x) \right| \Delta x \le K_1(a,\tau) \int_a^\tau r(x) \left| y^\Delta(x) \right|^2 \Delta x,$$

(3.1.23)

where

$$K_1(a,\tau) = \sqrt{2} \left(\int_a^\tau \frac{s^2(x)}{r(x)} \left(\int_a^x \frac{\Delta t}{r(t)} \right) \Delta x \right)^{\frac{1}{2}} + \sup_{a \le x \le \tau} \left(\mu(x) \frac{|s(x)|}{r(x)} \right).$$

(3.1.24)

Proof. Since $y^\Delta(t)$ does not change sign in $(a,\tau)_{\mathbb{T}}$, we have

$$|y(x)| = \int_a^x \left| y^\Delta(t) \right| \Delta t, \text{ for } x \in [a,\tau]_{\mathbb{T}}.$$

This implies that

$$|y(x)| = \int_a^x \frac{1}{\sqrt{r(t)}} \sqrt{r(t)} \left| y^\Delta(t) \right| \Delta t.$$

It follows from the Cauchy–Schwarz inequality with

$$f(t) = \frac{1}{(r(t))^{1/2}}, \quad g(t) = (r(t))^{\frac{1}{2}} \left| y^\Delta(t) \right|,$$

that

$$\int_a^x \left| y^\Delta(t) \right| \Delta t \le \left(\int_a^x \frac{1}{r(t)} \Delta t \right)^{\frac{1}{2}} \left(\int_a^x r(t) \left| y^\Delta(t) \right|^2 \Delta t \right)^{\frac{1}{2}}.$$

Then, for $a \le x \le \tau$, we get (note $y(a) = 0$)

$$|y(x)| \le \left(\int_a^x \frac{1}{r(t)} \Delta t \right)^{\frac{1}{2}} \left(\int_a^x r(t) \left| y^\Delta(t) \right|^2 \Delta t \right)^{\frac{1}{2}}.$$

(3.1.25)

Since $y^\sigma = y + \mu y^\Delta$, we have

$$y(x) + y^\sigma(x) = 2y(x) + \mu y^\Delta(x).$$

(3.1.26)

Setting

$$z(x) := \int_a^x r(t) \left| y^\Delta(t) \right|^2 \Delta t,$$

(3.1.27)

we see that $z(a) = 0$, and

$$z^\Delta(x) = r(x) \left| y^\Delta(x) \right|^2 > 0.$$

(3.1.28)

From this, we get that

$$\left|y^\Delta(x)\right|^2 = \frac{z^\Delta(x)}{r(x)}, \quad \text{and} \quad \left|y^\Delta(x)\right| = \left(\frac{z^\Delta(x)}{r(x)}\right)^{\frac{1}{2}}. \tag{3.1.29}$$

From (3.1.25)–(3.1.29), we have that

$$s(x)\left|y(x) + y^\sigma(x)\right|\left|y^\Delta(x)\right|$$

$$\leq 2|s(x)|\,|y(x)|\left|y^\Delta(x)\right| + \mu s(x)\left|y^\Delta\right|^2 \leq 2|s(x)|\left(\frac{1}{r(x)}\right)^{\frac{1}{2}} \times \left(\int_a^x \frac{1}{r(t)}\Delta t\right)^{\frac{1}{2}}$$

$$\times (z(x))^{\frac{1}{2}}\left(z^\Delta(x)\right)^{\frac{1}{2}} + \mu(x)\,|s(x)|\left(\frac{z^\Delta(x)}{r(x)}\right).$$

This implies that

$$\int_a^\tau s(x)\left|y(x) + y^\sigma(x)\right|\left|y^\Delta(x)\right|\Delta x \leq 2\int_a^\tau |s(x)|\left(\frac{1}{r(x)}\right)^{\frac{1}{2}}$$

$$\times \left(\int_a^x \frac{1}{r(t)}\Delta t\right)^{\frac{1}{2}}(z(x))^{\frac{1}{2}}\left(z^\Delta(x)\right)^{\frac{1}{2}}\Delta x + \int_a^\tau \left(\mu(x)\frac{|s(x)|}{r(x)}\right)z^\Delta(x)\Delta x$$

$$\leq 2\int_a^\tau |s(x)|\left(\frac{1}{r(x)}\right)^{\frac{1}{2}} \times \left(\int_a^x \frac{1}{r(t)}\Delta t\right)^{\frac{1}{2}}$$

$$\times (z(x))^{\frac{1}{2}}\left(z^\Delta(x)\right)^{\frac{1}{2}}\Delta x + \max_{a\leq x\leq\tau}\left(\mu\frac{|s(x)|}{r(x)}\right)\int_a^\tau z^\Delta(x)\Delta x. \tag{3.1.30}$$

Apply the Cauchy–Schwarz inequality and we have

$$\int_a^\tau s(x)\left|y(x) + y^\sigma(x)\right|\left|y^\Delta(x)\right|\Delta x$$

$$\leq 2\left(\int_a^\tau s^2(x)\left(\frac{1}{r(x)}\right)\left(\int_a^x \frac{1}{r(t)}\Delta t\right)\Delta x\right)^{\frac{1}{2}}\left(\int_a^\tau z(x)z^\Delta(x)\Delta x\right)^{\frac{1}{2}}$$

$$+ \sup_{a\leq x\leq\tau}\left(\mu(x)\frac{|s(x)|}{r(x)}\right)\int_a^\tau z^\Delta(x)\Delta x. \tag{3.1.31}$$

From (3.1.28), and the chain rule (1.1.7), we obtain

$$2z(x)z^\Delta(x) \leq \left(z^2(x)\right)^\Delta. \tag{3.1.32}$$

Substituting (3.1.32) into (3.1.31) and using the fact that $z(a) = 0$, we see that

$$\int_a^\tau s(x)\left|y(x) + y^\sigma(x)\right|\left|y^\Delta(x)\right|\Delta x$$

$$\leq 2\left(\int_a^\tau s^2(x)\frac{1}{r(x)}\left(\int_a^x \frac{1}{r(t)}\Delta t\right)^2\Delta x\right)^{\frac{1}{2}} \times \left(\frac{1}{2}\right)^{\frac{1}{2}}\left(\int_a^\tau \left(z^2(t)\right)^\Delta\Delta t\right)^{\frac{1}{2}}$$

$$= \sqrt{2}\left(\int_a^\tau s^2(x)\frac{1}{r(x)}\left(\int_a^x \frac{1}{r(t)}\Delta t\right)\Delta x\right)^{\frac{1}{2}}z(\tau) + \sup_{a\leq x\leq\tau}\left(\mu(x)\frac{|s(x)|}{r(x)}\right)z(\tau).$$

Using (3.1.27), we have from the last inequality that

$$\int_a^\tau s(x) |y(x) + y^\sigma(x)| |y^\Delta(x)| \Delta x \le K_1(a, \tau) \int_a^\tau r(x) |y^\Delta(x)|^2 \Delta x,$$

which is the desired inequality (3.1.23) where $K_1(a, \tau)$ is defined as in (3.1.24). The proof is complete. ■

Here, we only state the following theorem, since its proof is similar to the proof of Theorem 3.1.11, with $[a, \tau]$ replaced by $[\tau, b]$ and $|y(x)| = \int_x^b |y^\Delta(t)| \Delta t$.

Theorem 3.1.12 *Let $\mathbb{T}$ be a time scale with τ, $b \in \mathbb{T}$. Assume that $s \in C_{rd}([\tau, b]_\mathbb{T}, \mathbb{R})$ and r be a positive rd-continuous function on $(\tau, b)_\mathbb{T}$ such that $\int_\tau^b r^{-1}(t)\Delta t < \infty$. If $y : [\tau, b]_\mathbb{T} \to \mathbb{R}$ is delta differentiable with $y(b) = 0$ (and y^Δ does not change sign in $(\tau, b)_\mathbb{T}$), then we have*

$$\int_\tau^b s(x) |y(x) + y^\sigma(x)| |y^\Delta(x)| \Delta x \le K_2(\tau, b) \int_\tau^b r(x) |y^\Delta(x)|^2 \Delta x, \quad (3.1.33)$$

where

$$K_2(\tau, b) = \sqrt{2} \left(\int_\tau^b \frac{s^2(x)}{r(x)} \left(\int_x^b \frac{\Delta t}{r(t)} \right) \Delta x \right)^{\frac{1}{2}} + \sup_{\tau \le x \le b} \left(\mu(x) \frac{|s(x)|}{r(x)} \right). \quad (3.1.34)$$

In the following, we assume that there exists $\tau \in (a, b)$ which is the unique solution of the equation

$$K(a, b) = K_1(a, \tau) = K_2(\tau, b) < \infty, \quad (3.1.35)$$

where $K_1(a, \tau)$ and $K_2(\tau, b)$ are defined as in Theorems 3.1.11 and 3.1.12 and we establish an inequality when $y(a) = 0 = y(b)$.

Theorem 3.1.13 *Let $\mathbb{T}$ be a time scale with a, $b \in \mathbb{T}$. Assume that $s \in C_{rd}([a, b]_\mathbb{T}, \mathbb{R})$ and r be a positive rd-continuous function on $[a, b]_\mathbb{T}$ such that $\int_a^b r^{-1}(t)\Delta t < \infty$. If $y : [a, b]_\mathbb{T} \to \mathbb{R}$ is delta differentiable with $y(a) = 0 = y(b)$ (and y^Δ does not change sign in $(a, b)_\mathbb{T}$), then we have*

$$\int_a^b s(x) |y(x) + y^\sigma(x)| |y^\Delta(x)| \Delta x \le K(a, b) \int_a^b r(x) |y^\Delta(x)|^2 \Delta x, \quad (3.1.36)$$

where $K(a, b)$ is given as in (3.1.35).

Proof. Since

$$\int_a^b s(x) |y(x) + y^\sigma(x)| |y^\Delta(x)| \Delta x = \int_a^\tau s(x) |y(x) + y^\sigma(x)| |y^\Delta(x)| \Delta x$$

$$+ \int_\tau^b s(x) |y(x) + y^\sigma(x)| |y^\Delta(x)| \Delta x.$$

The rest of the proof is a combination of Theorems 3.1.11 and 3.1.12. ■

Corollary 3.1.3 *Let* $\mathbb{T}$ *be a time scale with* a, $\tau \in \mathbb{T}$, *and let* r *be a positive rd-continuous function on* $[a, \tau]_{\mathbb{T}}$ *such that* $\int_a^\tau r^{-1}(t)\Delta t < \infty$. *If* $y : [a, \tau]_{\mathbb{T}} \to \mathbb{R}$ *is delta differentiable with* $y(a) = 0$ *(and* y^Δ *does not change sign in* $(a, \tau)_{\mathbb{T}})$, *then we have*

$$\int_a^\tau r(x) \left| y(x) + y^\sigma(x) \right| \left| y^\Delta(x) \right| \Delta x \le K_1^*(a, \tau) \int_a^\tau r(x) \left| y^\Delta(x) \right|^2 \Delta x,$$

(3.1.37)

where

$$K_1^*(a, \tau) = \sqrt{2} \left(\int_a^\tau r(x) \left(\int_a^x \frac{\Delta t}{r(t)} \right) \Delta x \right)^{\frac{1}{2}} + \sup_{a \le x \le \tau} (\mu(x)).$$

(3.1.38)

Corollary 3.1.4 *Let* $\mathbb{T}$ *be a time scale with* τ, $b \in \mathbb{T}$, *and let* r *be a positive rd-continuous function on* $(\tau, b)_{\mathbb{T}}$ *such that* $\int_\tau^b r^{-1}(t)\Delta t < \infty$. *If* $y : [\tau, b]_{\mathbb{T}} \to \mathbb{R}$ *is delta differentiable with* $y(b) = 0$ *(and* y^Δ *does not change sign in* $(\tau, b)_{\mathbb{T}})$, *then we have*

$$\int_\tau^b r(x) \left| y(x) + y^\sigma(x) \right| \left| y^\Delta(x) \right| \Delta x \le K_2^*(\tau, b) \int_\tau^b r(x) \left| y^\Delta(x) \right|^2 \Delta x,$$

(3.1.39)

where

$$K_2^*(X, b) = \sqrt{2} \left(\int_\tau^b r(x) \left(\int_x^b \frac{\Delta t}{r(t)} \right) \Delta x \right)^{\frac{1}{2}} + \sup_{\tau \le x \le b} (\mu(x)).$$

(3.1.40)

In the following, we assume that there exists $\tau \in (a, b)$, which is the unique solution of the equation

$$K^*(a, b) = K_1^*(a, \tau) = K_2^*(\tau, b) < \infty,$$

where $K_1^*(a, \tau)$ and $K_2^*(\tau, b)$ are defined in Corollaries 3.1.3 and 3.1.4. Using this and Theorem 3.1.13 we obtain the following result.

Corollary 3.1.5 *Let* $\mathbb{T}$ *be a time scale with* a, $b \in \mathbb{T}$ *and let* r *be a positive rd-continuous function on* $(a, b)_{\mathbb{T}}$ *such that* $\int_a^b r^{-1}(t)\Delta t < \infty$. *If* $y : [a, b]_{\mathbb{T}} \to \mathbb{R}$ *is delta differentiable with* $y(a) = 0 = y(b)$ *(and* y^Δ *does not change sign in* $(a, b)_{\mathbb{T}})$, *then we have*

$$\int_a^b r(x) \left| y(x) + y^\sigma(x) \right| \left| y^\Delta(x) \right| \Delta x \le K^*(a, b) \int_a^b r(x) \left| y^\Delta(x) \right|^2 \Delta x. \quad (3.1.41)$$

On a time scale $\mathbb{T}$, we note from the chain rule (1.1.7) that

$$\left((t-a)^2 \right)^\Delta = 2 \int_0^1 [h(\sigma(t) - a) + (1-h)(t-a)] \, dh$$

$$\ge 2 \int_0^1 [h(t-a) + (1-h)(t-a)] \, dh = 2(t-a).$$

This implies that

$$\int_a^\tau (x-a)\Delta x \le \int_a^\tau \frac{1}{2}\left((x-a)^2\right)^\Delta \Delta x = \frac{(\tau-a)^2}{2}. \tag{3.1.42}$$

From this and (3.1.39) (by putting $r(t) = 1$), we get that

$$\begin{aligned} K_1^*(a,\tau) &= \sqrt{2}\left(\int_a^\tau (x-a)\,\Delta x\right)^{\frac{1}{2}} \le \sqrt{2}\left(\frac{(\tau-a)^2}{2}\right)^{\frac{1}{2}} + \max_{a\le x\le \tau}(\mu(x)) \\ &= \max_{a\le x\le \tau}(\mu(x)) + (\tau-a). \end{aligned} \tag{3.1.43}$$

Corollary 3.1.6 *Let $\mathbb{T}$ be a time scale with a, $\tau \in \mathbb{T}$. If $y : [a,\tau]_\mathbb{T} \to \mathbb{R}$ is delta differentiable with $y(a) = 0$ (and y^Δ does not change sign in $(a,\tau)_\mathbb{T}$), then we have*

$$\int_a^\tau |y(x) + y^\sigma(x)|\,|y^\Delta(x)|\,\Delta x \le \left((\tau-a) + \sup_{a\le x\le \tau}\mu(x)\right)\int_a^\tau |y^\Delta(x)|^2\,\Delta x. \tag{3.1.44}$$

In Corollary 3.1.5, we note that if $r(t) = 1$, then the unique solution of Eq. (3.1.35) is given by $h = (a+b)/2$. This gives us the following result.

Corollary 3.1.7 *Let $\mathbb{T}$ be a time scale with a, $b \in \mathbb{T}$. If $y : [a,b]_\mathbb{T} \to \mathbb{R}$ is delta differentiable with $y(a) = 0 = y(b)$ (and y^Δ does not change sign in $(a,b)_\mathbb{T}$), then we have*

$$\int_a^b |y(x) + y^\sigma(x)|\,|y^\Delta(x)|\,\Delta x \le \left(\frac{b-a}{2} + \sup_{a\le x\le b}\mu(x)\right)\int_a^b |y^\Delta(x)|^2\,\Delta x. \tag{3.1.45}$$

Remark 3.1.2 *In Corollary 3.1.7 if $\mathbb{T} = \mathbb{R}$ then $\mu(x) = 0$, $\sigma(x) = x$, $y(x) = y^\sigma(x)$ and the inequality (3.1.45) reduces to the original Opial inequality (3.0.1).*

3.2   Opial Type Inequalities II

In this section we give some other Opial type inequalities on time scales. The results are adapted from [124, 125, 138, 139].

Theorem 3.2.1 *Let $a,\tau \in \mathbb{T}$ and $r \in C_{rd}([a,\tau]_\mathbb{T}, \mathbb{R}^+)$ be such that $r(t)$ is nonincreasing on $[a,\tau]_\mathbb{T}$ and $p \ge 0$ and $q \ge 1$. Suppose that $x : [a,\tau]_\mathbb{T} \to \mathbb{R}$ is delta differentiable with $x(a) = 0$. Then*

$$\int_a^\tau r(t)\,|x(t)|^p\,|x^\Delta(t)|^q\,\Delta t \le \frac{q(\tau-a)^p}{p+q}\int_a^\tau r(t)\,|x^\Delta(t)|^{p+q}\,\Delta t. \tag{3.2.1}$$

Proof. Suppose that the function $g(t)$ is defined by

$$g(t) := \int_a^t r^{\frac{q}{p+q}}(s)\left|x^\Delta(s)\right|^q \Delta s,$$

so that

$$g(a) = 0, \quad \text{and} \quad g^\Delta(t) = r^{\frac{q}{p+q}}(t)\left|x^\Delta(t)\right|^q > 0. \tag{3.2.2}$$

In the case when $q > 1$, by using Hölder's inequality with indices q and $q/(q-1)$, we have

$$
\begin{aligned}
|x(t)| &\leq \int_a^t \left|x^\Delta(s)\right| \Delta s = \int_a^t r^{\frac{-1}{p+q}}(s) r^{\frac{1}{p+q}}(s)\left|x^\Delta(s)\right| \Delta s \\
&\leq \left(\int_a^t \left(r^{\frac{-1}{p+q}}(s)\right)^{\frac{q}{q-1}} \Delta s\right)^{\frac{q-1}{q}} \left(\int_a^t r^{\frac{q}{p+q}}(s)\left|x^\Delta(s)\right|^q\right)^{\frac{1}{q}} \Delta s \\
&\leq r^{\frac{-1}{p+q}}(t)(t-a)^{\frac{q-1}{q}} g^{\frac{1}{q}}(t),
\end{aligned}
$$

which yields that

$$r^{\frac{p}{p+q}}(t)|x(t)|^p \leq (t-a)^{\frac{p(q-1)}{q}} g^{\frac{p}{q}}(t). \tag{3.2.3}$$

In the case, when $q = 1$, we find that

$$
\begin{aligned}
|x(t)| &\leq \int_a^t \left|x^\Delta(s)\right| \Delta s = \int_a^t r^{\frac{-1}{p+1}}(s) r^{\frac{1}{p+1}}(s)\left|x^\Delta(s)\right| \Delta s \\
&\leq r^{\frac{-1}{p+1}}(t)\int_a^t r^{\frac{1}{p+1}}(s)\left|x^\Delta(s)\right| \Delta s = r^{\frac{-1}{p+1}}(t)g(t),
\end{aligned}
$$

which shows that the inequality (3.2.3) holds also when $q = 1$. Now, from (3.2.2) and (3.2.3), we see that

$$
\begin{aligned}
\int_a^\tau r(s)|x(s)|^p \left|x^\Delta(s)\right|^q \Delta s &= \int_a^\tau r^{\frac{p}{p+q}}(s)|x(s)|^p\, r^{\frac{q}{p+q}}(s)\left|x^\Delta(s)\right|^q \Delta s \\
&\leq \int_a^\tau (s-a)^{\frac{p(q-1)}{q}} g^{\frac{p}{q}}(s)g^\Delta(s)\Delta s \\
&\leq (\tau-a)^{\frac{p(q-1)}{q}} \int_a^\tau g^{\frac{p}{q}}(s)g^\Delta(s)\Delta s.
\end{aligned}
$$

From (3.2.2) and the chain rule (1.1.7), we obtain

$$g^{\frac{p}{q}}(s)g^\Delta(s) \leq \frac{q}{p+q}\left(g^{\frac{p+q}{q}}(s)\right)^\Delta.$$

This and the fact that $g(a) = 0$ imply that

$$
\begin{aligned}
\int_a^\tau r(s)|x(s)|^p \left|x^\Delta(s)\right|^q \Delta s &\leq \frac{q}{p+q}(\tau-a)^{\frac{p(q-1)}{q}} \int_a^\tau \left(g^{\frac{p+q}{q}}(s)\right)^\Delta \Delta s \\
&= \frac{q}{p+q}(\tau-a)^{\frac{p(q-1)}{q}} \left(g^{\frac{p+q}{q}}(\tau)\right). \tag{3.2.4}
\end{aligned}
$$

By Hölder's inequality with indices $(p+q)/p$ and $q/(p+q)$, we see that

$$
\begin{aligned}
g(\tau) &= \int_a^\tau r^{\frac{q}{p+q}}(s) \left|x^\Delta(s)\right|^q \Delta s \leq \left(\int_a^\tau 1 \Delta s\right)^{\frac{p}{p+q}} \\
&\quad \left(\int_a^\tau \left(r^{\frac{q}{p+q}}(s) \left|x^\Delta(s)\right|^q\right)^{\frac{p+q}{q}} \Delta s\right)^{\frac{q}{p+q}} \\
&= (\tau-a)^{\frac{p}{p+q}} \left(\int_a^\tau \left(r^{\frac{q}{p+q}}(s) \left|x^\Delta(s)\right|^q\right)^{\frac{p+q}{q}} \Delta s\right)^{\frac{q}{p+q}}. \quad (3.2.5)
\end{aligned}
$$

From (3.2.4) and (3.2.5), we have

$$
\int_a^\tau r(s) |x(s)|^p \left|x^\Delta(s)\right|^q \Delta s \leq \frac{q}{p+q}(\tau-a)^p \int_a^\tau r(s) \left|x^\Delta(s)\right|^{p+q} \Delta s,
$$

which is the desired inequality (3.2.1). The proof is complete. ∎

Remark 3.2.1 *When* $\mathbb{T} = \mathbb{R}$, *we see that the inequality (3.2.1) reduces to the Yang [153] inequality*

$$
\int_a^b r(t) |x(t)|^p \left|x'(t)\right|^q dt \leq \frac{q}{p+q}(b-a)^p \int_a^b r(t) \left|x'(t)\right|^{p+q} dt, \quad (3.2.6)
$$

where $r(t)$ *is a positive nonincreasing function and* x *is a continuous function on* $[a,b]$ *with* $x(a) = 0$, $p \geq 0$, $q \geq 1$. *When* $r(t) = 1$, *we get the Yang [152] inequality*

$$
\int_a^b |x(t)|^p \left|x'(t)\right|^q dt \leq \frac{q}{p+q}(b-a)^p \int_a^b \left|x'(t)\right|^{p+q} dt, \quad (3.2.7)
$$

where x *is a continuous function on* $[a,b]$ *with* $x(a) = 0$, $p \geq 0$, *and* $q \geq 1$.

Remark 3.2.2 *When* $q = 1$, *we get the Hua [86] inequality*

$$
\int_a^b |x(t)|^p \left|x'(t)\right| dt \leq \frac{(b-a)^p}{p+1} \int_a^b \left|x'(t)\right|^{p+1} dt, \quad (3.2.8)
$$

where p *is a positive integer and* x *is a continuous function with* $x(a) = 0$.

Remark 3.2.3 *Beesack and Das [42] showed that the inequalities (3.2.7) and (3.2.6) are sharp when* $q = 1$ *but are not sharp for* $q > 1$.

Theorem 3.2.2 *Let* $a, b \in \mathbb{T}$ *and* $r \in C_{rd}([\tau,b]_\mathbb{T}, \mathbb{R}^+)$ *be such that* $r(t)$ *is nonincreasing on* $[\tau,b]_\mathbb{T}$ *and* $p \geq 0$ *and* $q \geq 1$. *Suppose that* $x : [\tau,b]_\mathbb{T} \to \mathbb{R}$ *is delta differentiable with* $x(b) = 0$. *Then*

$$
\int_\tau^b r(t) |x(t)|^p \left|x^\Delta(t)\right|^q \Delta t \leq \frac{q}{p+q}(b-\tau)^p \int_\tau^b r(t) \left|x^\Delta(t)\right|^{p+q} \Delta t. \quad (3.2.9)
$$

Proof. Let

$$g(t) = \int_t^b r^{\frac{q}{p+q}}(s) \left| x^\Delta(s) \right|^q \Delta s.$$

Then

$$g(b) = 0, \quad \text{and} \quad g^\Delta(t) = -r^{\frac{q}{p+q}}(t) \left| x^\Delta(t) \right|^q > 0. \tag{3.2.10}$$

In the case when $q > 1$, by using Hölder's inequality with indices q and $q/(q-1)$, we have

$$
\begin{aligned}
|x(t)| &\leq \int_t^b \left| x^\Delta(s) \right| \Delta s = \int_t^b r^{\frac{-1}{p+q}}(s) r^{\frac{1}{p+q}}(s) \left| x^\Delta(s) \right| \Delta s \\
&\leq \left(\int_t^b \left(r^{\frac{-1}{p+q}}(s) \right)^{\frac{q}{q-1}} \Delta s \right)^{\frac{q-1}{q}} \left(\int_t^b r^{\frac{q}{p+q}}(s) \left| x^\Delta(s) \right|^q \right)^{\frac{1}{q}} \Delta s \\
&\leq r^{\frac{-1}{p+q}}(t) (b-t)^{\frac{q-1}{q}} g^{\frac{1}{q}}(t),
\end{aligned}
$$

which yields that

$$r^{\frac{p}{p+q}}(t) |x(t)|^p \leq (b-t)^{\frac{p(q-1)}{q}} g^{\frac{p}{q}}(t). \tag{3.2.11}$$

In the case, when $q = 1$, we find that

$$
\begin{aligned}
|x(t)| &\leq \int_t^b \left| x^\Delta(s) \right| \Delta s = \int_t^b r^{\frac{-1}{p+1}}(s) r^{\frac{1}{p+1}}(s) \left| x^\Delta(s) \right| \Delta s \\
&\leq r^{\frac{-1}{p+1}}(t) \int_t^b r^{\frac{1}{p+1}}(s) \left| x^\Delta(s) \right| \Delta s = r^{\frac{-1}{p+1}}(t) g(t),
\end{aligned}
$$

which shows that the inequality (3.2.11) holds also when $q = 1$. Thus from (3.2.10) and (3.2.11), we see that

$$
\begin{aligned}
\int_\tau^b r(s) |x(s)|^p \left| x^\Delta(s) \right|^q \Delta s &= \int_\tau^b r^{\frac{p}{p+q}}(s) |x(s)|^p \, r^{\frac{q}{p+q}}(s) \left| x^\Delta(s) \right|^q \Delta s \\
&\leq \int_\tau^b (b-s)^{\frac{p(q-1)}{q}} \left[-g^{\frac{p}{q}}(s) \right] g^\Delta(s) \Delta s \\
&\leq (b-\tau)^{\frac{p(q-1)}{q}} \int_\tau^b \frac{q}{p+q} \left(-g^{\frac{p+q}{q}}(s) \right)^\Delta \Delta s.
\end{aligned}
$$

This and the fact that $g(b) = 0$ imply that

$$\int_a^\tau r(s) |x(s)|^p \left| x^\Delta(s) \right|^q \Delta s \leq \frac{q}{p+q} (b-\tau)^{\frac{p(q-1)}{q}} \left(g^{\frac{p+q}{q}}(\tau) \right).$$

From Hölder's inequality we get

$$\int_\tau^b r(s) |x(s)|^p \left| x^\Delta(s) \right|^q \Delta s \leq \frac{q}{p+q} (b-\tau)^p \int_\tau^b r(s) \left| x^\Delta(s) \right|^{p+q} \Delta s,$$

which is the desired inequality (3.2.9). The proof is complete. ∎

The combination of Theorems 3.2.1 and 3.2.2 by choosing $\tau = (a + b)/2$ gives us the following result.

Theorem 3.2.3 *Let $a, b \in \mathbb{T}$ and $r \in C_{rd}([a, b]_{\mathbb{T}}, \mathbb{R}^+)$ be such that $r(t)$ is nonincreasing on $[a, b]_{\mathbb{T}}$ and nonincreasing on $[a, b]_{\mathbb{T}}$ and $p \geq 0$ and $q \geq 1$. Suppose that $x : [a, b]_{\mathbb{T}} \to \mathbb{R}$ is delta differentiable with $x(a) = x(b) = 0$. Then*

$$\int_a^b r(t) |x(t)|^p |x^\Delta(t)|^q \, \Delta t \leq \frac{q}{p+q} \left(\frac{b-a}{2} \right)^p \int_a^b r(t) |x^\Delta(t)|^{p+q} \, \Delta t.$$

Theorem 3.2.4 *Let $\mathbb{T}$ be a time scale with $a, \tau \in \mathbb{T}$ and $p \in C_{rd}([a, \tau), \mathbb{R})$ with $\int_a^\tau (p(t))^{1-\alpha} \, \Delta t < \infty$, $\alpha > 1$. Let the function $q(t)$ be positive, bounded, and nonincreasing on $[a, \tau]_{\mathbb{T}}$. If $x : [a, \tau] \cap \mathbb{T} \to \mathbb{R}$ is delta differentiable with $x(a) = 0$, then for $\gamma > 0$,*

$$\int_a^\tau q(t) |x(t)|^\gamma |x^\Delta(t)| \, \Delta t \ \leq \ \frac{1}{\gamma + 1} \left(\int_a^\tau \frac{1}{p^{\alpha-1}(t)} \Delta t \right)^{\frac{1+\gamma}{\alpha}}$$

$$\times \left(\int_a^\tau p(t) q^{\frac{\nu}{1+\gamma}}(t) |x^\Delta(t)|^\nu \, \Delta t \right)^{\frac{1+\gamma}{\nu}} (3.2.12)$$

where $\frac{1}{\alpha} + \frac{1}{\nu} = 1$.

Proof. Let

$$y(t) = \int_a^t q^{\frac{1}{1+\gamma}}(s) |x^\Delta(s)| \, \Delta s.$$

Then

$$\begin{aligned} y^\gamma(t) &= \left(\int_a^t q^{\frac{1}{1+\gamma}}(s) |x^\Delta(s)| \, \Delta s \right)^\gamma \geq q^{\frac{\gamma}{1+\gamma}}(t) \left(\int_a^t |x^\Delta(s)| \, \Delta s \right)^\gamma \\ &= q^{\frac{\gamma}{1+\gamma}}(t) \left(|x^\Delta(t)| \right)^\gamma, \end{aligned}$$

and

$$y^\Delta(t) = q^{\frac{1}{1+\gamma}}(t) |x^\Delta(t)|.$$

Therefore, we have

$$\begin{aligned} \int_a^\tau q(t) |x(t)|^\gamma |x^\Delta(t)| \, \Delta t &= \int_a^\tau q^{\frac{1}{1+\gamma}}(t) q^{\frac{\gamma}{1+\gamma}}(t) |x(t)|^\gamma |x^\Delta(t)| \, \Delta t \\ &\leq \int_a^\tau q^{\frac{1}{1+\gamma}}(t) q^{\frac{\gamma}{1+\gamma}}(t) |x(t)|^\gamma |x^\Delta(t)| \, \Delta t \\ &\leq \int_a^\tau y^\gamma(t) y^\Delta(t) \Delta t = \frac{1}{\gamma + 1} y^{\gamma+1}(\tau) \\ &= \frac{1}{\gamma + 1} \left(\int_a^\tau q^{\frac{1}{1+\gamma}}(s) |x^\Delta(s)| \, \Delta s \right)^{\gamma+1}. \end{aligned}$$

In the case when $\alpha > 1$, by using the Hölder inequality with indices α and ν, we have that

$$\int_a^\tau q^{\frac{1}{1+\gamma}}(s)\left|x^\Delta(s)\right|\Delta s = \int_a^\tau p^{\frac{-1}{\nu}}(s)q^{\frac{1}{1+\gamma}}(s)p^{\frac{1}{\nu}}(s)\left|x^\Delta(s)\right|\Delta s$$

$$= \left(\int_a^\tau p^{\frac{-\alpha}{\nu}}(s)\Delta s\right)^{\frac{1}{\alpha}}\left(\int_a^\tau q^{\frac{\nu}{1+\gamma}}(s)p(s)\left|x^\Delta(s)\right|^\nu\Delta s\right)^{\frac{1}{\nu}}.$$

This implies that

$$\int_a^\tau q(t)\,|x(t)|^\gamma\left|x^\Delta(t)\right|\Delta t \le \frac{1}{\gamma+1}\left(\int_a^\tau p^{1-\alpha}(s)\Delta s\right)^{\frac{\gamma+1}{\alpha}}$$

$$\times\left(\int_a^\tau q^{\frac{\nu}{1+\gamma}}(s)p(s)\left|x^\Delta(s)\right|^\nu\Delta s\right)^{\frac{\gamma+1}{\nu}},$$

which is the desired inequality (3.2.12). The proof is complete. ∎

A slight modification of the argument above yields the following result.

Theorem 3.2.5 *Let* $\mathbb{T}$ *be a time scale with* $a,\ \tau \in \mathbb{T}$ *and* $p \in C_{rd}([\tau,b),\mathbb{R})$ *with* $\int_\tau^b (p(t))^{1-\alpha}\,\Delta t < \infty,\ \alpha > 1.$ *Let the function* $q(t)$ *be positive, bounded, and nondecreasing on* $[\tau,b]_\mathbb{T}.$ *If* $x : [\tau,b]\cap\mathbb{T} \to \mathbb{R}$ *is delta differentiable with* $x(b) = 0,$ *then for* $\gamma > 0,$

$$\int_\tau^b q(t)\,|x(t)|^\gamma\left|x^\Delta(t)\right|\Delta t \le \frac{1}{\gamma+1}\left(\int_\tau^b \frac{1}{p^{\alpha-1}(t)}\Delta t\right)^{\frac{1+\gamma}{\alpha}}$$

$$\times\left(\int_\tau^b p(t)q^{\frac{\nu}{1+\gamma}}(t)\left|x^\Delta(t)\right|^\nu\Delta t\right)^{\frac{1+\gamma}{\nu}} \quad (3.2.13)$$

where $\frac{1}{\alpha} + \frac{1}{\nu} = 1.$

Theorem 3.2.6 *Let* $\mathbb{T}$ *be a time scale with* $a,\ b \in \mathbb{T}$ *and* $\tau \in [a,b]_\mathbb{T}.$ *Let* $p \in C_{rd}([a,\tau),\mathbb{R})$ *with* $\int_a^\tau (p(t))^{1-\alpha}\,\Delta t < \infty,$ *and* $\int_\tau^b (p(t))^{1-\alpha}\,\Delta t < \infty,\ \alpha \ge 1+\gamma.$ *Let the function* $q(t)$ *be positive, bounded, and nonincreasing on* $[a,\tau]_\mathbb{T}$ *and nondecreasing in* $[\tau,b]_\mathbb{T}.$ *Suppose that*

$$\chi := \left(\int_a^\tau \frac{1}{p^{\alpha-1}(t)}\Delta t\right)^{\frac{1+\gamma}{\alpha}} = \left(\int_\tau^b \frac{1}{p^{\alpha-1}(t)}\Delta t\right)^{\frac{1+\gamma}{\alpha}}.$$

If $x : [a,b]_\mathbb{T} \to \mathbb{R}$ *is delta differentiable with* $x(a) = x(b) = 0,$ *then for* $\gamma > 0,$

$$\int_a^b q(t)\,|x(t)|^\gamma\left|x^\Delta(t)\right|\Delta t \le \frac{\chi}{\gamma+1}\left(\int_a^b p(t)q^{\frac{\nu}{1+\gamma}}(t)\left|x^\Delta(t)\right|^\nu\Delta t\right)^{\frac{1+\gamma}{\nu}},$$

$$(3.2.14)$$

where $\frac{1}{\alpha} + \frac{1}{\nu} = 1.$

Proof. Since

$$\int_a^b q(t) \, |x(t)|^\gamma \, \left|x^\Delta(t)\right| \Delta t = \int_a^\tau q(t) \, |x(t)|^\gamma \, \left|x^\Delta(t)\right| \Delta t + \int_\tau^b q(t) \, |x(t)|^\gamma \, \left|x^\Delta(t)\right| \Delta t,$$

we have by Theorems 3.2.4, 3.2.5 and the inequality

$$a^\gamma + b^\gamma \le (a+b)^\gamma, \text{ for } a,\, b > 0 \text{ and } \gamma \ge 1,$$

that

$$\int_a^b q(t) \, |x(t)|^\gamma \, \left|x^\Delta(t)\right| \Delta t$$

$$\le \frac{1}{\gamma+1} \left(\int_a^\tau \frac{1}{p^{\alpha-1}(t)} \Delta t \right)^{\frac{1+\gamma}{\alpha}} \times \left(\int_a^\tau p(t) q^{\frac{\nu}{1+\gamma}}(t) \left|x^\Delta(t)\right|^\nu \Delta t \right)^{\frac{1+\gamma}{\nu}}$$

$$+ \frac{1}{\gamma+1} \left(\int_\tau^b p^{1-\alpha}(s) \Delta s \right)^{\frac{\gamma+1}{\alpha}} \left(\int_\tau^b q^{\frac{\nu}{1+\gamma}}(s) p(s) \left|x^\Delta(s)\right|^\nu \Delta s \right)^{\frac{\gamma+1}{\nu}}$$

$$= \frac{\chi}{\gamma+1} \left[\left(\int_a^\tau p(t) q^{\frac{\nu}{1+\gamma}}(t) \left|x^\Delta(t)\right|^\nu \Delta t \right)^{\frac{1+\gamma}{\nu}} + \left(\int_\tau^b q^{\frac{\nu}{1+\gamma}}(s) p(s) \left|x^\Delta(s)\right|^\nu \Delta s \right)^{\frac{\gamma+1}{\nu}} \right]$$

$$\le \frac{\chi}{\gamma+1} \left(\int_a^b p(t) q^{\frac{\nu}{1+\gamma}}(t) \left|x^\Delta(t)\right|^\nu \Delta t \right)^{\frac{1+\gamma}{\nu}},$$

which is the desired inequality (3.2.14). The proof is complete. ∎

Remark 3.2.4 *As a special case of Theorem 3.2.6 when $\mathbb{T} = \mathbb{R}$, we have the Maroni [107] inequality*

$$\int_a^b |x(t)| \, \left|x'(t)\right| dt \le \frac{\chi}{2} \left(\int_a^b p(t) \left|x'(t)\right|^\nu dt \right)^{\frac{2}{\nu}},$$

where $\alpha \ge 1$, $\frac{1}{\alpha} + \frac{1}{\nu} = 1$ and x is an absolutely continuous function on $[a,b]$ with $x(a) = x(b) = 0$ and

$$\chi = \left(\int_a^\tau \left(\frac{1}{p(t)} \right)^{\alpha-1} dt \right)^{\frac{2}{\alpha}} = \left(\int_\tau^b \left(\frac{1}{p(t)} \right)^{\alpha-1} dt \right)^{\frac{2}{\alpha}},$$

where τ is the unique solution of the equation

$$\left(\int_a^\tau \left(\frac{1}{p(t)} \right)^{\alpha-1} dt \right)^{\frac{2}{\alpha}} = \left(\int_\tau^b \left(\frac{1}{p(t)} \right)^{\alpha-1} dt \right)^{\frac{2}{\alpha}}.$$

In the following, we establish an inequality of Opial type which depends on the smallest eigenvalue of a boundary value problems on time scales. We will assume that the boundary value problem

$$\left. \begin{array}{l} (r(t)(u^\Delta(t))^p)^\Delta = \beta s^\Delta(t) u^p(t), \\ u(0) = 0, \; r(b) \left(u^\Delta(b)\right)^p = \beta s(b) u^p(b), \end{array} \right\} \tag{3.2.15}$$

has a solution $u(t)$ such that $u^\Delta(t) \geq 0$ on the interval $[0, b]_{\mathbb{T}}$, where r, s are nonnegative rd-continuous functions on $(0, b)_{\mathbb{T}}$. Let

$$w(t) = \left(\frac{u^\Delta(t)}{u(t)}\right)^p.$$

Theorem 3.2.7 *Let $\mathbb{T}$ be a time scale with 0, $b \in \mathbb{T}$ and let r, s be nonnegative rd-continuous functions on $(0, b)_{\mathbb{T}}$ such that $(3.2.15)$ has a solution for some $\beta > 0$. If $x : [0, b] \cap \mathbb{T} \to \mathbb{R}$ is delta differentiable with $x(0) = 0$, then for $p > 0$,*

$$\int_0^b s(t) |x(t)|^p |x^\Delta(t)| \, \Delta t \leq \frac{1}{(p+1)\beta} \int_0^b r(t) |x^\Delta(t)|^{p+1} \, \Delta t + \frac{p}{(p+1)\beta}$$

$$\times \int_0^b \left[r(t) w^{\frac{p+1}{p}}(t) - r^\sigma(t) w^\sigma(t) w^{\frac{1}{p}}(t) \right] |x^\sigma(t)|^{p+1} \, \Delta t. \qquad (3.2.16)$$

Proof. Let

$$f(t) = |x^\Delta(t)| \quad \text{and} \quad F(t) = \int_0^t f(t)\Delta t.$$

Using the inequality

$$A^{p+1} + pB^{p+1} - (p+1)AB^p \geq 0, \text{ for all } A \neq B > 0 \text{ and } p > 0, \qquad (3.2.17)$$

and substituting f for A and $w^{\frac{1}{p}} F^\sigma$ for B, we obtain

$$f^{p+1} + pw^\lambda (F^\sigma)^{p+1} - (p+1)fw (F^\sigma)^p \geq 0, \text{ where } \lambda = \frac{(p+1)}{p}.$$

Multiplying this inequality by $r(t)$ and integrating from 0 to b and using the fact that $F^\Delta(t) = f(t) > 0$, we have

$$\int_0^b r(t) f^{p+1}(t) \Delta t + p \int_0^b r(t) w^\lambda(t) (F^\sigma(t))^{p+1} \, \Delta t$$

$$\geq (p+1) \int_0^b r(t) w(t) f(t) (F^\sigma(t))^p \, \Delta t$$

$$= (p+1) \int_0^b r(t) w(t) (F^\sigma(t))^p F^\Delta(t) \Delta t. \qquad (3.2.18)$$

By the chain rule $(1.1.7)$ and the fact that $F^\Delta(t) > 0$, we obtain

$$\left(F^{p+1}(t)\right)^\Delta = (p+1) \int_0^1 [(1-h) F(t) + hF^\sigma(t)]^p \, dh F^\Delta(t). \qquad (3.2.19)$$

Also note

$$\int_t^{\sigma(t)} f(s)\Delta s = F(\sigma(t)) - F(t) = \mu(t)F^\Delta(t) = \mu(t)f(t) > 0.$$

From the definition of $F(t)$, we see that

$$F^\sigma(t) \geq F(t). \tag{3.2.20}$$

Substituting this into (3.2.19), we see that

$$(p+1)\left[F(t)\right]^p F^\Delta(t) \leq \left(F^{p+1}(t)\right)^\Delta \leq (p+1)\left[F^\sigma(t)\right]^p F^\Delta(t). \tag{3.2.21}$$

Substituting (3.2.21) into (3.2.18), we have

$$\int_0^b r(t)f^{p+1}(t)\Delta t + p\int_0^b r(t)w^\lambda(t)\left(F^\sigma(t)\right)^{p+1}\Delta t$$

$$\geq \quad \int_0^b r(t)w(t)\left(F^{p+1}(t)\right)^\Delta \Delta t. \tag{3.2.22}$$

Integrating by parts and using the assumption $F(0) = 0$, we see that

$$\int_0^b r(t)w(t)\left(F^{p+1}(t)\right)^\Delta \Delta t$$

$$= \quad r(t)w(t)F^{p+1}(t)\big|_0^b - \int_0^b (r(t)w(t))^\Delta \left(F^\sigma(t)\right)^{p+1}\Delta t$$

$$= \quad r(b)w(b)F^{p+1}(b) - \int_0^b (r(t)w(t))^\Delta \left(F^\sigma(t)\right)^{p+1}\Delta t. \tag{3.2.23}$$

From (3.2.22) and (3.2.23), we see that

$$\int_0^b r(t)f^{p+1}(t)\Delta t + p\int_0^b r(t)w^\lambda(t)\left(F^\sigma(t)\right)^{p+1}\Delta t$$

$$\geq \quad r(b)w(b)F^{p+1}(b) - \int_0^b (r(t)w(t))^\Delta \left(F^\sigma(t)\right)^{p+1}\Delta t. \tag{3.2.24}$$

From the definition of the function $w(t)$, we see that

$$r(t)w(t) = \frac{r(t)\left(u^\Delta(t)\right)^p}{u^p(t)}. \tag{3.2.25}$$

From this, we obtain that

$$(r(t)w(t))^\Delta = \frac{1}{u^p(t)}\left(r(t)\left(u^\Delta(t)\right)^p\right)^\Delta + \left(r\left(u^\Delta\right)^p\right)^\sigma\left[\frac{-(u^p(t))^\Delta}{u^p(t)u^p(\sigma(t))}\right]. \tag{3.2.26}$$

In view of (3.2.15) and (3.2.26), we get that

$$(r(t)w(t))^\Delta = \beta s^\Delta(t) - \frac{\left(r\left(u^\Delta\right)^p\right)^\sigma (u^p(t))^\Delta}{u^p(t)u^p(\sigma(t))}. \tag{3.2.27}$$

Using the fact that $u^\Delta(t) \geq 0$ and the chain rule (1.1.7), we see that

$$
\begin{aligned}
(u^p(t))^\Delta &= p \int_0^1 [hu^\sigma + (1-h)u]^{p-1} u^\Delta(t) dh \\
&\geq p \int_0^1 [hu + (1-h)u]^{p-1} u^\Delta(t) dh \\
&= p(u(t))^{p-1} u^\Delta(t).
\end{aligned}
\tag{3.2.28}
$$

It follows from (3.2.27) and (3.2.28) that

$$
\begin{aligned}
(r(t)w(t))^\Delta &\leq \beta s^\Delta(t) - \frac{\left(r\left(u^\Delta\right)^p\right)^\sigma p(u(t))^{p-1} u^\Delta(t)}{u^p(t) u^p(\sigma(t))} \\
&= \beta s^\Delta(t) - \frac{p\left(r\left(u^\Delta\right)^p\right)^\sigma u^\Delta(t)}{u(t) u^p(\sigma(t))} \\
&= \beta s^\Delta(t) - \frac{p r^\sigma(t) \left(\left(u^\Delta\right)^p\right)^\sigma u^\Delta(t)}{u(t) u^p(\sigma(t))} \\
&= \beta s^\Delta(t) - p r^\sigma(t) w^\sigma(t) w^{\frac{1}{p}}(t).
\end{aligned}
\tag{3.2.29}
$$

From (3.2.29) and (3.2.24), we have

$$
\begin{aligned}
&\int_0^b r(t) f^{p+1}(t) \Delta t + p \int_0^b r(t) w^\lambda(t) \left(F^\sigma(t)\right)^{p+1} \Delta t \\
&\geq r(b) w(b) F^{p+1}(b) - \int_0^b \beta s^\Delta(t) \left(F^\sigma(t)\right)^{p+1} \Delta t \\
&+ p \int_0^b r^\sigma(t) w^\sigma(t) w^{\frac{1}{p}}(t) \left(F^\sigma(t)\right)^{p+1} \Delta t.
\end{aligned}
$$

This implies that

$$
\begin{aligned}
&\int_0^b r(t) f^{p+1}(t) \Delta t \\
&+ p \int_0^b \left[r(t) w^\lambda(t) - r^\sigma(t) w^\sigma(t) w^{\frac{1}{p}}(t)\right] \left(F^\sigma(t)\right)^{p+1} \Delta t \\
&\geq r(b) w(b) F^{p+1}(b) - \int_0^b \beta s^\Delta(t) \left(F^\sigma(t)\right)^{p+1} \Delta t.
\end{aligned}
\tag{3.2.30}
$$

Using integration by parts again and using (3.2.21), we see that

$$-\beta \int_0^b s^\Delta(t)\, (F^\sigma(t))^{p+1}\, \Delta t$$

$$= -\beta\, s(t)\, (F(t))^{p+1}\Big|_0^b + \int_0^b s(t)\, (F^{p+1}(t))^\Delta\, \Delta t$$

$$= -\beta s(b)\, (F(b))^{p+1} + \int_0^b s(t)\, (F^{p+1}(t))^\Delta\, \Delta t$$

$$\geq -\beta s(b)\, (F(b))^{p+1} + \beta(p+1) \int_0^b s(t)\, [F(t)]^p\, F^\Delta(t)\Delta t \quad (3.2.31)$$

Substituting (3.2.31) into (3.2.30), we have

$$\int_0^b r(t) f^{p+1}(t)\Delta t + p \int_0^b \left[r(t) w^\lambda(t) - r^\sigma(t) w^\sigma(t) w^{\frac{1}{p}}(t) \right] (F^\sigma(t))^{p+1}\, \Delta t$$

$$\geq [r(b)w(b - \beta s(b)] F^{p+1}(b) + (p+1)\beta \int_0^b s(t)\, [F(t)]^p\, F^\Delta(t)\Delta t.$$

From this, we obtain

$$\int_0^b r(t) f^{p+1}(t)\Delta t + p \int_0^b \left[r(t) w^\lambda(t) - r^\sigma(t) w^\sigma(t) w^{\frac{1}{p}}(t) \right] (F^\sigma(t))^{p+1}\, \Delta t$$

$$\geq (p+1)\beta \int_0^b s(t) F^p(t) F^\Delta(t)\Delta t.$$

This implies that

$$\int_0^b r(t) f^{p+1}(t)\Delta t + p \int_0^b \left[r(t) w^\lambda(t) - r^\sigma(t) w^\sigma(t) w^{\frac{1}{p}}(t) \right] (F^\sigma(t))^{p+1}\, \Delta t$$

$$\geq (p+1)\beta \int_0^b s(t) F^p(t) F^\Delta(t)\Delta t,$$

which is the desired inequality (3.2.16) after replacing f by $x^\Delta(t)$ and F by $x(t)$. The proof is complete. ∎

Remark 3.2.5 *Note that when $\mathbb{T} = \mathbb{R}$, we have $r(t) = r^\sigma(t)$ and $w(t) = w^\sigma(t)$. Then (3.2.16) reduces to the inequality*

$$\int_0^a s(t)\, |x(t)|^p \left| x^{'}(t) \right| dt \leq \frac{1}{\lambda_0(p+1)} \int_0^a r(t) \left| x^{'}(t) \right|^{p+1} dt, \qquad (3.2.32)$$

due to Boyd and Wong [54], where $p > 0$ and x is an absolutely continuous function defined on $[a, b]$ with $x(0) = 0$, and r and s are nonnegative functions in $C^1[0, a]$, λ_0 is the smallest eigenvalue of the boundary value problem

$$\left(r(t) \left(u^{'}(t) \right)^p \right)^{'} = \lambda s^{'}(t) u^p(t),$$

with $u(0) = 0$ and $r(a) \left(u^{'}(a) \right)^p = \lambda s^{'}(a) u^p(a)$ for which $u^{'} > 0$ in $[0, a]$.

The following are results motivated by Beesack and Das [42] which are easy to apply in practice.

Theorem 3.2.8 *Let* $\mathbb{T}$ *be a time scale with* $a, \tau \in \mathbb{T}$ *and* p, q *be positive real numbers such that* $p + q > 1$, *and let* r, s *be nonnegative rd-continuous functions on* $(a, \tau)_{\mathbb{T}}$ *such that* $\int_a^\tau r^{\frac{-1}{p+q-1}}(t)\Delta t < \infty$. *If* $y : [a, \tau] \cap \mathbb{T} \to \mathbb{R}$ *is delta differentiable with* $y(a) = 0$ *(and* y^Δ *does not change sign in* $(a, \tau)_{\mathbb{T}}$*), then*

$$\int_a^\tau s(x) |y(x)|^p \left|y^\Delta(x)\right|^q \Delta x \le K_1(a, \tau, p, q) \int_a^\tau r(x) \left|y^\Delta(x)\right|^{p+q} \Delta x,$$
$$(3.2.33)$$

where

$$K_1(a, \tau, p, q) = \left(\frac{q}{p+q}\right)^{\frac{q}{p+q}}$$
$$\times \left(\int_a^\tau \frac{s^{\frac{p+q}{p}}(x)}{r^{\frac{q}{p}}(x)} \left(\int_a^x r^{\frac{-1}{p+q-1}}(t)\Delta t\right)^{p+q-1} \Delta x\right)^{\frac{p}{p+q}} (3.2.34)$$

Proof. Let

$$|y(x)| = \int_a^x \left|y^\Delta(t)\right| \Delta t = \int_a^x \frac{1}{(r(t))^{\frac{1}{p+q}}} (r(t))^{\frac{1}{p+q}} \left|y^\Delta(t)\right| \Delta t.$$

Now, since r is nonnegative on $(a, \tau)_{\mathbb{T}}$, it follows from the Hölder inequality with

$$f(t) = \frac{1}{(r(t))^{\frac{1}{p+q}}}, \quad g(t) = (r(t))^{\frac{1}{p+q}} \left|y^\Delta(t)\right|, \quad \gamma = \frac{p+q}{p+q-1} \text{ and } \nu = p+q,$$

that

$$\int_a^x \left|y^\Delta(t)\right| \Delta t \le \left(\int_a^x \frac{1}{(r(t))^{\frac{1}{p+q-1}}} \Delta t\right)^{\frac{p+q-1}{p+q}} \left(\int_a^x r(t) \left|y^\Delta(t)\right|^{p+q} \Delta t\right)^{\frac{1}{p+q}}.$$

Then, for $a \le x \le \tau$, we get that

$$|y(x)|^p \le \left(\int_a^x \frac{1}{(r(t))^{\frac{1}{p+q-1}}} \Delta t\right)^{p\left(\frac{p+q-1}{p+q}\right)} \left(\int_a^x r(t) \left|y^\Delta(t)\right|^{p+q} \Delta t\right)^{\frac{p}{p+q}}.$$
$$(3.2.35)$$

Setting

$$z(x) := \int_a^x r(t) \left|y^\Delta(t)\right|^{p+q} \Delta t \qquad (3.2.36)$$

we see that $z(a) = 0$, and

$$z^\Delta(x) = r(x) \left| y^\Delta(x) \right|^{p+q} > 0. \tag{3.2.37}$$

This gives us

$$\left| y^\Delta(x) \right|^q = \left(\frac{z^\Delta(x)}{r(x)} \right)^{\frac{q}{p+q}}. \tag{3.2.38}$$

From (3.2.35) and (3.2.38), since s is a nonnegative on $(a, \tau)_\mathbb{T}$, we have

$$s(x) \left| y(x) \right|^p \left| y^\Delta(x) \right|^q \le s(x) \left(\frac{1}{r(x)} \right)^{\frac{q}{p+q}}$$

$$\times \left(\int_a^x \frac{1}{r^{\frac{1}{p+q-1}}(t)} \Delta t \right)^{p\left(\frac{p+q-1}{p+q} \right)} (z(x))^{\frac{p}{p+q}} \left(z^\Delta(x) \right)^{\frac{q}{p+q}}.$$

This implies that

$$\int_a^\tau s(x) \left| y(x) \right|^p \left| y^\Delta(x) \right|^q \Delta x \le \int_a^\tau s(x) \left(\frac{1}{r(x)} \right)^{\frac{q}{p+q}}$$

$$\times \left(\int_a^x \frac{1}{r^{\frac{1}{p+q-1}}(t)} \Delta t \right)^{p\left(\frac{p+q-1}{p+q} \right)} (z(x))^{\frac{p}{p+q}} \left(z^\Delta(x) \right)^{\frac{q}{p+q}} \Delta x. \tag{3.2.39}$$

Next note

$$\int_a^\tau s(x) \left| y(x) \right|^p \left| y^\Delta(x) \right|^q \Delta x$$

$$\le \left(\int_a^\tau s^{\frac{p+q}{p}}(x) \left(\frac{1}{r(x)} \right)^{\frac{q}{p}} \left(\int_a^x \frac{1}{r^{\frac{1}{p+q-1}}(t)} \Delta t \right)^{(p+q-1)} \Delta x \right)^{\frac{p}{p+q}}$$

$$\times \left(\int_a^\tau z^{\frac{p}{q}}(x) z^\Delta(x) \Delta x \right)^{\frac{q}{p+q}}. \tag{3.2.40}$$

From (3.2.37), the chain rule (1.1.7) and the fact that $z^\Delta(t) > 0$, we obtain

$$z^{\frac{p}{q}}(x) z^\Delta(x) \le \frac{q}{p+q} \left(z^{\frac{p+q}{q}}(x) \right)^\Delta. \tag{3.2.41}$$

Substituting (3.2.41) into (3.2.40) and using the fact that $z(a) = 0$, we have

$$\int_a^\tau s(x)\,|y(x)|^p\,\left|y^\Delta(x)\right|^q \Delta x$$

$$\leq \left(\int_a^\tau s^{\frac{p+q}{p}}(x)\left(\frac{1}{r(x)}\right)^{\frac{q}{p}}\left(\int_a^x \frac{1}{r^{\frac{1}{p+q-1}}(t)}\Delta t\right)^{(p+q-1)}dx\right)^{\frac{p}{p+q}}$$

$$\times \left(\frac{q}{p+q}\right)^{\frac{q}{p+q}}\left(\int_a^\tau \left(z^{\frac{p+q}{q}}(t)\right)^\Delta \Delta t\right)^{\frac{q}{p+q}}$$

$$= \left(\int_a^\tau s^{\frac{p+q}{p}}(x)\left(\frac{1}{r(x)}\right)^{\frac{q}{p}}\left(\int_a^x \frac{1}{r^{\frac{1}{p+q-1}}(t)}\Delta t\right)^{(p+q-1)}\Delta x\right)^{\frac{p}{p+q}}$$

$$\times \left(\frac{q}{p+q}\right)^{\frac{q}{p+q}} z(\tau).$$

Using (3.2.36), we have from the last inequality that

$$\int_a^\tau s(x)\,|y(x)|^p\,\left|y^\Delta(x)\right|^q \Delta x \leq K_1(a,\tau,p,q)\int_a^\tau r(x)\,\left|y^\Delta(x)\right|^{p+q}\Delta x,$$

which is the desired inequality (3.2.33). The proof is complete. ∎

Similar reasoning yields the following result.

Theorem 3.2.9 *Let $\mathbb{T}$ be a time scale with a, $b \in \mathbb{T}$ and p, q be positive real numbers such that $p + q > 1$, and let r, s be nonnegative rd-continuous functions on $(\tau, b)_\mathbb{T}$ such that $\int_\tau^b r^{\frac{-1}{p+q-1}}(t)\Delta t < \infty$. If $y : [\tau, b] \cap \mathbb{T} \to \mathbb{R}$ is delta differentiable with $y(b) = 0$, (and y^Δ does not change sign in $(\tau, b)_\mathbb{T}$), then we have*

$$\int_\tau^b s(x)\,|y(x)|^p\,\left|y^\Delta(x)\right|^q \Delta x \leq K_2(\tau,b,p,q)\int_\tau^b r(x)\,\left|y^\Delta(x)\right|^{p+q}\Delta x,$$

$$(3.2.42)$$

where

$$K_2(\tau,b,p,q) = \left(\frac{q}{p+q}\right)^{\frac{q}{p+q}}$$

$$\times \left(\int_\tau^b \frac{(s(x))^{\frac{p+q}{p}}}{(r(x))^{\frac{q}{p}}}\left(\int_x^b r^{\frac{-1}{p+q-1}}(t)\Delta t\right)^{(p+q-1)}\Delta x\right)^{\frac{p}{p+q}} \quad (3.2.43)$$

In the following, we assume that

$$K(p,q) = K_1(a,\tau,p,q) = K_2(\tau,b,p,q) < \infty,$$

where $K_1(a,\tau,p,q)$ and $K_2(\tau,b,p,q)$ are defined as in Theorems 3.2.8 and 3.2.9 and τ is the unique solution of the equation $K_1(a,\tau,p,q) = K_2(\tau,b,p,q)$.

Note that since

$$\int_a^b s(x) \, |y(x)|^p \, \left|y^\Delta(x)\right|^q \Delta x \; = \; \int_a^\tau s(x) \, |y(x)|^p \, \left|y^\Delta(x)\right|^q \Delta x$$
$$+ \int_\tau^b s(x) \, |y(x)|^p \, \left|y^\Delta(x)\right|^q \Delta x,$$

so combining Theorems 3.2.8 and 3.2.9 will give us the following result.

Theorem 3.2.10 *Let* $\mathbb{T}$ *be a time scale with* a, $b \in \mathbb{T}$ *and* p, q *be positive real numbers such that* $pq > 0$ *and* $p + q > 1$, *and let* r, s *be nonnegative rd-continuous functions on* $(a, b)_\mathbb{T}$ *such that* $\int_a^b r^{\frac{-1}{p+q-1}}(t)\Delta t < \infty$. *If* $y : [a, b] \cap \mathbb{T} \to \mathbb{R}$ *is delta differentiable with* $y(a) = 0 = y(b)$, *(and* y^Δ *does not change sign in* (a, b)), *then we have*

$$\int_a^b s(x) \, |y(x)|^p \, \left|y^\Delta(x)\right|^q \Delta x \le K(p, q) \int_a^b r(x) \, \left|y^\Delta(x)\right|^{p+q} \Delta x. \qquad (3.2.44)$$

Corollary 3.2.1 *Let* $\mathbb{T}$ *be a time scale with* a, $\tau \in \mathbb{T}$ *and* p, q *be positive real numbers such that* $p + q > 1$, *and let* r *be a nonnegative rd-continuous function on* $(a, \tau)_\mathbb{T}$ *such that* $\int_a^\tau r^{\frac{-1}{p+q-1}}(x)\Delta x < \infty$. *If* $y : [a, \tau] \cap \mathbb{T} \to \mathbb{R}$ *is delta differentiable with* $y(a) = 0$, *(and* y^Δ *does not change sign in* $(a, \tau)_\mathbb{T}$), *then we have*

$$\int_a^\tau r(x) \, |y(x)|^p \, \left|y^\Delta(x)\right|^q \Delta x \le K_1^*(a, \tau, p, q) \int_a^\tau r(x) \, \left|y^\Delta(x)\right|^{p+q} \Delta x,$$
$$(3.2.45)$$

where

$$K_1^*(a, \tau, p, q) = \left(\frac{q}{p+q}\right)^{\frac{q}{p+q}} \left(\int_a^\tau r(x) \left(\int_a^x r^{\frac{-1}{p+q-1}}(t)\Delta t\right)^{(p+q-1)} \Delta x\right)^{\frac{p}{p+q}}.$$
$$(3.2.46)$$

On a time scale $\mathbb{T}$, we note as a consequence of the chain rule (1.1.7) that

$$\left((t - a)^{p+q}\right)^\Delta \; = \; (p+q) \int_0^1 [h(\sigma(t) - a) + (1 - h)(t - a)]^{p+q-1} \, dh$$
$$\ge \; (p+q) \int_0^1 [h(t - a) + (1 - h)(t - a)]^{p+q-1} \, dh$$
$$= \; (p+q)(t - a)^{p+q-1}.$$

This implies that

$$\int_a^\tau (x - a)^{(p+q-1)} \Delta x \le \int_a^\tau \frac{1}{(p+q)} \left((x - a)^{p+q}\right)^\Delta \Delta x = \frac{(\tau - a)^{p+q}}{(p+q)}.$$
$$(3.2.47)$$

From this and (3.2.46) with $r(t) = 1$, we get that

$$K_1^*(a, \tau, p, q) \leq \left(\frac{q}{p+q}\right)^{\frac{q}{p+q}} \left(\frac{(\tau - a)^{p+q}}{(p+q)}\right)^{\frac{p}{p+q}} = \frac{q^{\frac{q}{p+q}}}{p+q}(\tau - a)^p. \quad (3.2.48)$$

Thus setting $r = 1$ in (3.2.45) and using (3.2.48), we have the following inequality.

Corollary 3.2.2 *Let* $\mathbb{T}$ *be a time scale with* a, $\tau \in \mathbb{T}$ *and* p, q *be positive real numbers such that* $p + q > 1$. *If* $y : [a, \tau] \cap \mathbb{T} \to \mathbb{R}$ *is delta differentiable with* $y(a) = 0$, *(and* y^Δ *does not change sign in* $(a, \tau)_{\mathbb{T}})$ *then we have*

$$\int_a^\tau |y(x)|^p \left|y^\Delta(x)\right|^q \Delta x \leq \frac{q^{\frac{q}{p+q}}}{p+q}(\tau - a)^p \int_a^\tau \left|y^\Delta(x)\right|^{p+q} \Delta x. \quad (3.2.49)$$

Choose $c = (a + b)/2$ and applying (3.2.46) to $[a, c]$ and $[c, b]$ and then add we obtain the following inequality.

Corollary 3.2.3 *Let* $\mathbb{T}$ *be a time scale with* a, $b \in \mathbb{T}$ *and* p, q *be positive real numbers such that* $p + q > 1$. *If* $y : [a, b] \cap \mathbb{T} \to \mathbb{R}$ *is delta differentiable with* $y(a) = 0 = y(b)$, *then we have*

$$\int_a^b |y(x)|^p \left|y^\Delta(x)\right|^q \Delta x \leq \frac{q^{\frac{q}{p+q}}}{p+q} \left(\frac{b-a}{2}\right)^p \int_a^b \left|y^\Delta(x)\right|^{p+q} \Delta x. \quad (3.2.50)$$

3.3 Opial Type Inequalities III

The main results in this section will be proved by employing the inequality (see [110, page 500])

$$|a + b|^r \leq 2^{r-1} \left(|a|^r + |b|^r\right), \, for \, r \geq 1, \quad (3.3.1)$$

and the inequality (see [9, page 51])

$$2^{r-1} \left(a^r + b^r\right) \leq (a + b)^r \leq (a^r + b^r), \, 0 \leq r \leq 1, \quad (3.3.2)$$

where a, b are positive real numbers. The results are adapted from [91].

Theorem 3.3.1 *Let* $\mathbb{T}$ *be a time scale with* a, $\tau \in \mathbb{T}$ *and* p, q *be positive real numbers such that* $p \geq 1$, *and let* r, s *be nonnegative rd-continuous functions* on $(a, \tau)_{\mathbb{T}}$ such that $\int_a^\tau r^{\frac{-1}{p+q-1}}(t)\Delta t < \infty$. If $y : [a, \tau] \cap \mathbb{T} \to \mathbb{R}^+$ is delta differentiable with $y(a) = 0$, (and y^Δ does not change sign in $(a, \tau)_{\mathbb{T}}$), then we have

$$\int_a^\tau s(x) |y(x) + y^\sigma(x)|^p |y^\Delta(x)|^q \Delta x \le K_1(a, \tau, p, q) \int_a^\tau r(x) |y^\Delta(x)|^{p+q} \Delta x,$$

(3.3.3)

where

$$
K_1(a, \tau, p, q) = 2^{2p-1} \left(\frac{q}{p+q} \right)^{\frac{q}{p+q}}
$$

$$
\times \left(\int_a^\tau \frac{(s(x))^{\frac{p+q}{p}}}{(r(x))^{\frac{q}{p}}} \left(\int_a^x r^{\frac{-1}{p+q-1}}(t)\Delta t \right)^{(p+q-1)} \Delta x \right)^{\frac{p}{p+q}}
$$

$$
+ 2^{p-1} \sup_{a \le x \le \tau} \left(\mu^p(x) \frac{s(x)}{r(x)} \right).
$$

(3.3.4)

Proof. Since $y^\Delta(t)$ does not change sign in $(a, \tau)_{\mathbb{T}}$, we have

$$|y(x)| = \int_a^x |y^\Delta(t)| \Delta t, \text{ for } x \in [a, \tau]_{\mathbb{T}}.$$

This implies that

$$|y(x)| = \int_a^x \frac{1}{(r(t))^{\frac{1}{p+q}}} (r(t))^{\frac{1}{p+q}} |y^\Delta(t)| \Delta t.$$

Now, since r is nonnegative on $(a, \tau)_{\mathbb{T}}$, then it follows from the Hölder inequality with

$$f(t) = \frac{1}{(r(t))^{\frac{1}{p+q}}}, \; g(t) = (r(t))^{\frac{1}{p+q}} |y^\Delta(t)|, \; \gamma = \frac{p+q}{p+q-1} \text{ and } \nu = p+q,$$

that

$$\int_a^x |y^\Delta(t)| \Delta t \le \left(\int_a^x \frac{1}{(r(t))^{\frac{1}{p+q-1}}} \Delta t \right)^{\frac{p+q-1}{p+q}} \left(\int_a^x r(t) |y^\Delta(t)|^{p+q} \Delta t \right)^{\frac{1}{p+q}}.$$

Then, for $a \le x \le \tau$, we get (note that $y(a) = 0$) that

$$|y(x)|^p \le \left(\int_a^x \frac{1}{(r(t))^{\frac{1}{p+q-1}}} \Delta t \right)^{p\left(\frac{p+q-1}{p+q}\right)} \left(\int_a^x r(t) |y^\Delta(t)|^{p+q} \Delta t \right)^{\frac{p}{p+q}}.$$

(3.3.5)

Since $y^\sigma = y + \mu y^\Delta$, we have

$$y(x) + y^\sigma(x) = 2y(x) + \mu y^\Delta(x).$$

Applying the inequality (3.3.1), we get (note $p \geq 1$) that

$$|y + y^\sigma|^p \leq 2^{p-1}(2^p |y|^p + \mu^p |y^\Delta|^p) = 2^{2p-1} |y|^p + 2^{p-1} \mu^p |y^\Delta|^p. \quad (3.3.6)$$

Setting

$$z(x) := \int_a^x r(t) |y^\Delta(t)|^{p+q} \Delta t, \quad (3.3.7)$$

we see that $z(a) = 0$, and

$$z^\Delta(x) = r(x) |y^\Delta(x)|^{p+q} > 0. \quad (3.3.8)$$

From this, we get that

$$|y^\Delta(x)|^{p+q} = \frac{z^\Delta(x)}{r(x)}, \quad \text{and} \quad |y^\Delta(x)|^q = \left(\frac{z^\Delta(x)}{r(x)}\right)^{\frac{q}{p+q}}. \quad (3.3.9)$$

From (3.3.6) and (3.3.9), since s is nonnegative on $(a, \tau)_\mathbb{T}$, we have

$$s(x) |y(x) + y^\sigma(x)|^p |y^\Delta(x)|^q$$
$$\leq 2^{2p-1} s(x) |y(x)|^p |y^\Delta(x)|^q + 2^{p-1} \mu^p(x) s(x) |y^\Delta|^{p+q}$$
$$\leq 2^{2p-1} s(x) \left(\frac{1}{r(x)}\right)^{\frac{q}{p+q}} \times \left(\int_a^x \frac{1}{r^{\frac{1}{p+q-1}}(t)} \Delta t\right)^{p(\frac{p+q-1}{p+q})}$$
$$\times (z(x))^{\frac{p}{p+q}} (z^\Delta(x))^{\frac{q}{p+q}} + 2^{p-1} \mu^p(x) s(x) \left(\frac{z^\Delta(x)}{r(x)}\right).$$

This implies that

$$\int_a^\tau s(x) |y(x) + y^\sigma(x)|^p |y^\Delta(x)|^q \Delta x$$
$$\leq 2^{2p-1} \int_a^\tau s(x) \left(\frac{1}{r(x)}\right)^{\frac{q}{p+q}} \times \left(\int_a^x \frac{1}{r^{\frac{1}{p+q-1}}(t)} \Delta t\right)^{p(\frac{p+q-1}{p+q})}$$
$$\times (z(x))^{\frac{p}{p+q}} (z^\Delta(x))^{\frac{q}{p+q}} \Delta x + 2^{p-1} \int_a^\tau \left(\mu^p \frac{s(x)}{r(x)}\right) z^\Delta(x) \Delta x$$
$$\leq 2^{2p-1} \int_a^\tau s(x) \left(\frac{1}{r(x)}\right)^{\frac{q}{p+q}} \times \left(\int_a^x \frac{1}{r^{\frac{1}{p+q-1}}(t)} \Delta t\right)^{p(\frac{p+q-1}{p+q})}$$
$$\times (z(x))^{\frac{p}{p+q}} (z^\Delta(x))^{\frac{q}{p+q}} \Delta x$$
$$+ 2^{p-1} \max_{a \leq x \leq \tau} \left(\mu^p \frac{s(x)}{r(x)}\right) \int_a^\tau z^\Delta(x) \Delta x. \quad (3.3.10)$$

Applying the Hölder inequality with indices $(p+q)/p$ and $(p+q)/q$ on the first integral on the right-hand side, we have

$$
\int_a^\tau s(x)\,|y(x)+y^\sigma(x)|^p\,\left|y^\Delta(x)\right|^q \Delta x
$$

$$
\leq\; 2^{2p-1}\left(\int_a^\tau s^{\frac{p+q}{p}}(x)\left(\frac{1}{r(x)}\right)^{\frac{q}{p}}\left(\int_a^x \frac{1}{r^{\frac{1}{p+q-1}}(t)}\Delta t\right)^{(p+q-1)}\Delta x\right)^{\frac{p}{p+q}}
$$

$$
\times\left(\int_a^\tau z^{\frac{p}{q}}(x)z^\Delta(x)\Delta x\right)^{\frac{q}{p+q}}+2^{p-1}\sup_{a\leq x\leq\tau}\mu^p\frac{s(x)}{r(x)}\int_a^\tau z^\Delta(x)\Delta x \qquad(3.3.11)
$$

From (3.3.8), and the chain rule (1.1.7), we obtain

$$
z^{\frac{p}{q}}(x)z^\Delta(x)\leq\frac{q}{p+q}\left(z^{\frac{p+q}{q}}(x)\right)^\Delta. \qquad(3.3.12)
$$

Substituting (3.3.12) into (3.3.11) and using the fact that $z(a)=0$, we have that

$$
\int_a^\tau s(x)\,|y(x)+y^\sigma(x)|^p\,\left|y^\Delta(x)\right|^q \Delta x
$$

$$
\leq\; 2^{2p-1}\left(\int_a^\tau s^{\frac{p+q}{p}}(x)\left(\frac{1}{r(x)}\right)^{\frac{q}{p}}\left(\int_a^x \frac{1}{r^{\frac{1}{p+q-1}}(t)}\Delta t\right)^{(p+q-1)}\Delta x\right)^{\frac{p}{p+q}}
$$

$$
\times\left(\frac{q}{p+q}\right)^{\frac{q}{p+q}}\left(\int_a^\tau\left(z^{\frac{p+q}{q}}(t)\right)^\Delta\Delta t\right)^{\frac{q}{p+q}}
$$

$$
+2^{p-1}\sup_{a\leq x\leq\tau}\left(\mu^p\frac{s(x)}{r(x)}\right)\int_a^\tau z^\Delta(x)\Delta x
$$

$$
=\;\left(\int_a^\tau s^{\frac{p+q}{p}}(x)\left(\frac{1}{r(x)}\right)^{\frac{q}{p}}\left(\int_a^x \frac{1}{r^{\frac{1}{p+q-1}}(t)}\Delta t\right)^{(p+q-1)}\Delta x\right)^{\frac{p}{p+q}}
$$

$$
\times 2^{2p-1}\left(\frac{q}{p+q}\right)^{\frac{q}{p+q}}z(\tau)+2^{p-1}\sup_{a\leq x\leq\tau}\left(\mu^p\frac{s(x)}{r(x)}\right)z(\tau).
$$

Using (3.3.7), we have from the last inequality that

$$
\int_a^\tau s(x)\,|y(x)+y^\sigma(x)|^p\,\left|y^\Delta(x)\right|^q \Delta x\leq K_1(a,\tau,p,q)\int_a^\tau r(x)\left|y^\Delta(x)\right|^{p+q}\Delta x,
$$

which is the desired inequality (3.3.3). The proof is complete. ∎

Similar reasoning as in Theorem 3.3.1, with $[a,\tau]_\mathbb{T}$ replaced by $[\tau,b]_\mathbb{T}$ and $|y(x)|=\int_x^b\left|y^\Delta(t)\right|\Delta t$, yields the following result.

Theorem 3.3.2 *Let* $\mathbb{T}$ *be a time scale with* $\tau, b \in \mathbb{T}$ *and* p, q *be positive real numbers such that* $p \geq 1$, *and let* r, s *be nonnegative rd-continuous functions on* $(\tau, b)_{\mathbb{T}}$ *such that* $\int_{\tau}^{b} r^{\frac{-1}{p+q-1}}(t)\Delta t < \infty$. *If* $y : [\tau, b] \cap \mathbb{T} \to \mathbb{R}^{+}$ *is delta differentiable with* $y(b) = 0$, *(and* y^{Δ} *does not change sign in* $(\tau, b)_{\mathbb{T}}$), *then we have*

$$\int_{\tau}^{b} s(x) \left| y(x) + y^{\sigma}(x) \right|^{p} \left| y^{\Delta}(x) \right|^{q} \Delta x \leq K_{2}(\tau, b, p, q) \int_{\tau}^{b} r(x) \left| y^{\Delta}(x) \right|^{p+q} \Delta x,$$

(3.3.13)

where

$$K_{2}(\tau, b, p, q) = 2^{2p-1} \left(\frac{q}{p+q} \right)^{\frac{q}{p+q}}$$

$$\times \left(\int_{\tau}^{b} \frac{(s(x))^{\frac{p+q}{p}}}{(r(x))^{\frac{q}{p}}} \left(\int_{x}^{b} r^{\frac{-1}{p+q-1}}(t)\Delta t \right)^{(p+q-1)} \Delta x \right)^{\frac{p}{p+q}}$$

$$+ 2^{p-1} \sup_{\tau \leq x \leq b} \left(\mu^{p}(x) \frac{s(x)}{r(x)} \right).$$

(3.3.14)

In the following, we assume that

$$K(p, q) = K_{1}(a, \tau, p, q) = K_{2}(\tau, b, p, q) < \infty,$$

where $K_{1}(a, \tau, p, q)$ and $K_{2}(\tau, b, p, q)$ are defined as in Theorems 3.3.1 and 3.3.2 and τ is the unique solution of the equation $K_{1}(a, \tau, p, q) = K_{2}(\tau, b, p, q)$. Note that,

$$\int_{a}^{b} s(x) \left| y(x) + y^{\sigma}(x) \right|^{p} \left| y^{\Delta}(x) \right|^{q} \Delta x$$

$$= \int_{a}^{\tau} s(x) \left| y(x) + y^{\sigma}(x) \right|^{p} \left| y^{\Delta}(x) \right|^{q} \Delta x + \int_{\tau}^{b} s(x) \left| y(x) + y^{\sigma}(x) \right|^{p} \left| y^{\Delta}(x) \right|^{q} \Delta x,$$

so combining Theorems 3.3.1 and 3.3.2 gives us the following result.

Theorem 3.3.3 *Let* $\mathbb{T}$ *be a time scale with* $a, b \in \mathbb{T}$ *and* p, q *be positive real numbers such that* $p \geq 1$, *and let* r, s *be nonnegative rd-continuous functions on* $(a, b)_{\mathbb{T}}$ *such that* $\int_{a}^{b} r^{\frac{-1}{p+q-1}}(t)\Delta t < \infty$. *If* $y : [a, b] \cap \mathbb{T} \to \mathbb{R}^{+}$ *is delta differentiable with* $y(a) = 0 = y(b)$, *(and* y^{Δ} *does not change sign in* $(a, b)_{\mathbb{T}}$), *then we have*

$$\int_{a}^{b} s(x) \left| y(x) + y^{\sigma}(x) \right|^{p} \left| y^{\Delta}(x) \right|^{q} \Delta x \leq K(p, q) \int_{a}^{b} r(x) \left| y^{\Delta}(x) \right|^{p+q} \Delta x.$$

(3.3.15)

For $r = s$ in Theorem 3.3.1, we obtain the following result.

Corollary 3.3.1 *Let* $\mathbb{T}$ *be a time scale with* a, $\tau \in \mathbb{T}$ *and* p, q *be positive real numbers such that* $p \geq 1$, *and let* r *be a nonnegative rd-continuous function on* $(a, \tau)_{\mathbb{T}}$ *such that* $\int_a^\tau r^{\frac{-1}{p+q-1}}(t)\Delta t < \infty$. *If* $y : [a, \tau] \cap \mathbb{T} \to \mathbb{R}^+$ *is delta differentiable with* $y(a) = 0$, *(and* y^Δ *does not change sign in* $(a, \tau)_{\mathbb{T}}$*) then we have*

$$\int_a^\tau r(x) |y(x) + y^\sigma(x)|^p |y^\Delta(x)|^q \, \Delta x \leq K_1^*(a, \tau, p, q) \int_a^\tau r(x) |y^\Delta(x)|^{p+q} \, \Delta x,$$

$$(3.3.16)$$

where

$$K_1^*(a, \tau, p, q) \;=\; 2^{2p-1} \left(\frac{q}{p+q} \right)^{\frac{q}{p+q}}$$

$$\times \left(\int_a^\tau r(x) \left(\int_a^x r^{\frac{-1}{p+q-1}}(t)\Delta t \right)^{(p+q-1)} \Delta x \right)^{\frac{p}{p+q}}$$

$$+ 2^{p-1} \sup_{a \leq x \leq \tau} (\mu^p(x)). \qquad (3.3.17)$$

From Theorems 3.3.2 and 3.3.3 one can derive similar results by setting $r = s$. From (3.2.47) and (3.3.17) (by putting $r(t) = 1$), we get

$$K_1^*(a, \tau, p, q) \;=\; 2^{2p-1} \left(\frac{q}{p+q} \right)^{\frac{q}{p+q}} \times \left(\int_a^\tau (x - a)^{(p+q-1)} \Delta x \right)^{\frac{p}{p+q}}$$

$$\leq \; 2^{2p-1} \left(\frac{q}{p+q} \right)^{\frac{q}{p+q}} \left(\frac{(\tau - a)^{p+q}}{(p+q)} \right)^{\frac{p}{p+q}}$$

$$+ 2^{p-1} \max_{a \leq x \leq \tau} (\mu^p(x))$$

$$= \; 2^{p-1} \max_{a \leq x \leq \tau} (\mu^p(x)) + 2^{2p-1} \frac{q^{\frac{q}{p+q}}}{p+q} (\tau - a)^p. \qquad (3.3.18)$$

Setting $r = 1$ in (3.3.16) and using (3.3.18), we have the following result.

Corollary 3.3.2 *Let* $\mathbb{T}$ *be a time scale with* a, $\tau \in \mathbb{T}$ *and* p, q *be positive real numbers such that* $p \geq 1$. *If* $y : [a, \tau] \cap \mathbb{T} \to \mathbb{R}^+$ *is delta differentiable with* $y(a) = 0$, *(and* y^Δ *does not change sign in* $(a, \tau)_{\mathbb{T}}$*), then we have*

$$\int_a^\tau |y(x) + y^\sigma(x)|^p |y^\Delta(x)|^q \, \Delta x \leq L(a, \tau, p, q) \int_a^\tau |y^\Delta(x)|^{p+q} \, \Delta x, \quad (3.3.19)$$

where

$$L(a, \tau, p, q) := \left(2^{2p-1} \frac{q^{\frac{q}{p+q}}}{p+q} (\tau - a)^p + 2^{p-1} \sup_{a \leq x \leq \tau} \mu^p(x) \right).$$

Choose $\tau = (a + b)/2$ and apply (3.3.17) to $[a, \tau]$ and $[\tau, b]$ and then add to obtain the following inequality.

Corollary 3.3.3 *Let* $\mathbb{T}$ *be a time scale with* $a, b \in \mathbb{T}$ *and* p, q *be positive real numbers such that* $p \geq 1$. *If* $y : [a, b] \cap \mathbb{T} \to \mathbb{R}^+$ *is delta differentiable with* $y(a) = 0 = y(b)$, *then we have*

$$\int_a^b |y(x) + y^\sigma(x)|^p \left| y^\Delta(x) \right|^q \Delta x \leq F(a, b, p, q) \int_a^b \left| y^\Delta(x) \right|^{p+q} \Delta x, \quad (3.3.20)$$

where

$$F(a, b, p, q) := 2^{2p-1} \frac{q^{\frac{q}{p+q}}}{p+q} \left(\frac{b-a}{2} \right)^p + 2^{p-1} \sup_{a \leq x \leq b} (\mu^p(x)).$$

Setting $p = q = 1$ in (3.3.20) we have the following Opial type inequality on a time scale.

Corollary 3.3.4 *Let* $\mathbb{T}$ *be a time scale with* $a, b \in \mathbb{T}$. *If* $y : [a, \tau] \cap \mathbb{T} \to \mathbb{R}^+$ *is delta differentiable with* $y(a) = 0 = y(b)$, *then we have*

$$\int_a^b |y(x) + y^\sigma(x)| \left| y^\Delta(x) \right| \Delta x \leq \left(\frac{b-a}{2} + \sup_{a \leq x \leq b} \mu(x) \right) \int_a^b \left| y^\Delta(x) \right|^2 \Delta x. \quad (3.3.21)$$

In the following, we establish some dynamic inequalities of Opial type on time scales of the form

$$\int_a^\tau s(x) |y(x) + y^\sigma(x)|^p \left| y^\Delta(x) \right|^q \Delta x \leq K(a, \tau, p, q) \int_a^\tau r(x) \left| y^\Delta(x) \right|^{p+q} \Delta x,$$

where p, q be positive real numbers such that $p \leq 1, p + q > 1$.

The proof of our next result is similar to that in Theorem 3.3.1 except here we use the inequality (3.3.2).

Theorem 3.3.4 *Let* $\mathbb{T}$ *be a time scale with* $a, \tau \in \mathbb{T}$ *and* p, q *be positive real numbers such that* $p \leq 1, p + q > 1$ *and let* r, s *be nonnegative rd-continuous functions on* $(a, \tau)_\mathbb{T}$ *such that* $\int_a^\tau r^{\frac{-1}{p+q-1}}(t) \Delta t < \infty$. *If* $y : [a, \tau] \cap \mathbb{T} \to \mathbb{R}^+$ *is delta differentiable with* $y(a) = 0$, *(and* y^Δ *does not change sign in* $(a, \tau)_\mathbb{T})$, *then we have*

$$\int_a^\tau s(x) |y(x) + y^\sigma(x)|^p \left| y^\Delta(x) \right|^q \Delta x \leq K_1(a, \tau, p, q) \int_a^\tau r(x) \left| y^\Delta(x) \right|^{p+q} \Delta x, \quad (3.3.22)$$

where

$$K_1(a, \tau, p, q) = \sup_{a \leq x \leq \tau} \left(\mu^p(x) \frac{s(x)}{r(x)} \right) + 2^p \left(\frac{q}{p+q} \right)^{\frac{q}{p+q}}$$

$$\left(\int_a^\tau \frac{s^{\frac{p+q}{p}}(x)}{r^{\frac{q}{p}}(x)} \left(\int_a^x r^{\frac{-1}{p+q-1}}(t) \Delta t \right)^{p+q-1} \Delta x \right)^{\frac{p}{p+q}} \quad (3.3.23)$$

Proof. Since $y^\Delta(t)$ does not change sign in $(a, \tau)_\mathbb{T}$, we have

$$|y(x)| = \int_a^x |y^\Delta(t)| \, \Delta t, \text{ for } x \in [a, \tau]_\mathbb{T}.$$

This implies that

$$|y(x)| = \int_a^x \frac{1}{(r(t))^{\frac{1}{p+q}}} (r(t))^{\frac{1}{p+q}} |y^\Delta(t)| \, \Delta t.$$

Now, since r is nonnegative on $(a, \tau)_\mathbb{T}$, then it follows from the Hölder inequality with

$$f(t) = \frac{1}{(r(t))^{\frac{1}{p+q}}}, \quad g(t) = (r(t))^{\frac{1}{p+q}} |y^\Delta(t)|, \quad \gamma = \frac{p+q}{p+q-1}, \text{ and } \nu = p+q,$$

that

$$\int_a^x |y^\Delta(t)| \, \Delta t \le \left(\int_a^x \frac{1}{(r(t))^{\frac{1}{p+q-1}}} \Delta t \right)^{\frac{p+q-1}{p+q}} \left(\int_a^x r(t) |y^\Delta(t)|^{p+q} \, \Delta t \right)^{\frac{1}{p+q}}.$$

Then, for $a \le x \le \tau$, we get (note $y(a) = 0$) that

$$|y(x)|^p \le \left(\int_a^x \frac{1}{(r(t))^{\frac{1}{p+q-1}}} \Delta t \right)^{p\left(\frac{p+q-1}{p+q}\right)} \left(\int_a^x r(t) |y^\Delta(t)|^{p+q} \, \Delta t \right)^{\frac{p}{p+q}}.$$
$$(3.3.24)$$

Since $y^\sigma = y + \mu y^\Delta$, we have

$$y(x) + y^\sigma(x) = 2y(x) + \mu y^\Delta(x).$$

Applying the inequality (3.3.2), we get (note $p \le 1$) that

$$|y + y^\sigma|^p = |2y(x) + \mu y^\Delta(x)|^p \le 2^p |y|^p + \mu^p |y^\Delta|^p. \tag{3.3.25}$$

Setting

$$z(x) := \int_a^x r(t) |y^\Delta(t)|^{p+q} \, \Delta t, \tag{3.3.26}$$

we see that $z(a) = 0$, and

$$z^\Delta(x) = r(x) |y^\Delta(x)|^{p+q} > 0. \tag{3.3.27}$$

From this, we get that

$$|y^\Delta(x)|^{p+q} = \frac{z^\Delta(x)}{r(x)}, \quad \text{and } |y^\Delta(x)|^q = \left(\frac{z^\Delta(x)}{r(x)} \right)^{\frac{q}{p+q}}. \tag{3.3.28}$$

Thus, since s is nonnegative on $(a, \tau)_{\mathbb{T}}$, we have from (3.3.25) and (3.3.28) that

$$s(x) \left| y(x) + y^\sigma(x) \right|^p \left| y^\Delta(x) \right|^q$$

$$\leq 2^p s(x) \left| y(x) \right|^p \left| y^\Delta(x) \right|^q + \mu^p(x) s(x) \left| y^\Delta \right|^{p+q}$$

$$\leq 2^p s(x) \left(\frac{1}{r(x)} \right)^{\frac{q}{p+q}} \times \left(\int_a^x \frac{1}{r^{\frac{1}{p+q-1}}(t)} \Delta t \right)^{p\left(\frac{p+q-1}{p+q} \right)}$$

$$\times (z(x))^{\frac{p}{p+q}} \left(z^\Delta(x) \right)^{\frac{q}{p+q}} + \mu^p(x) s(x) \left(\frac{z^\Delta(x)}{r(x)} \right).$$

The rest of the proof is similar to the proof of Theorem 3.3.1. The proof is complete. ∎

Theorem 3.3.5 *Let $\mathbb{T}$ be a time scale with $\tau, b \in \mathbb{T}$ and p, q be positive real numbers such that $p \leq 1$, $p + q > 1$ and let r, s be nonnegative rd-continuous functions on $(\tau, b)_{\mathbb{T}}$ such that $\int_\tau^b r^{\frac{-1}{p+q-1}}(t)\Delta t < \infty$. If $y : [\tau, b] \cap \mathbb{T} \to \mathbb{R}^+$ is delta differentiable with $y(b) = 0$, (and y^Δ does not change sign in $(\tau, b)_{\mathbb{T}}$), then we have*

$$\int_\tau^b s(x) \left| y(x) + y^\sigma(x) \right|^p \left| y^\Delta(x) \right|^q \Delta x \leq K_2(\tau, b, p, q) \int_\tau^b r(x) \left| y^\Delta(x) \right|^{p+q} \Delta x,$$
$$(3.3.29)$$

where

$$K_2(\tau, b, p, q) = 2^p \left(\frac{q}{p+q} \right)^{\frac{q}{p+q}} \left(\int_\tau^b \frac{(s(x))^{\frac{p+q}{p}}}{(r(x))^{\frac{q}{p}}} \left(\int_x^b (r(t))^{\frac{-1}{p+q-1}} \Delta t \right)^{(p+q-1)} \Delta x \right)^{\frac{p}{p+q}}$$

$$+ \sup_{\tau \leq x \leq b} \left(\mu^p(x) \frac{s(x)}{r(x)} \right). \tag{3.3.30}$$

In the following, we assume

$$K(p, q) = K_1(a, \tau, p, q) = K_2(\tau, b, p, q) < \infty,$$

where $K_1(a, \tau, p, q)$ and $K_2(\tau, b, p, q)$ are defined as in Theorems 3.3.4 and 3.3.5 and τ is the unique solution of the equation $K_1(a, \tau, p, q) = K_2(\tau, b, p, q)$.

Theorem 3.3.6 *Let $\mathbb{T}$ be a time scale with $a, b \in \mathbb{T}$ and p, q be positive real numbers such that $p \leq 1$, $p + q > 1$ and let r, s be nonnegative rd−continuous functions on $(a, b)_{\mathbb{T}}$ such that $\int_a^b (r(t))^{\frac{-1}{p+q-1}}\Delta t < \infty$. If $y : [a, b] \cap \mathbb{T} \to \mathbb{R}^+$ is delta differentiable with $y(a) = 0 = y(b)$, (and y^Δ does not change sign in $(a, b)_{\mathbb{T}}$), then we have*

$$\int_a^b s(x) \left| y(x) + y^\sigma(x) \right|^p \left| y^\Delta(x) \right|^q \Delta x \leq K(p, q) \int_a^b r(x) \left| y^\Delta(x) \right|^{p+q} \Delta x.$$
$$(3.3.31)$$

Proof. Since

$$\int_a^b s(x) \left|y(x) + y^\sigma(x)\right|^p \left|y^\Delta(x)\right|^q \Delta x = \int_a^\tau s(x) \left|y(x) + y^\sigma(x)\right|^p \left|y^\Delta(x)\right|^q \Delta x$$
$$+ \int_\tau^b s(x) \left|y(x) + y^\sigma(x)\right|^p \left|y^\Delta(x)\right|^q \Delta x,$$

the rest of the proof is a combination of Theorems 3.3.4 and 3.3.5. ∎

For $r = s$ in Theorem 3.3.4, we obtain the following result.

Corollary 3.3.5 *Let $\mathbb{T}$ be a time scale with $a, \tau \in \mathbb{T}$ and p, q be positive real numbers such that $p \le 1$, $p + q > 1$ and let r be a nonnegative rd-continuous function on $(a, \tau)_{\mathbb{T}}$ such that $\int_a^\tau (r(t))^{\frac{-1}{p+q-1}} \Delta t < \infty$. If $y : [a, \tau] \cap \mathbb{T} \to \mathbb{R}^+$ is delta differentiable with $y(a) = 0$, (and y^Δ does not change sign in $(a, \tau)_{\mathbb{T}}$), then we have*

$$\int_a^\tau r(x) \left|y(x) + y^\sigma(x)\right|^p \left|y^\Delta(x)\right|^q \Delta x \le K_1^*(a, \tau, p, q) \int_a^\tau r(x) \left|y^\Delta(x)\right|^{p+q} \Delta x,$$
$$(3.3.32)$$

where

$$K_1^*(a, \tau, p, q) = \sup_{a \le x \le \tau} (\mu^p(x)) + 2^p \left(\frac{q}{p+q}\right)^{\frac{q}{p+q}}$$
$$\times \left(\int_a^\tau r(x) \left(\int_a^x r^{\frac{-1}{p+q-1}}(t)\Delta t\right)^{(p+q-1)} \Delta x\right)^{\frac{p}{p+q}}. \quad (3.3.33)$$

Using the inequality (3.2.47) and (3.3.33) (by putting $r(t) = 1$), we get that

$$K_1^*(a, \tau, p, q) = 2^p \left(\frac{q}{p+q}\right)^{\frac{q}{p+q}} \times \left(\int_a^\tau (x-a)^{(p+q-1)} \Delta x\right)^{\frac{p}{p+q}}$$
$$\le 2^p \left(\frac{q}{p+q}\right)^{\frac{q}{p+q}} \left(\frac{(\tau-a)^{p+q}}{(p+q)}\right)^{\frac{p}{p+q}} + \max_{a \le x \le \tau} (\mu^p(x))$$
$$= \max_{a \le x \le \tau} (\mu^p(x)) + 2^p \frac{q^{\frac{q}{p+q}}}{p+q}(\tau-a)^p. \quad (3.3.34)$$

Setting $r = 1$ in (3.3.32) and using (3.3.34), we have the following result.

Corollary 3.3.6 *Let $\mathbb{T}$ be a time scale with $a, \tau \in \mathbb{T}$ and p, q be positive real numbers such that $p \le 1$ and $p + q > 1$. If $y : [a, \tau] \cap \mathbb{T} \to \mathbb{R}^+$ is delta differentiable with $y(a) = 0$, (and y^Δ does not change sign in $(a, \tau)_{\mathbb{T}}$), then we have*

$$\int_a^\tau \left|y(x) + y^\sigma(x)\right|^p \left|y^\Delta(x)\right|^q \Delta x \le L(a, \tau, p, q) \int_a^\tau \left|y^\Delta(x)\right|^{p+q} \Delta x, \quad (3.3.35)$$

where

$$L(a, \tau, p, q) := \left(2^p \frac{q^{\frac{q}{p+q}}}{p+q} (\tau - a)^p + \sup_{a \leq x \leq \tau} \mu^p(x) \right).$$

Choose $\tau = (a + b)/2$ and apply (3.3.33) to $[a, \tau]$ and $[\tau, b]$ and then add to obtain the following inequality.

Corollary 3.3.7 *Let* $\mathbb{T}$ *be a time scale with* $a, b \in \mathbb{T}$ *and* p, q *be positive real numbers such that* $p \leq 1$ *and* $p + q > 1$. *If* $y : [a, b] \cap \mathbb{T} \to \mathbb{R}^+$ *is delta differentiable with* $y(a) = 0 = y(b)$, *then we have*

$$\int_a^b |y(x) + y^\sigma(x)|^p \, |y^\Delta(x)|^q \, \Delta x \leq F(a, b, p, q) \int_a^b |y^\Delta(x)|^{p+q} \, \Delta x, \quad (3.3.36)$$

where

$$F(a, b, p, q) := \frac{q^{\frac{q}{p+q}}}{p+q} (b - a)^p + \sup_{a \leq x \leq b} (\mu^p(x)).$$

Next various types of Opial's inequality involving several functions are presented on arbitrary time scales. The well-known Muirhead's inequality will be employed to obtain the results.

Theorem 3.3.7 ([111]) *Let* S^n *be the symmetry group of the set* $[1, n]_\mathbb{N}$, *and* $A := (\alpha_1, \alpha_2, \dots, \alpha_n)$, $B := (\beta_1, \beta_2, \dots, \beta_n)$ *be two vectors with nonnegative entries and* $\sum_{j=1}^k \alpha_j \geq \sum_{j=1}^k \beta_j$ *for all* $k \in [1, n-1]_\mathbb{N}$ *and* $\sum_{j=1}^n \alpha_j = \sum_{j=1}^n \beta_j$, *then it is said that* A *majorizes* B *(we prefer the notation* $A \triangleright B$*), and the following inequality is true:*

$$\sum_{\pi \in S^n} \prod_{j=1}^n x_{\pi_j}^{\alpha_j} \geq \sum_{\pi \in S^n} \prod_{j=1}^n x_{\pi_j}^{\beta_j},$$

where π_j *denotes the* j-th *component of the permutation* π, *and* $x_j \in \mathbb{R}_0^+$ *holds for all* $j \in [1, n]_\mathbb{N}$.

One can easily see that for $(2, 0) \triangleright (1, 1)$ Theorem 3.3.7 gives us the following well-known inequality

$$x_1^2 + x_2^2 \geq 2x_1 x_2 \quad \text{with} \quad x_1, x_2 \geq 0.$$

This inequality gives us the well-known inequality between arithmetic and geometric means by letting $y_1 := 2x_1^2$ and $y_2 := 2x_2^2$, i.e.,

$$\sqrt{y_1 y_2} \leq \frac{y_1 + y_2}{2}.$$

Throughout, for convenience, the empty sum and the empty product are assumed to be 0 and 1, respectively, i.e., for $\alpha, \beta \in \mathbb{Z}$ with $\beta < \alpha$, $\sum_{j=\alpha}^\beta f_j = 0$ and $\prod_{j=\alpha}^\beta f_j = 1$.

Theorem 3.3.8 *Let* $n \in \mathbb{N}$ *and* $f_j : \mathbb{T} \to \mathbb{R}$ *be differentiable functions for* $j \in [1, n]_{\mathbb{N}}$, *then we have*

$$\left[\prod_{j=1}^{n} f_j(t) \right]^{\Delta} = \sum_{j=1}^{n} \left\{ \left[\prod_{i=1}^{j-1} f_i^{\sigma}(t) \right] f_j^{\Delta}(t) \left[\prod_{i=j+1}^{n} f_i(t) \right] \right\}, \quad \text{for } t \in \mathbb{T}.$$

Now, we are ready to establish some generalized Opial inequalities.

Theorem 3.3.9 *Let* $n \in \mathbb{N}$, $a, b \in \mathbb{T}$ *and* $f_j \in C_{\mathrm{rd}}^1(\mathbb{T}, \mathbb{R})$ *for all* $j \in [1, n+1]_{\mathbb{N}}$ *with* $f_j(a) = 0$ *for all* $j \in [1, n + 1]_{\mathbb{N}}$. *Then*

$$\int_a^b \sum_{j=1}^{n+1} \left| \left[\prod_{i=1}^{j-1} f_i^{\sigma}(\xi) \right] f_j^{\Delta}(\xi) \left[\prod_{i=j+1}^{n+1} f_i(\xi) \right] \right| \Delta \xi \qquad (3.3.37)$$

$$\leq \quad \frac{(b-a)^n}{n+1} \int_a^b \sum_{j=1}^{n+1} |f_j^{\Delta}(\xi)|^{n+1} \Delta \xi,$$

with equality when $f_j(t) = c(t-a)$ *for all* $j \in [1, n+1]_{\mathbb{N}}$, *where* c *is a constant.*

Proof. The proof of this theorem follows similar steps to that in the following one, so we skip it here. ∎

Theorem 3.3.10 *Let* $m, n \in \mathbb{N}$, $a, b \in \mathbb{T}$ *and* $f_j \in C_{\mathrm{rd}}^1(\mathbb{T}, \mathbb{R})$ *with* $f_j(a) = 0$ *for all* $j \in [1, n+1]_{\mathbb{N}}$. *Then*

$$\int_a^b \sum_{j=0}^{m} \left| \left(\prod_{i=1}^{n+1} f_i^{\sigma}(\xi) \right)^j \left(\prod_{i=1}^{n+1} f_i(\xi) \right)^{m-j} \sum_{j=1}^{n+1} \left(\prod_{i=1}^{j-1} f_i^{\sigma}(\xi) \right) f_j^{\Delta}(\xi) \left(\prod_{i=j+1}^{n+1} f_i(\xi) \right) \right| \Delta \xi$$

$$\leq \frac{(b-a)^{(n+1)(m+1)-1}}{n+1} \int_a^b \left[\sum_{j=1}^{n+1} |f_j^{\Delta}(\xi)| \right]^{(n+1)(m+1)} \Delta \xi.$$

$$(3.3.38)$$

with equality when $f_j(t) = c(t - a)$ *for all* $j \in [1, n + 1]_{\mathbb{N}}$, *where* c *is a constant.*

Proof. Set $F_j(t) := \int_a^t |f_j^{\Delta}(\xi)| \Delta \xi$ for $t \in [a, b]_{\mathbb{T}}$ and all $j \in [1, n, +1]_{\mathbb{N}}$. Then, on $[a, b]_{\mathbb{T}}$, we have $F_j^{\Delta} = |f_j^{\Delta}|$ and $F_j \geq |f_j|$ for all $j \in [1, n + 1]_{\mathbb{N}}$. Now, set $F(t) := \prod_{j=1}^{n+1} F_j(t)$ for $t \in [a, b]_{\mathbb{T}}$. It follows that

$$\int_a^b \sum_{j=0}^m \left| \left[\prod_{i=1}^{n+1} f_i^\sigma(\xi) \right]^j \left(\prod_{i=1}^{n+1} f_i(\xi) \right)^{m-j} \sum_{j=1}^{n+1} \left[\prod_{i=1}^{j-1} f_i^\sigma(\xi) \right] f_j^\Delta(\xi) \prod_{i=j+1}^{n+1} f_i(\xi) \right| \Delta\xi$$

$$\leq \int_a^b \sum_{j=0}^m \left[\prod_{i=1}^{n+1} F_i^\sigma(\xi) \right]^j \left[\prod_{i=1}^{n+1} F_i(\xi) \right]^{m-j} \times \sum_{j=1}^{n+1} \prod_{i=1}^{j-1} \left(F_i^\sigma(\xi) F_j^\Delta(\xi) \prod_{i=j+1}^{n+1} F_i(\xi) \right) \Delta\xi$$

$$= \int_a^b \sum_{j=0}^m \left[F^\sigma(\xi) \right]^j \left[F(\xi) \right]^{m-j} F^\Delta(\xi) \Delta\xi = \int_a^b \left[(F(\xi))^{(m+1)} \right]^\Delta \Delta\xi$$

$$= (F(b))^{m+1} = \prod_{j=1}^{n+1} \left[F_j(b) \right]^{m+1} \leq \frac{1}{n+1} \sum_{j=1}^{n+1} \left[F_j(b) \right]^{(n+1)(m+1)} \tag{3.3.39}$$

is true, where we have applied the arithmetic mean and geometric mean inequalities in the last step. Also for $j \in [1, n+1]_\mathbb{N}$, we have

$$\left[F_j(b) \right]^{(n+1)(m+1)} = \left(\int_a^b \left| f_j^\Delta(\xi) \right| \Delta\xi \right)^{(n+1)(m+1)}$$

$$\leq (b-a)^{(n+1)(m+1)-1} \int_a^b \left| f_j^\Delta(\xi) \right|^{(n+1)(m+1)} \Delta\xi, \tag{3.3.40}$$

by applying Hölder's inequality. Also by letting $f_j(t) = c(t-a)$ for $t \in [a,b]_\mathbb{T}$ and all $j \in [1, n+1]_\mathbb{N}$ for some constant c, one can easily see that (3.3.38) holds with equality. The proof is complete. ∎

Theorem 3.3.11 Let $n \in \mathbb{N}$, $a, b \in \mathbb{T}$ and $f_j \in C^1_{\mathrm{rd}}(\mathbb{T}, \mathbb{R})$ with $f_j(a) = 0$ for all $j \in [1, n+1]_\mathbb{N}$, and that $p \in C_{\mathrm{rd}}(\mathbb{T}, \mathbb{R}^+)$ with $\int_a^b \left[p(\xi) \right]^{-1/n} \Delta\xi < \infty$ and $q \in C_{\mathrm{rd}}(\mathbb{T}, \mathbb{R}^+)$ be a nonincreasing function. Then

$$\int_a^b q^\sigma(\xi) \sum_{j=1}^{n+1} \left| \left[\prod_{i=1}^{j-1} f_i^\sigma(\xi) \right] f_j^\Delta(\xi) \left(\prod_{i=j+1}^{n+1} f_i(\xi) \right) \right| \Delta\xi$$

$$\leq \frac{1}{n+1} \left(\int_a^b \frac{1}{(p(\xi))^{\frac{1}{n}}} \Delta\xi \right)^n \left(\int_a^b p(\xi) q^\sigma(\xi) \sum_{j=1}^{n+1} \left(f_j^\Delta(\xi) \right)^{n+1} \Delta\xi \right). \tag{3.3.41}$$

Proof. Set

$$F_j(t) := \int_a^t \left[q^\sigma(\xi) \right]^{\frac{1}{n+1}} \left| f_j^\Delta(\xi) \right| \Delta\xi,$$

and then

$$F_j^\Delta(t) = (q^\sigma(t))^{\frac{1}{n+1}} \left| f_j^\Delta(t) \right|$$

for $t \in [a,b]_\mathbb{T}$ and all $j \in [1, n+1]_\mathbb{N}$. Then we have

$$F_j(t) \geq q^{\frac{1}{n+1}}(t) \int_a^t |f_j^\Delta(\xi)| \Delta\xi \geq q^{\frac{1}{n+1}}(t) \left| \int_a^t f_j^\Delta(\xi) \Delta\xi \right| \geq q^{\frac{1}{n+1}}(t) |f_j(t)|,$$

for $t \in [a,b]_\mathbb{T}$ and all $j \in [1, n+1]_\mathbb{N}$ (note here that for any $t \in [a,b]_\mathbb{T}$, $\xi \in [a,t)_\mathbb{T}$ implies $\sigma(\xi) \leq t$ and thus $q^\sigma(\xi) \leq q(t)$). Now, set

$$F(t) := \prod_{j=1}^{n+1} F_j(t), \quad \text{for} \quad t \in [a,b]_\mathbb{T}.$$

Similar reasoning as in the proof of Theorem 3.3.10 yields

$$\int_a^b q^\sigma(\xi) \sum_{j=1}^{n+1} \left| \left[\prod_{i=1}^{j-1} f_i^\sigma(\xi) \right] f_j^\Delta(\xi) \left[\prod_{i=j+1}^{n+1} f_i(\xi) \right] \right| \Delta\xi$$

$$\leq \int_a^b \sum_{j=1}^{n+1} \left\{ \left[\prod_{i=1}^{j-1} F_i^\sigma(\xi) \right] F_j^\Delta(\xi) \left[\prod_{i=j+1}^{n+1} F_i(\xi) \right] \right\} \Delta\xi$$

$$= \int_a^b F^\Delta(\xi) \Delta\xi = F(b) = \prod_{j=1}^{n+1} F_j(b) \leq \frac{1}{n+1} \sum_{j=1}^{n+1} \left[F_j(b) \right]^{n+1}$$

$$= \frac{1}{n+1} \sum_{j=1}^{n+1} \left\{ \int_a^b \frac{1}{(p(\xi))^{\frac{1}{n+1}}} \left[p(\xi) q^\sigma(\xi) \right]^{\frac{1}{n+1}} |f_j^\Delta(\xi)| \Delta\xi \right\}^{n+1}$$

$$\leq \frac{1}{n+1} \left(\int_a^b \frac{1}{[p(\xi)]^{\frac{1}{n}}} \Delta\xi \right)^n \tag{3.3.42}$$

$$\times \sum_{j=1}^{n} \left\{ \int_a^b p(\xi) q^\sigma(\xi) |f_j^\Delta(\xi)|^{n+1} \Delta\xi \right\}.$$

The proof is complete. ∎

Theorem 3.3.12 *Let $n \in \mathbb{N}$, $a,b \in \mathbb{T}$ and $f_j \in C_{rd}^1(\mathbb{T}, \mathbb{R})$ with $f_j(a) = 0$ for all $j \in [1, n+1]_\mathbb{N}$, and that $p \in C_{rd}(\mathbb{T}, \mathbb{R})$ with $\int_a^b [p(\xi)]^{-1/(m+n)} \Delta\xi < \infty$ and $q \in C_{rd}(\mathbb{T}, \mathbb{R})$ be a nonincreasing function. Then*

$$\int_a^b q^\sigma(\xi) \sum_{j=0}^{m} \left| \left[\prod_{i=1}^{n+1} f_i^\sigma(\xi) \right]^j \left[\prod_{i=1}^{n+1} f_i(\xi) \right]^{m-j} \right|$$

$$\times \sum_{j=1}^{n+1} \left| \left[\prod_{i=1}^{j-1} f_i^\sigma(\xi) \right] f_j^\Delta(\xi) \left[\prod_{i=j+1}^{n+1} f_i(\xi) \right] \right| \Delta\xi$$

$$\leq \frac{1}{n+1} \left(\int_a^b \frac{1}{[p(\xi)]^{\frac{1}{m+n}}} \Delta\xi \right)^{\frac{(m+n)(m+1)(n+1)}{m+n+1}}$$

$$\times \sum_{j=1}^{n+1} \left(\int_a^b p(\xi) q^\sigma(\xi) |f_j^\Delta(\xi)|^{m+n+1} \Delta\xi \right)^{\frac{(m+1)(n+1)}{m+n+1}}.$$

The results can be extended by applying the Muirhead inequality. For example the result of Theorem 3.3.10 can be arranged as follows.

Corollary 3.3.8 *In addition to the assumptions of Theorem 3.3.10, suppose that there exists $(\alpha_1, \alpha_2, \ldots, \alpha_{n+1})$ such that $\sum_{j=1}^{n+1} \alpha_j = n+1$ and $\sum_{j=1}^{k} \alpha_j \geq k$ for all $k \in [1, n]_{\mathbb{N}}$, then the right-hand side of (3.3.38) can be replaced by the following one*

$$\frac{1}{(n+1)!} \sum_{\pi \in S^{n+1}} \left\{ \prod_{j=1}^{n+1} (b-a)^{\alpha_j(m+1)} \int_a^b \left| \left[\prod_{j=1}^{n+1} f_{\pi_j}^{\Delta}(\xi) \right]^{\alpha_j(m+1)} \right| \Delta\xi \right\},$$

where S^{n+1} is the set of all permutations of the set $[1, n+1]_{\mathbb{N}}$, and π_j stands for the j-th component of the permutation π.

Proof. From the second term in (3.3.39) and Muirhead's inequality, we have

$$\prod_{j=1}^{n+1} \left[F_j(b) \right]^{m+1} = \frac{1}{(n+1)!} \sum_{\pi \in S^{n+1}} \prod_{j=1}^{n+1} \left[F_{\pi_j}(b) \right]^{m+1}$$

$$\leq \frac{1}{(n+1)!} \sum_{\pi \in S^{n+1}} \prod_{j=1}^{n+1} \left[F_{\pi_j}(b) \right]^{\alpha_j(m+1)},$$

where $(\alpha_1, \alpha_2, \ldots, \alpha_{n+1}) \rhd (1, 1, \ldots, 1)$. The rest of the proof is similar to that of Theorem 3.3.10. ■

3.4 Higher Order Opial Type Inequalities

In this section, we present some Opial type inequalities involving higher order derivatives. The results in this section are adapted from [49, 91, 127, 136, 139, 149]. To prove the results we need the following theorem [111, p. 338].

Theorem 3.4.1 *Let $k \geq 2$ be an integer and $x_i \geq 0$ be reals for all $i \in [1, k]_{\mathbb{N}}$. Then*

$$\left(\sum_{i=1}^{k} x_i \right)^{\alpha} \leq \left\{ \begin{array}{ll} 1, & \text{if } 0 \leq \alpha \leq 1 \\ k^{\alpha-1}, & \text{if } \alpha \geq 1 \end{array} \right\} \sum_{i=1}^{k} x_i^{\alpha}. \tag{3.4.1}$$

We recall the definition of the generalized Taylor Monomials which is given as follows:

$$h_k(t, s) := \left\{ \begin{array}{ll} 1, & k = 0, \\ \displaystyle\int_s^t h_{k-1}(\xi, s)\Delta\xi, & k \in \mathbb{N}, \end{array} \right.$$

for all $s, t \in \mathbb{T}$. For convenience, for y^{Δ^n} we mean the (nth) delta derivative of y which is equivalent to $\left(y^{\Delta^{n-1}} \right)^{\Delta}$ for $n \in \mathbb{N}$.

Theorem 3.4.2 *Let* $\mathbb{T}$ *be a time scale with* $a, b \in \mathbb{T}$ *and* $y \in C_{rd}^{(n)}([a,b] \cap \mathbb{T})$.
If $y^{\Delta^i}(a) = 0$, *for* $i = 0, 1, \ldots, n-1$, *then*

$$\int_a^b |y(t)| \left| y^{\Delta^n}(t) \right| \Delta t \; \leq \; \sqrt{\frac{1}{2}} \left(\int_a^b \left(\int_a^t |h_{n-1}(t, \sigma(s))|^2 \, \Delta s \right) \Delta t \right)^{\frac{1}{2}}$$

$$\times \int_a^b \left| y^{\Delta^n}(t) \right|^2 \Delta t. \tag{3.4.2}$$

Proof. From the Taylor formula (1.4.6), since $y^{\Delta^i}(a) = 0$, for $i = 0, 1, \ldots, n-1$, we have

$$y(t) := \int_a^t h_{n-1}(t, \sigma(s)) y^{\Delta^n}(s) \Delta s. \tag{3.4.3}$$

This implies that

$$\left| y^{\Delta^n}(t) \right| |y(t)| \leq \left| y^{\Delta^n}(t) \right| \int_a^t |h_{n-1}(t, \sigma(s))| \left| y^{\Delta^n}(s) \right| \Delta s.$$

Applying the Schwartz inequality, we have

$$\left| y^{\Delta^n}(t) \right| |y(t)| \leq \left| y^{\Delta^n}(t) \right| \left(\int_a^t |h_{n-1}(t, \sigma(s))|^2 \, \Delta s \right)^{\frac{1}{2}} \left(\int_a^t \left| y^{\Delta^n}(s) \right|^2 \Delta s \right)^{\frac{1}{2}}.$$

Then

$$\int_a^b \left| y^{\Delta^n}(t) \right| |y(t)| \, \Delta t \; \leq \; \int_a^b \left(\int_a^t |h_{n-1}(t, \sigma(s))|^2 \, \Delta s \right)^{\frac{1}{2}} \left| y^{\Delta^n}(t) \right|$$

$$\times \left(\int_a^t \left| y^{\Delta^n}(s) \right|^2 \Delta s \right)^{\frac{1}{2}} \Delta t. \tag{3.4.4}$$

Let

$$z(t) := \int_a^t \left| y^{\Delta^n}(s) \right|^2 \Delta s.$$

Then $z(a) = 0$ and $\left| y^{\Delta^n}(t) \right|^2 = z^{\Delta}(t)$. From this and (3.4.4), we have

$$\int_a^b \left| y^{\Delta^n}(t) \right| |y(t)| \, \Delta t \leq \int_a^b \left(\int_a^t |h_{n-1}(t, \sigma(s))|^2 \, \Delta s \right)^{\frac{1}{2}} \left(z(t) z^{\Delta}(t) \right)^{\frac{1}{2}} \Delta t. \tag{3.4.5}$$

Applying the Schwartz inequality we obtain

$$\int_a^b \left| y^{\Delta^n}(t) \right| |y(t)| \, \Delta t \; \leq \; \left(\int_a^b \left(\int_a^t |h_{n-1}(t, \sigma(s))|^2 \, \Delta s \right) \Delta t \right)^{\frac{1}{2}}$$

$$\times \left(\int_a^b z(t) z^{\Delta}(t) \Delta t \right)^{\frac{1}{2}}. \tag{3.4.6}$$

By the chain rule (1.1.7), we see (note that $z(t) > 0$ and $z^\Delta(t) > 0$), that

$$\left(z^2(t)\right)^\Delta = 2z^\Delta(t) \int_0^1 [hz^\sigma + (1-h)z]\, dh \geq 2z^\Delta(t)z(t).$$

Then, since $z(a) = 0$, we have that

$$\int_a^b z(t)z^\Delta(t)\Delta t \leq \frac{1}{2}\int_a^b \left(z^2(t)\right)^\Delta \Delta t = \frac{1}{2}z^2(b), \qquad (3.4.7)$$

Substituting (3.4.7) into (3.4.6), we have

$$\int_a^b \left|y^{\Delta^n}(t)\right| |y(t)|\, \Delta t \leq \sqrt{\frac{1}{2}}\left(\int_a^b \left(\int_a^t |h_{n-1}(t,\sigma(s))|^2\, \Delta s\right)\Delta t\right)^{\frac{1}{2}} z(b)$$

$$= \sqrt{\frac{1}{2}}\left(\int_a^b \left(\int_a^t |h_{n-1}(t,\sigma(s))|^2\, \Delta s\right)\Delta t\right)^{\frac{1}{2}} \int_a^b \left|y^{\Delta^n}(t)\right|^2 \Delta t,$$

which is the desired inequality (3.4.2). The proof is complete. ∎

Remark 3.4.1 *Let $0 \leq k < n$, be fixed, and let $x \in C_{rd}^{(n-k)}([a,b] \cap \mathbb{T})$ be such that $x^{\Delta^i}(a) = 0$, $0 \leq i \leq n-k-1$. Then from (3.4.2) it follows that*

$$\int_a^b |x(t)|\left|x^{\Delta^{n-k}}(t)\right|\Delta t \quad \leq \quad \sqrt{\frac{1}{2}}\left(\int_a^b \left(\int_a^t |h_{n-k-1}(t,\sigma(s))|^2\, \Delta s\right)\Delta t\right)^{\frac{1}{2}}$$

$$\times \int_a^b \left|x^{\Delta^n}(t)\right|^2 \Delta t.$$

Thus for $x = y^{\Delta^k}$, where $y \in C_{rd}^{(n-k)}[a,b]$, $y^{\Delta^i} = 0$, $k \leq i \leq n-1$, we have the following result.

Corollary 3.4.1 *Let $\mathbb{T}$ be a time scale with $a, b \in \mathbb{T}$ and $y \in C_{rd}^{(n)}([a,b] \cap \mathbb{T})$. If $y^{\Delta^i}(a) = 0$, $k \leq i \leq n-1$, then*

$$\int_a^b \left|y^{\Delta^k}(t)\right|\left|y^{\Delta^n}(t)\right|\Delta t \quad \leq \quad \sqrt{\frac{1}{2}}\left(\int_a^b \left(\int_a^t |h_{n-k-1}(t,\sigma(s))|^2\, \Delta s\right)\Delta t\right)^{\frac{1}{2}}$$

$$\times \int_a^b \left|y^{\Delta^n}(t)\right|^2 \Delta t.$$

Theorem 3.4.2 can be extended to a general inequality with two different constants by applying the Hölder inequality with indices p and q satisfying $1/p + 1/q = 1$.

Theorem 3.4.3 *Let* $\mathbb{T}$ *be a time scale with* a, $b \in \mathbb{T}$, p *and* q *are real numbers such that* $1/p + 1/q = 1$ *and* $y \in C_{rd}^{(n)}([a,b] \cap \mathbb{T})$. *If* $y^{\Delta^i}(a) = 0$, *for* $i = 0, 1, \ldots, n-1$, *then*

$$\int_a^b |y(t)| \left| y^{\Delta^n}(t) \right| \Delta t \leq \left(\frac{1}{2} \right)^{\frac{1}{q}} \left(\int_a^b \left(\int_a^t |h_{n-1}(t, \sigma(s))|^p \Delta s \right) \Delta t \right)^{\frac{1}{p}}$$

$$\times \int_a^b \left| y^{\Delta^n}(t) \right|^q \Delta t.$$

Theorem 3.4.4 *Let* $\mathbb{T}$ *be a time scale with* a, $b \in \mathbb{T}$ *and* l, m *be positive real numbers such that* $l + m > 1$, *and* $y \in C_{rd}^{(n)}([a,b] \cap \mathbb{T})$. *If* $y^{\Delta^i}(a) = 0$, *for* $i = 0, 1, \ldots, n-1$, *then*

$$\int_a^b |y(t)|^l \left| y^{\Delta^n}(t) \right|^m \Delta t \leq \left(\frac{m}{l+m} \right)^{\frac{m}{l+m}} \left(\int_a^b H^{(l+m-1)}(t, s) \Delta t \right)^{\frac{l}{l+m}}$$

$$\times \int_a^b \left| y^{\Delta^n}(t) \right|^{l+m} \Delta t, \qquad (3.4.8)$$

where

$$H(t, s) := \int_a^t (h_{n-1}(t, \sigma(s)))^{\frac{l+m}{l+m-1}} \Delta s.$$

Proof. From the Taylor formula (1.4.6) and since $y^{\Delta^i}(a) = 0$, for $i = 0, 1, \ldots, n-1$, we have

$$|y(t)| \leq \int_a^t h_{n-1}(t, \sigma(s)) \left| y^{\Delta^n}(s) \right| \Delta s.$$

Applying the Hölder inequality with $\gamma = l + m$ and $\nu = \frac{l+m}{l+m-1}$, we have

$$|y(t)| \leq \left(\int_a^t (h_{n-1}(t, \sigma(s)))^{\frac{l+m}{l+m-1}} \Delta s \right)^{\frac{l+m-1}{l+m}} \left(\int_a^t \left| y^{\Delta^n}(s) \right|^{l+m} \Delta s \right)^{\frac{1}{l+m}}.$$

This implies that

$$|y(t)|^l \left| y^{\Delta^n}(t) \right|^m \leq \left| y^{\Delta^n}(t) \right|^m \left(\int_a^t (h_{n-1}(t, \sigma(s)))^{\frac{l+m}{l+m-1}} \Delta s \right)^{l \left(\frac{l+m-1}{l+m} \right)}$$

$$\left(\int_a^t \left| y^{\Delta^n}(s) \right|^{l+m} \Delta s \right)^{\frac{l}{l+m}}.$$

Then

$$\int_a^b \left| y^{\Delta^n}(t) \right|^m |y(t)|^l \Delta t \leq \int_a^b \left(\int_a^t (h_{n-1}(t, \sigma(s)))^{\frac{l+m}{l+m-1}} \Delta s \right)^{l \left(\frac{l+m-1}{l+m} \right)}$$

$$\times \left| y^{\Delta^n}(t) \right|^m \left(\int_a^t \left| y^{\Delta^n}(s) \right|^{l+m} \Delta s \right)^{\frac{l}{l+m}} \Delta t. \quad (3.4.9)$$

Let $z(t) := \int_a^t \left|y^{\Delta^n}(s)\right|^{l+m} \Delta s$. This implies that $z(a) = 0$, and

$$\left|y^{\Delta^n}(t)\right|^m = \left(z^\Delta(t)\right)^{\frac{m}{l+m}} > 0.$$

From this and (3.4.9), we have

$$\int_a^b |y(t)|^l \left|y^{\Delta^n}(t)\right|^m \Delta t \leq \int_a^b \left(\int_a^t (h_{n-1}(t, \sigma(s)))^{\frac{l+m}{l+m-1}} \Delta s\right)^{l\left(\frac{l+m-1}{l+m}\right)}$$
$$\times \left(z^\Delta(t)\right)^{\frac{m}{l+m}} (z(t))^{\frac{l}{l+m}} \Delta t. \qquad (3.4.10)$$

Applying the Hölder inequality with $\gamma = (l+m)/l$ and $\nu = (l+m)/m$, we have

$$\int_a^b |y(t)|^l \left|y^{\Delta^n}(t)\right|^m \Delta t \leq \left(\int_a^b \left(\int_a^t (h_{n-1}(t, \sigma(s)))^{\frac{l+m}{l+m-1}} \Delta s\right)^{l\left(\frac{l+m-1}{l+m}\right)\left(\frac{l+m}{l}\right)} \Delta t\right)^{\frac{l}{l+m}}$$
$$\times \left(\int_a^b z^\Delta(t) (z(t))^{(l/m)} \Delta t\right)^{\frac{m}{m+l}}. \qquad (3.4.11)$$

From (1.1.7), we have (note that $z(t)$ and $z^\Delta(t) > 0$) that

$$\left(z^{\frac{l+m}{m}}(t)\right)^\Delta \geq \frac{l+m}{m} \int_0^1 [hz^\sigma + (1-h)z]^{\frac{l+m}{m}-1} z^\Delta(t)$$
$$\geq \frac{l+m}{m} (z(t))^{(l/m)} z^\Delta(t).$$

Then, since $z(a) = 0$, we have

$$\int_a^b z^{l/m}(t) z^\Delta(t) \Delta t \leq \frac{m}{l+m} \int_a^b \left(z^{\frac{l+m}{m}}(t)\right)^\Delta \Delta t = \frac{m}{l+m} z^{\frac{l+m}{m}}(b). \qquad (3.4.12)$$

Substituting (3.4.12) into (3.4.11) yields

$$\int_a^b |y(t)|^l \left|y^{\Delta^n}(t)\right|^m \Delta t$$
$$\leq \left(\frac{m}{l+m}\right)^{\frac{m}{l+m}} \left(\int_a^b \left(\int_a^t (h_{n-1}(t, \sigma(s)))^{\frac{l+m}{l+m-1}} \Delta s\right)^{l\left(\frac{l+m-1}{l+m}\right)\left(\frac{l+m}{l}\right)} \Delta t\right)^{\frac{l}{l+m}} z(b)$$
$$= \left(\frac{m}{l+m}\right)^{\frac{m}{l+m}} \left(\int_a^b \left(\int_a^t (h_{n-1}(t, \sigma(s)))^{\frac{l+m}{l+m-1}} \Delta s\right)^{(l+m-1)} \Delta t\right)^{\frac{l}{l+m}}$$
$$\times \int_a^b \left|y^{\Delta^n}(t)\right|^{l+m} \Delta t,$$

which is the desired inequality (3.4.8). The proof is complete. ∎

Using the ideas in Remark 3.4.1 we obtain the following result.

Corollary 3.4.2 *Let* $\mathbb{T}$ *be a time scale with* a, $b \in \mathbb{T}$ *and* l, m *be positive real numbers such that* $l + m > 1$ *and* $y \in C_{rd}^{(n)}([a,b] \cap \mathbb{T})$. *If* $y^{\Delta^i}(a) = 0$, $k \le i \le n - 1$, *then*

$$\int_a^b \left|y^{\Delta^k}(t)\right|^l \left|y^{\Delta^n}(t)\right|^m \Delta t \; \le \; \left(\frac{m}{l+m}\right)^{\frac{m}{l+m}} \left(\int_a^b H^{(l+m-1)}(t,s)\Delta t\right)^{\frac{l}{l+m}}$$

$$\times \int_a^b \left|y^{\Delta^n}(t)\right|^{l+m} \Delta t,$$

where

$$H(t,s) := \int_a^t \left(h_{n-k-1}(t,\sigma(s))\right)^{\frac{l+m}{l+m-1}} \Delta s.$$

Note that Theorem 3.4.4 cannot be applied when $l + m = 1$. In the following theorem we prove an inequality which can be applied in this case.

Theorem 3.4.5 *Let* $\mathbb{T}$ *be a time scale with* a, $b \in \mathbb{T}$ *and* l, m *be positive real numbers such that* $l + m = 1$ *and* $y \in C_{rd}^{(n)}([a,b] \cap \mathbb{T})$. *If* $y^{\Delta^i}(a) = 0$, *for* $i = 0, 1, \ldots, n - 1$, *then*

$$\int_a^b |y(t)|^l \left|y^{\Delta^n}(t)\right|^m \Delta t \le m^m \left(\int_a^b h_{n-1}(t,a)\Delta t\right)^l \int_a^b \left|y^{\Delta^n}(t)\right| \Delta t.$$

$$(3.4.13)$$

Proof. Using the fact that $|h_n(t,s)|$ is increasing with respect to its first component for $t \ge \sigma(s) > a$, we have from the Taylor formula (1.4.6) and $y^{\Delta^i}(a) = 0$, for $i = 0, 1, \ldots, n - 1$, that

$$|y(t)| \le h_{n-1}(t,a) \int_a^t \left|y^{\Delta^n}(s)\right| \Delta s.$$

This implies that

$$|y(t)|^l \left|y^{\Delta^n}(t)\right|^m \le (h_{n-1}(t,a))^l \left|y^{\Delta^n}(t)\right|^m \left(\int_a^t \left|y^{\Delta^n}(s)\right| \Delta s\right)^l.$$

Now applying the Hölder inequality with indices $1/l$ and $1/m$, we obtain

$$\int_a^b |y(t)|^l \left|y^{\Delta^n}(t)\right|^m \Delta t \; \le \; \left(\int_a^b h_{n-1}(t,a)\Delta t\right)^l$$

$$\times \left(\int_a^b \left|y^{\Delta^n}(t)\right| \left(\int_a^t \left|y^{\Delta^n}(s)\right| \Delta s\right)^{l/m} \Delta t\right)^m.$$

Let $z(t) = \int_a^t \left| y^{\Delta^n}(s) \right| \Delta s$. Then $z(a) = 0$ and $z^\Delta(t) = \left| y^{\Delta^n}(t) \right|$, so

$$\int_a^b \left| y^{\Delta^n}(t) \right| \left(\int_a^t \left| y^{\Delta^n}(s) \right| \Delta s \right)^{l/m} \Delta t = \int_a^b z^\Delta(t) \left(z(t) \right)^{l/m} \Delta t,$$

and hence

$$\int_a^b |y(t)|^l \left| y^{\Delta^n}(t) \right|^m \Delta t \le \left(\int_a^b h_{n-1}(t,a) \Delta t \right)^l \left(\int_a^b z^\Delta(t) \left(z(t) \right)^{l/m} \Delta t \right)^m.$$

$$(3.4.14)$$

As in the proof of Theorem 3.4.4, we have that

$$\int_a^b z^\Delta(t) \left(z(t) \right)^{l/m} \Delta t \;\le\; \int_a^b z^{l/m}(t) z^\Delta(t) \Delta t \le m \int_a^b \left(z^{\frac{l+m}{m}}(t) \right)^\Delta \Delta t$$

$$= \; m z^{\frac{1}{m}}(b) = m \left(\int_a^b \left| y^{\Delta^n}(t) \right| \Delta t \right)^{1/m}.$$

Substituting into (3.4.14), we have

$$\int_a^b |y(t)|^l \left| y^{\Delta^n}(t) \right|^m \Delta t \le m^m \left(\int_a^b h_{n-1}(t,a) \Delta t \right)^l \left(\int_a^b \left| y^{\Delta^n}(t) \right| \Delta t \right),$$

which is the desired inequality (3.4.13). The proof is complete. ∎

Using the ideas in Remark 3.4.1 we obtain the following result.

Corollary 3.4.3 *Let* $\mathbb{T}$ *be a time scale with* $a, b \in \mathbb{T}$ *and* l, m *be positive real numbers such that* $l + m = 1$, *and* $y \in C_{rd}^{(n)}([a,b] \cap \mathbb{T})$. *If* $y^{\Delta^i}(a) = 0$, $k \le i \le n-1$, *then*

$$\int_a^b \left| y^{\Delta^k}(t) \right|^l \left| y^{\Delta^n}(t) \right|^m \Delta t \le m^m \left(\int_a^b |h_{n-k-1}(t,a)| \Delta t \right)^l \left(\int_a^b \left| y^{\Delta^n}(t) \right| \Delta t \right).$$

Theorem 3.4.6 *Let* $\mathbb{T}$ *be a time scale with* $a, b \in \mathbb{T}$ *and* p, q *be positive real numbers such that* $p \ge 0$ *and* $q \ge 1$. *Suppose that* h *is a positive rd-continuous function and is nonincreasing on* $[a, \tau]_{\mathbb{T}}$ *and* $y \in C_{rd}^{(n)}([a,b] \cap \mathbb{T})$. *If* $y^{\Delta^i}(a) = 0$, *for* $i = 0, 1, \ldots, n-1$, *then*

$$\int_a^b h(t) |y(t)|^p \left| y^{\Delta^n}(t) \right|^q \Delta t \le \frac{q(b-a)^{pn}}{p+q} \int_a^b h(t) \left| y^{\Delta^n}(t) \right|^{p+q} \Delta t. \quad (3.4.15)$$

Proof. Let

$$z(t) = \int_a^t \int_a^{t_{n-1}} \cdots \int_a^{t_1} \left| y^{\Delta^n}(s) \right| \Delta s \Delta t_1 \ldots \Delta t_{n-1}, \qquad (3.4.16)$$

Then

$$z^{\Delta}(t), \ldots, z^{\Delta^n}(t) \geq 0, \; z^{\Delta^n}(t) = \left| y^{\Delta^n}(s) \right| \geq 0, \quad \text{and} \quad z(t) \geq |y(t)|.$$
$$\text{(3.4.17)}$$

and

$$z^{\Delta^i}(t) = \int_a^t z^{\Delta^{i+1}}(s)\Delta s \leq (t-a)z^{\Delta^{i+1}}(t), \quad i = 0,1,\ldots,n-2. \quad \text{(3.4.18)}$$

Applying Theorem 3.2.1 in conjunction with (3.4.16)–(3.4.18), we have

$$\int_a^{\tau} h(t)\,|y(t)|^p \left| y^{\Delta^n}(t) \right|^q \Delta t$$

$$\leq \int_a^{\tau} h(t)\,|y(t)|^p \left| z^{\Delta^n}(t) \right| \Delta t$$

$$\leq \int_a^{\tau} h(t)\,|y(t)|^p \left| z^{\Delta^n}(t) \right| \Delta t \leq \int_a^{\tau} h(t)\left[(t-a)z^{\Delta}(t) \right]^p \left| z^{\Delta^n}(t) \right|^q \Delta t$$

$$\vdots$$

$$\leq \int_a^{\tau} h(t)\left[(t-a)^{n-1} z^{\Delta^{n-1}}(t) \right]^p \left| z^{\Delta^n}(t) \right|^q \Delta t$$

$$\leq (\tau - a)^{p(n-1)}\frac{q}{p+q}(\tau - a)^p \int_a^{\tau} h(t)\left| z^{\Delta^n}(t) \right|^{p+q} \Delta t$$

$$= \frac{q(\tau - a)^{pn}}{p+q} \int_a^{\tau} \left| y^{\Delta^n}(t) \right|^{p+q} \Delta t,$$

which is the desired inequality (3.4.15). The proof is complete. ∎

Theorem 3.4.7 *Let $\mathbb{T}$ be a time scale with a, $b \in \mathbb{T}$ and p, q be positive real numbers such that $p \geq 0$ and $q \geq 1$. Suppose that h is a positive rd-continuous function and is nonincreasing on $[a, \tau]_{\mathbb{T}}$ and x, $y \in C_{rd}^{(n)}([a,b]\cap\mathbb{T})$. If $x^{\Delta^i}(a) = y^{\Delta^i}(a) = 0$, for $i = 0,1,\ldots,n-1$, then*

$$\int_a^b h(t)\left[|x(t)|^p \left| y^{\Delta^n}(t) \right|^q + |y(t)|^p \left| x^{\Delta^n}(t) \right|^q \right] \Delta t$$

$$\leq \frac{2q(b-a)^{pn}}{p+q} \int_a^b h(t)\left[\left| x^{\Delta^n}(t) \right|^{p+q} + \left| y^{\Delta^n}(t) \right|^{p+q} \right] \Delta t. \quad \text{(3.4.19)}$$

Proof. Let

$$z(t) = \int_a^t \int_a^{t_{n-1}} \cdots \int_a^{t_1} \left[\left| x^{\Delta^n}(s) \right|^{p+q} + \left| y^{\Delta^n}(s) \right|^{p+q} \right]^{\frac{1}{p+q}} \Delta s \Delta t_1 \ldots \Delta t_{n-1},$$

Then

$$z^{\Delta}(t), \ldots, z^{\Delta^{n-1}}(t) \geq 0, \ z^{\Delta^n}(t) = \left[\left| x^{\Delta^n}(s) \right|^{p+q} + \left| y^{\Delta^n}(s) \right|^{p+q} \right]^{\frac{1}{p+q}}$$

$$\geq \max\{ \left| x^{\Delta^n}(s) \right|, \left| y^{\Delta^n}(s) \right| \} \geq 0,$$

$$z(t) \geq |x(t)| \ \text{and} \ z(t) \geq |y(t)|.$$

Applying Theorem 3.4.6, we have

$$\int_a^\tau h(t) \left[|x(t)|^p \left| y^{\Delta^n}(t) \right|^q + |y(t)|^p \left| x^{\Delta^n}(t) \right|^q \right] \Delta t$$

$$\leq \int_a^\tau h(t) |z(t)|^p \left[\left| x^{\Delta^n}(t) \right|^q + \left| y^{\Delta^n}(t) \right|^q \right] \Delta t$$

$$\leq 2 \int_a^\tau h(t) |z(t)|^p \left| z^{\Delta^n}(t) \right|^q \Delta t$$

$$\leq (\tau - a)^{pn} \frac{2q}{p+q} \int_a^\tau h(t) \left| z^{\Delta^n}(t) \right|^{p+q} \Delta t$$

$$= \frac{2q(\tau - a)^{pn}}{p+q} \int_a^\tau h(t) \left[\left| x^{\Delta^n}(t) \right|^{p+q} + \left| y^{\Delta^n}(t) \right|^{p+q} \right] \Delta t,$$

which is the desired inequality (3.4.19). The proof is complete. ∎

Theorem 3.4.8 *Let* $\mathbb{T}$ *be a time scale with* $a, b \in \mathbb{T}$ *and* p, q *be positive real numbers such that* $p \geq 0$ *and* $q \geq 1$. *Suppose that* h *is a positive rd-continuous function and is nonincreasing on* $[a, \tau]_{\mathbb{T}}$ *and* $x, y \in C_{rd}^{(n)}$ *(*$[a, b] \cap \mathbb{T}$*). If* $x^{\Delta^i}(a) = y^{\Delta^i}(a) = 0$ *and* $x^{\Delta^i}(b) = y^{\Delta^i}(b) = 0$, *for* $i = 0, 1, \ldots, n-1$, *then*

$$\int_a^b h(t) \left[|x(t)|^p \left| y^{\Delta^n}(t) \right|^q + |y(t)|^p \left| x^{\Delta^n}(t) \right|^q \right] \Delta t$$

$$\leq \frac{2q}{p+q} (\frac{b-a}{2})^{pn} \int_a^b h(t) \left[\left| x^{\Delta^n}(t) \right|^{p+q} + \left| y^{\Delta^n}(t) \right|^{p+q} \right] \Delta t. \quad (3.4.20)$$

In the following, we prove some inequalities with two different weight functions.

Theorem 3.4.9 *Let* $\mathbb{T}$ *be a time scale with* $a, \tau \in \mathbb{T}$, *let* $p \in C_{rd}([a, \tau]_{\mathbb{T}}, \mathbb{R})$ *with*

$$\int_a^\tau (p(t))^{1-\alpha} \Delta t < \infty, \quad (\alpha > 1).$$

Suppose that q *is a positive bounded rd-continuous function and is nonincreasing on* $[a, \tau]_{\mathbb{T}}$. *Let* $y \in C_{rd}^{(n)}([a, \tau]_{\mathbb{T}})$. *If* $y^{\Delta^i}(a) = 0$, *for* $i = 0, 1, \ldots, n-1$, *then for* $r > 0$, *we have*

$$\int_a^\tau q(t) |y(t)|^r \left| y^{\Delta^n}(t) \right| \Delta t \leq \Lambda_1(r, \alpha) \left(\int_a^\tau p(t) q^{\frac{r}{r+1}}(t) \left| y^{\Delta^n}(t) \right|^\nu \Delta t \right)^{\frac{1+r}{\nu}},$$

$$(3.4.21)$$

where

$$\Lambda_1(r, \alpha) := \frac{(\tau - a)^{r(n-1)}}{r + 1} \left(\int_a^\tau p^{1-\alpha}(t)\Delta t \right)^{\frac{1+r}{\alpha}}, \quad \frac{1}{\alpha} + \frac{1}{\nu} = 1. \quad (3.4.22)$$

Proof. Let

$$z(t) = \int_a^t \int_a^{t_{n-1}} \cdots \int_a^{t_1} \left| y^{\Delta^n}(s) \right| \Delta s \Delta t_1 \ldots \Delta t_{n-1}. \quad (3.4.23)$$

Then

$$z^r(t) = \left(\int_a^t \int_a^{t_{n-1}} \cdots \int_a^{t_1} \left| y^{\Delta^n}(s) \right| \Delta s \Delta t_1 \ldots \Delta t_{n-1} \right)^r, \quad (3.4.24)$$

$$z^{\Delta^n}(t) = \left| y^{\Delta^n}(s) \right| \geq 0, \quad \text{and} \quad z(t) \geq |y(t)|, \quad (3.4.25)$$

and

$$z^{\Delta^i}(t) = \int_a^t z^{\Delta^{i+1}}(s)\Delta s \leq (t - a)z^{\Delta^i}(t), \quad i = 0, 1, \ldots, n - 2. \quad (3.4.26)$$

Applying Theorem 3.2.4 in conjunction with (3.4.23)–(3.4.26), we have

$$\int_a^\tau q(t) |y(t)|^r \left| y^{\Delta^n}(t) \right| \Delta t$$

$$\leq \int_a^\tau q(t) |y(t)|^r \left| z^{\Delta^{n-1}}(t) \right| \Delta t$$

$$\leq \int_a^\tau q(t) |y(t)|^r \left| z^{\Delta^{n-1}}(t) \right| \Delta t \leq \int_a^\tau q(t) \left[(t - a)z^\Delta(t) \right]^r \left| z^{\Delta^{n-1}}(t) \right| \Delta t$$

$$\vdots$$

$$\leq \int_a^\tau q(t) \left[(t - a)^{n-1} z^{\Delta_{n-1}}(t) \right]^r \left| z^{\Delta^n}(t) \right| \Delta t$$

$$\leq (\tau - a)^{r(n-1)} \int_a^\tau q(t) \left[z^{\Delta^{n-1}}(t) \right]^r \left| z^{\Delta^n}(t) \right| \Delta t$$

$$\leq \frac{(\tau - a)^{r(n-1)}}{r + 1} \left(\int_a^\tau p^{1-\alpha}(t)\Delta t \right)^{\frac{1+r}{\alpha}} \left(\int_a^\tau p(t)q^{\frac{r}{r+1}}(t) \left| y^{\Delta^n}(t) \right|^\nu \Delta t \right)^{\frac{1+r}{\nu}},$$

which is the desired inequality (3.4.21). The proof is complete. ∎

Theorem 3.4.10 *Let* $\mathbb{T}$ *be a time scale with* $a, b \in \mathbb{T}$ *and* l, m, r *be positive real numbers such that* $l + m > 1$ *and* $r > 1$. *Furthermore, let* p *and* q *be positive rd-continuous functions defined on* $[a, b] \cap \mathbb{T}$ *and* $y \in C_{rd}^{(n)}([a, b] \cap \mathbb{T})$. *If* $y^{\Delta^i}(a) = 0$, *for* $i = 0, 1, \ldots, n - 1$, *then*

$$\int_a^b q(t) |y(t)|^l \left| y^{\Delta^n}(t) \right|^m \Delta t \leq \Lambda_1(l, m, r, p, q) \left(\int_a^b p(s) \left| y^{\Delta^n}(s) \right|^r \Delta s \right)^{\frac{l+m}{r}}, \quad (3.4.27)$$

where

$$\Lambda_1(l,m,p,q,r) := \left(\frac{m}{l+m}\right)^{m/r} \left(\int_a^b q^{\frac{r}{r-m}}(t) p^{\frac{-m}{r-m}}(t) \left(P(t)\right)^{l\left(\frac{r-1}{r-m}\right)} \Delta t\right)^{\frac{r-m}{r}},$$

(3.4.28)

and

$$P(t) := \int_a^t p^{\frac{-1}{r-1}}(s) \left(h_{n-1}(t,\sigma(s))\right)^{\frac{r}{r-1}} \Delta s.$$

Proof. From the Taylor formula (1.4.6), we see that

$$|y(t)| \leq \int_a^t p^{\frac{-1}{r}}(s) \left(h_{n-1}(t,\sigma(s))\right) p^{\frac{1}{r}}(s) \left|y^{\Delta^n}(s)\right| \Delta s.$$

Applying the Hölder inequality on the right-hand side with indices r and $r/(r-1)$, we have

$$\begin{aligned}
|y(t)| &\leq \int_a^t p^{\frac{-1}{r}}(s) h_{n-1}(t,\sigma(s)) p^{\frac{1}{r}}(s) \left|y^{\Delta^n}(s)\right| \Delta s \\
&\leq \left(\int_a^t p^{\frac{-1}{r-1}}(s) \left(h_{n-1}(t,\sigma(s))\right)^{\frac{r}{r-1}} \Delta s\right)^{\frac{r-1}{r}} \\
&\quad \times \left(\int_a^t p(s) \left|y^{\Delta^n}(s)\right|^r \Delta s\right)^{1/r}.
\end{aligned}$$

This implies that

$$q(t)\,|y(t)|^l \left|y^{\Delta^n}(t)\right|^m \leq q(t) P^{l\left(\frac{r-1}{r}\right)}(t) \left|y^{\Delta^n}(t)\right|^m \left(\int_a^t p(s) \left|y^{\Delta^n}(s)\right|^r \Delta s\right)^{l/r}.$$

Integrating from a to b, we have

$$\begin{aligned}
&\int_a^b q(t)\,|y(t)|^l \left|y^{\Delta^n}(t)\right|^m \Delta t \\
&\leq \int_a^b q(t) P^{l\left(\frac{r-1}{r}\right)}(t) \left|y^{\Delta^n}(t)\right|^m \left(\int_a^t p(s) \left|y^{\Delta^n}(s)\right|^r \Delta s\right)^{l/r} \Delta t.
\end{aligned}$$

Let

$$z(t) := \int_a^t p(s) \left|y^{\Delta^n}(s)\right|^r \Delta s.$$

Then $z(a) = 0$, $z^\Delta(t) = p(t) \left|y^{\Delta^n}(t)\right|^r$ and $\left|y^{\Delta^n}(t)\right|^m = \left(z^\Delta(t)\right)^{\frac{m}{r}} p^{\frac{-m}{r}}(t)$. This implies that

$$\begin{aligned}
&\int_a^b q(t)\,|y(t)|^l \left|y^{\Delta^n}(t)\right|^m \Delta t \\
&\leq \int_a^b q(t) P^{l\left(\frac{r-1}{r}\right)}(t) p^{\frac{-m}{r}}(t) \left(z^\Delta(t)\right)^{\frac{m}{r}} \left(z(t)\right)^{l/r} \Delta t.
\end{aligned}$$

(3.4.29)

Applying the Hölder inequality with indices r/m and $r/(r-m)$, we obtain

$$\int_a^b q(t)P^{l(\frac{r-1}{r})}(t)p^{\frac{-m}{r}}(t)\left(z^\Delta(t)\right)^{\frac{m}{r}}(z(t))^{l/r}\Delta t$$

$$\leq \left(\int_a^b \left(z^\Delta(t)\right)(z(t))^{l/m}\Delta t\right)^{m/r}$$

$$\times \left(\int_a^b q^{\frac{r}{r-m}}(t)P^{l(\frac{r-1}{r-m})}(t)p^{\frac{-m}{r-m}}(t)\Delta t\right)^{\frac{r-m}{r}}.$$

Substituting into (3.4.29), we have

$$\int_a^b q(t)\,|y(t)|^l\,\left|y^{\Delta^n}(t)\right|^m\Delta t$$

$$\leq \left(\int_a^b q^{\frac{r}{r-m}}(t)P^{l(\frac{r-1}{r-m})}(t)p^{\frac{-m}{r-m}}(t)\Delta t\right)^{\frac{r-m}{r}}\left(\int_a^b \left(z^\Delta(t)\right)(z(t))^{l/m}\Delta t\right)^{m/r}.$$

Also we have

$$\left(\int_a^b z^\Delta(t)\,(z(t))^{l/m}\,\Delta t\right)^{m/r} \leq \left(\frac{m}{l+m}\int_a^b \left(z^{\frac{l+m}{m}}(t)\right)^\Delta\Delta t\right)^{m/r}$$

$$= \left(\frac{m}{l+m}\right)^{m/r}(z(b))^{\frac{l+m}{r}} = \left(\frac{m}{l+m}\right)^{m/r}\left(\int_a^b p(s)\left|y^{\Delta^n}(s)\right|^r\Delta s\right)^{\frac{l+m}{r}}.$$

This implies that

$$\int_a^b q(t)\,|y(t)|^l\,\left|y^{\Delta^n}(t)\right|^m\Delta t \leq \Lambda_1(l,m,p,q,r)\left(\int_a^b p(s)\left|y^{\Delta^n}(s)\right|^r\Delta s\right)^{\frac{l+m}{r}},$$

which is the desired inequality (3.4.27) where $\Lambda_1(l,m,p,q,r)$ is defined as in (3.4.28). The proof is complete. ∎

Using the ideas in Remark 3.4.1 we obtain the following result.

Theorem 3.4.11 *Let $\mathbb{T}$ be a time scale with $a,\,b\in\mathbb{T}$ and $l,\,m$ be positive real numbers such that $l+m>1$ and $r>1$. Furthermore, let p and q be positive rd-continuous functions defined on $[a,b]\cap\mathbb{T}$ and $y\in C_{rd}^{(n)}([a,b]\cap\mathbb{T})$. If $y^{\Delta^i}(a)=0,\ k\leq i\leq n-1$, then*

$$\int_a^b q(t)\left|y^{\Delta^k}(t)\right|^l\left|y^{\Delta^n}(t)\right|^m\Delta t \leq \Lambda_2(l,m,r,p,q)\left(\int_a^b p(s)\left|y^{\Delta^n}(s)\right|^r\Delta s\right)^{\frac{l+m}{r}},$$

where

$$\Lambda_2(l,m,p,q,r) := \left(\frac{m}{l+m}\right)^{m/r}\left(\int_a^b q^{\frac{r}{r-m}}(t)p^{\frac{-m}{r-m}}(t)\,(P(t))^{l(\frac{r-1}{r-m})}\Delta t\right)^{\frac{r-m}{r}},$$

and

$$P(t) := \int_a^t p^{\frac{-1}{r-1}}(s) h_{n-k-1}^{\frac{r}{r-1}}(t, \sigma(s)) \Delta s.$$

Theorem 3.4.12 *Let $\mathbb{T}$ be a time scale with $a, b \in \mathbb{T}$ and α, β be positive real numbers such that $\alpha + \beta > 1$, and let p, q be nonnegative rd-continuous functions on $(a, b)_{\mathbb{T}}$ and $y \in C_{rd}^{(n)}([a, b] \cap \mathbb{T})$. If $y^{\Delta^i}(a) = 0$, for $i = 0, 1, \ldots, n-1$, then*

$$\int_a^b q(t) |y(t)|^\alpha \left| y^{\Delta^n}(t) \right|^\beta \Delta t \le \Lambda_3(a, b, \alpha, \beta) \int_a^b p(t) \left| y^{\Delta^n}(t) \right|^{\alpha + \beta} \Delta t,$$

$$(3.4.30)$$

where

$$\Lambda_3(a, b, \alpha, \beta) = \left(\frac{\beta}{\alpha + \beta} \right)^{\frac{\beta}{\alpha+\beta}}$$

$$\left(\int_a^b \frac{q^{\frac{\alpha+\beta}{\alpha}}(t)}{p^{\frac{\beta}{\alpha}}(t)} \left(\int_a^t \frac{h_{n-1}^{\frac{\alpha+\beta}{\alpha+\beta-1}}(t, \sigma(s))}{p^{\frac{1}{\alpha+\beta-1}}(s)} \Delta s \right)^{\alpha+\beta-1} \Delta t \right)^{\frac{\alpha}{\alpha+\beta}} (3.4.31)$$

Proof. From the Taylor formula (1.4.6), we see that

$$|y(t)| \le \int_a^t h_{n-1}(t, \sigma(s)) \left| y^{\Delta^n}(s) \right| \Delta s$$

$$= \int_a^t \frac{h_{n-1}(t, \sigma(s))}{(p(s))^{\frac{1}{\alpha+\beta}}} (p(s))^{\frac{1}{\alpha+\beta}} \left| y^{\Delta^n}(s) \right| \Delta s.$$

Now, since p is nonnegative on $(a, b)_{\mathbb{T}}$, it follows from the Hölder inequality with

$$f(s) = \frac{|h_{n-1}(t, \sigma(s))|}{(p(s))^{\frac{1}{\alpha+\beta}}}, \quad g(s) = (p(s))^{\frac{1}{\alpha+\beta}} \left| y^{\Delta^n}(s) \right|,$$

$$\gamma = \frac{\alpha + \beta}{\alpha + \beta - 1} \quad \text{and } \nu = \alpha + \beta,$$

that

$$|y(t)| \le \left(\int_a^t \frac{h_{n-1}^{\frac{\alpha+\beta}{\alpha+\beta-1}}(t, \sigma(s))}{(p(s))^{\frac{1}{\alpha+\beta-1}}} \Delta s \right)^{\frac{\alpha+\beta-1}{\alpha+\beta}} \left(\int_a^t p(s) \left| y^{\Delta^n}(s) \right|^{\alpha+\beta} \Delta s \right)^{\frac{1}{\alpha+\beta}}.$$

Then

$$|y(t)|^\alpha \le \left(\int_a^t \frac{h_{n-1}^{\frac{\alpha+\beta}{\alpha+\beta-1}}(t, \sigma(s))}{(p(s))^{\frac{1}{\alpha+\beta-1}}} \Delta s \right)^{\alpha\left(\frac{\alpha+\beta-1}{\alpha+\beta}\right)} \left(\int_a^t p(s) \left| y^{\Delta^n}(s) \right|^{\alpha+\beta} \Delta s \right)^{\frac{\alpha}{\alpha+\beta}}.$$

$$(3.4.32)$$

Setting $z(t) := \int_a^t p(s) \left| y^{\Delta^n}(s) \right|^{\alpha+\beta} \Delta s$, we see that $z(a) = 0$, and

$$z^\Delta(t) = p(t) \left| y^{\Delta^n}(t) \right|^{\alpha+\beta} > 0. \tag{3.4.33}$$

This gives us

$$\left| y^{\Delta^n}(t) \right|^\beta = \left(\frac{z^\Delta(t)}{p(t)} \right)^{\frac{\beta}{\alpha+\beta}}. \tag{3.4.34}$$

Since q is nonnegative on $(a, b)_{\mathbb{T}}$, we have from (3.4.32) and (3.4.34) that

$$q(t) |y(t)|^\alpha \left| y^{\Delta^n}(t) \right|^\beta \le q(t) \left(\frac{1}{p(t)} \right)^{\frac{\beta}{\alpha+\beta}}$$

$$\times \left(\int_a^t \frac{(h_{n-1}(t, \sigma(s)))^{\frac{\alpha+\beta}{\alpha+\beta-1}}}{p^{\frac{1}{\alpha+\beta-1}}(s)} \Delta s \right)^{\alpha\left(\frac{\alpha+\beta-1}{\alpha+\beta}\right)} (z(t))^{\frac{\alpha}{\alpha+\beta}} \left(z^\Delta(t) \right)^{\frac{\beta}{\alpha+\beta}}.$$

This implies that

$$\int_a^b q(t) |y(t)|^\alpha \left| y^{\Delta^n}(t) \right|^\beta \Delta t \le \int_a^b q(t) \left(\frac{1}{p(t)} \right)^{\frac{\beta}{\alpha+\beta}}$$

$$\times \left(\int_a^t \frac{(h_{n-1}(t, \sigma(s)))^{\frac{\alpha+\beta}{\alpha+\beta-1}}}{p^{\frac{1}{\alpha+\beta-1}}(s)} \Delta s \right)^{\alpha\left(\frac{\alpha+\beta-1}{\alpha+\beta}\right)} (z(t))^{\frac{\alpha}{\alpha+\beta}} \left(z^\Delta(t) \right)^{\frac{\beta}{\alpha+\beta}} \Delta t \tag{3.4.35}$$

so applying the Hölder inequality with indices $(\alpha+\beta)/\alpha$ and $(\alpha+\beta)/\beta$, we have

$$\int_a^b q(t) |y(t)|^\alpha \left| y^{\Delta^n}(t) \right|^\beta \Delta t$$

$$\le \left(\int_a^b \frac{s^{\frac{\alpha+\beta}{\alpha}}(t)}{p^{\frac{\beta}{\alpha}}(t)} \left(\int_a^t \frac{(h_{n-1}(t, \sigma(s)))^{\frac{\alpha+\beta}{\alpha+\beta-1}}}{p^{\frac{1}{\alpha+\beta-1}}(s)} \Delta s \right)^{(\alpha+\beta-1)} \Delta t \right)^{\frac{\alpha}{\alpha+\beta}}$$

$$\times \left(\int_a^b z^{\frac{\alpha}{\beta}}(t) z^\Delta(t) \Delta t \right)^{\frac{\beta}{\alpha+\beta}}. \tag{3.4.36}$$

From (3.4.33), the chain rule (1.1.7) and the fact that $z^\Delta(s) > 0$, we obtain

$$z^{\frac{\alpha}{\beta}}(t) z^\Delta(t) \le \frac{\beta}{\alpha+\beta} \left(z^{\frac{\alpha+\beta}{\beta}}(t) \right)^\Delta. \tag{3.4.37}$$

Substituting (3.4.37) into (3.4.36) and using the fact that $z(a) = 0$, we have

$$\int_a^b q(t) \left|y(t)\right|^\alpha \left|y^{\Delta^n}(t)\right|^\beta \Delta t$$

$$\leq \left(\int_a^b \frac{q^{\frac{\alpha+\beta}{\alpha}}(t)}{p^{\frac{\beta}{\alpha}}(t)} \left(\int_a^t \frac{(h_{n-1}(t,\sigma(s)))^{\frac{\alpha+\beta}{\alpha+\beta-1}}}{p^{\frac{1}{\alpha+\beta-1}}(s)} \Delta s\right)^{(\alpha+\beta-1)} dt\right)^{\frac{\alpha}{\alpha+\beta}}$$

$$\times \left(\frac{\alpha}{\alpha+\beta}\right)^{\frac{\beta}{\alpha+\beta}} \left(\int_a^b \left(z^{\frac{\alpha+\beta}{\beta}}(s)\right)^\Delta \Delta s\right)^{\frac{\beta}{\alpha+\beta}}$$

$$= \left(\int_a^b \frac{q^{\frac{\alpha+\beta}{\alpha}}(t)}{p^{\frac{\beta}{\alpha}}(t)} \left(\int_a^t \frac{(h_{n-1}(t,\sigma(s)))^{\frac{\alpha+\beta}{\alpha+\beta-1}}}{p^{\frac{1}{\alpha+\beta-1}}(s)} \Delta s\right)^{(\alpha+\beta-1)} \Delta t\right)^{\frac{\alpha}{\alpha+\beta}}$$

$$\times \left(\frac{\beta}{\alpha+\beta}\right)^{\frac{\beta}{\alpha+\beta}} z(b).$$

Using $z(b) := \int_a^b p(s) \left|y^{\Delta^n}(s)\right|^{\alpha+\beta} \Delta s$, we have from the last inequality that

$$\int_a^b q(t) \left|y(t)\right|^\alpha \left|y^{\Delta^n}(t)\right|^\beta \Delta t \leq \Lambda_3(a,b,\alpha,\beta) \int_a^b p(t) \left|y^{\Delta^n}(t)\right|^{\alpha+\beta} \Delta t,$$

which is the desired inequality (3.4.30). The proof is complete. ∎

Using the ideas in Remark 3.4.1 we obtain the following result.

Theorem 3.4.13 *Let $\mathbb{T}$ be a time scale with a, $b \in \mathbb{T}$ and α, β be positive real numbers such that $\alpha + \beta > 1$, and let p, q be nonnegative rd-continuous functions on $(a,b)_\mathbb{T}$ and $y \in C_{rd}^{(n)}([a,b] \cap \mathbb{T})$. If $y^{\Delta^i}(a) = 0$, $k \leq i \leq n-1$, then*

$$\int_a^b q(t) \left|y^{\Delta^k}(t)\right|^\alpha \left|y^{\Delta^n}(t)\right|^\beta \Delta t \leq \Lambda_4(a,b,\alpha,\beta) \int_a^b p(t) \left|y^{\Delta^n}(t)\right|^{\alpha+\beta} \Delta t,$$

where

$$\Lambda_4(a,b,\alpha,\beta) = \left(\frac{\beta}{\alpha+\beta}\right)^{\frac{\beta}{\alpha+\beta}}$$

$$\times \left(\int_a^b \frac{q^{\frac{\alpha+\beta}{\alpha}}(t)}{p^{\frac{\beta}{\alpha}}(t)} \left(\int_a^t \frac{h_{n-1}^{\frac{\alpha+\beta}{\alpha+\beta-1}}(t,\sigma(s))}{p^{\frac{1}{\alpha+\beta-1}}(s)} \Delta s\right)^{(\alpha+\beta-1)} \Delta t\right)^{\frac{\alpha}{\alpha+\beta}}.$$

Next instead of (3.4.3) we use the relation between g_n and h_n and define

$$y(t) := (-1)^n \int_t^b g_{n-1}(\sigma(s),t)y^{\Delta^n}(s)\Delta s. \tag{3.4.38}$$

Proceeding as above and using (3.4.38) one can obtain some results when $y^{\Delta^i}(b) = 0$, *for* $i = 0, 1, \ldots, n - 1$. For example one can get the following results.

Theorem 3.4.14   *Let* $\mathbb{T}$ *be a time scale with* a, $b \in \mathbb{T}$ *and* $y \in C_{rd}^{(n)}([a, b] \cap \mathbb{T})$. *If* $y^{\Delta^i}(b) = 0$, *for* $i = 0, 1, \ldots, n - 1$, *then*

$$\int_a^b |y(t)| \left| y^{\Delta^n}(t) \right| \Delta t \leq \sqrt{\frac{1}{2}} \left(\int_a^b \left(\int_t^b g_{n-1}^2(\sigma(s), t) \Delta s \right) \Delta t \right)^{\frac{1}{2}} \int_a^b \left| y^{\Delta^n}(t) \right|^2 \Delta t.$$

$$(3.4.39)$$

Theorem 3.4.15   *Let* $\mathbb{T}$ *be a time scale with* a, $b \in \mathbb{T}$ *and* l, m *be positive real numbers such that* $l + m > 1$. *Let* $y \in C_{rd}^{(n)}([a, b] \cap \mathbb{T})$. *If* $y^{\Delta^i}(b) = 0$, *for* $i = 0, 1, \ldots, n - 1$, *then*

$$\int_a^b |y(t)|^l \left| y^{\Delta^n}(t) \right|^m \Delta t \quad \leq \quad \left(\int_a^b \left(\int_t^b g_{n-1}^{\frac{l+m}{l+m-1}}(\sigma(s), t) \Delta s \right)^{(l+m-1)} \Delta t \right)^{\frac{l}{l+m}}$$

$$\times \int_a^b \left| y^{\Delta^n}(t) \right|^{l+m} \Delta t.$$

Theorem 3.4.16 .   *Let* $\mathbb{T}$ *be a time scale with* 0, $h \in \mathbb{T}$ *and* l, n *be positive integers. If* $y \in C_{rd}^{(n)}([0, h] \cap \mathbb{T})$ *with* $y^{\Delta^i}(0) = 0$, *for* $i = 0, 1, 2, \ldots, n - 1$, *then we have*

$$\int_0^h \left| \left\{ \sum_{k=0}^l y^k(t)(y^\sigma(t))^{l-k} \right\} y^{\Delta^n}(t) \right| \Delta t \leq h^{nl} \int_0^h \left| y^{\Delta^n}(t) \right|^{l+1} \Delta t. \quad (3.4.40)$$

Proof. We consider

$$z(t) = \int_0^t \int_0^{\tau_{n-1}} \cdots \int_0^{\tau_2} \left\{ \int_0^{\tau_1} \left| y^{\Delta^n}(s) \right| \Delta s \right\} \Delta \tau_1 \Delta \tau_2 \ldots \Delta \tau_{n-1}.$$

Hence, we have

$$z^\Delta(t) \quad = \quad \int_0^t \int_0^{\tau_{n-1}} \cdots \int_0^{\tau_2} \left\{ \int_0^{\tau_1} \left| y^{\Delta^n}(s) \right| \Delta s \right\} \Delta \tau_1 \Delta \tau_2 \ldots \Delta \tau_{n-2}, \ldots,$$

$$z^{\Delta^n}(t) \quad = \quad \int_0^t \left| y^{\Delta^n}(s) \right| \Delta s, \quad z^{\Delta^n}(t) = \left| y^{\Delta^n}(t) \right|,$$

and for $0 \leq t \leq h$,

$$y(t) \quad \leq \quad \int_0^t \left| y^\Delta(t_1) \right| \Delta t_1 \leq \int_0^t \int_0^{t_1} \left| y^{\Delta\Delta}(t_2) \right| \Delta t_2 \Delta t_1 \leq \ldots \leq z(t)$$

$$= \quad \int_0^t z^\Delta(s) \Delta s \leq \int_0^t z^\Delta(t) \Delta s \leq \int_0^h z^\Delta(t) \Delta s \leq h z^\Delta(t)$$

$$\leq \quad h^2 z^{\Delta\Delta}(t) \leq \ldots \leq h^{n-1} z^{\Delta^{n-1}}(t) = h^{n-1} f(t),$$

where we put $f(t) = z^{\Delta^{n-1}}(t)$. Therefore,

$$\int_0^h \left| \left\{ \sum_{k=0}^l y^k(t)(y^\sigma(t))^{l-k} \right\} y^{\Delta^n}(t) \right| \Delta t$$

$$\leq \int_0^h \left\{ \sum_{k=0}^l |y(t)|^k |y^\sigma(t)|^{l-k} \left| y^{\Delta^n}(t) \right| \right\} \Delta t$$

$$\leq \int_0^h \left\{ (h^{n-1} f(t))^k \left| h^{n-1} f^\sigma(t) \right|^{l-k} \left| y^\Delta(t) \right| \right\} \Delta t$$

$$= h^{(n-1)l} \int_0^h \left\{ (f(t))^k (f^\sigma(t))^{l-k} f^\Delta(t) \right\} \Delta t = h^{(n-1)l} \int_0^h (f^{l+1}(t))^\Delta \Delta t$$

$$= h^{(n-1)l} f^{l+1}(h) = h^{(n-1)l} \left[\int_0^h \left| y^{\Delta^n}(t) \right| \Delta t \right]^{l+1}.$$

Applying the Hölder inequality with indices $(l+1)/l$ and $l+1$, we see that

$$\int_0^h \left| \left\{ \sum_{k=0}^l y^k(t)(y^\sigma(t))^{l-k} \right\} y^{\Delta^n}(t) \right| \Delta t \quad \leq \quad h^{(n-1)l} h^l \int_0^h \left| y^{\Delta^n}(t) \right|^{l+1} \Delta t$$

$$= h^{nl} \int_0^h \left| y^{\Delta^n}(t) \right|^{l+1} \Delta t,$$

which is the desired inequality (3.4.40). The proof is complete. ∎

Similarly using Taylor formula (1.4.6) one can prove the following result.

Theorem 3.4.17 *Let* $\mathbb{T}$ *be a time scale with* $0, h \in \mathbb{T}$ *and* l, n *be positive integers. If* $y \in C_{ld}^{(n)}([0, h] \cap \mathbb{T})$ *with* $y^{\Delta^i}(0) = 0$, *for* $i = 0, 1, 2, \ldots, n-1$, *then we have*

$$\int_0^h \left| \left\{ \sum_{k=0}^l y^k(t)(y^\rho(t))^{l-k} \right\} y^{\nabla^n}(t) \right| \Delta t \leq h^{nl} \int_0^h \left| y^{\nabla^n}(t) \right|^{l+1} \nabla t. \quad (3.4.41)$$

Theorem 3.4.18 . *Let* $\mathbb{T}$ *be a time scale with* $a, \tau \in \mathbb{T}$ *and* p, q *be positive real numbers such that* $p > 1$, $1/p + 1/q = 1$, *and let* r, s *be nonnegative rd-continuous functions on* $(a, \tau)_{\mathbb{T}}$. *If* $y \in C_{rd}^{(n)}([a, \tau] \cap \mathbb{T})$ *with* $y^{\Delta^i}(a) = 0$, *for* $i = 0, 1, 2, \ldots, n-1$, *then*

$$\int_a^\tau s(t) |y(t) + y^\sigma(t)|^p \left| y^{\Delta^n}(t) \right|^q \Delta t \leq K(a, \tau, p, q) \int_a^\tau r(t) \left| y^{\Delta^n}(t) \right|^{p+q} \Delta t,$$

$$(3.4.42)$$

where $K(a,\tau,p,q) = K_1(a,\tau,p,q) + K_2(a,\tau,p,q),$

$$K_1(a,\tau,p,q) = 2^{2p-1}\left(\frac{q}{p+q}\right)^{\frac{q}{p+q}}$$

$$\times \left(\int_a^\tau \frac{(s(t))^{\frac{p+q}{p}}}{(r(t))^{\frac{q}{p}}}\left(\int_a^t \frac{h_{n-1}(t,\sigma(s))}{r^{\frac{1}{p+q-1}}(s)}\Delta s\right)^{(p+q-1)}\Delta t\right)^{\frac{p}{p+q}},$$

and

$$K_2(a,\tau,p,q) = 2^{p-1}\left(\frac{q}{p+q}\right)^{\frac{q}{p+q}}$$

$$\times \left(\int_a^\tau \frac{\mu^{p+q}(t)(s(t))^{\frac{p+q}{p}}}{(r(t))^{\frac{q}{p}}}\left(\int_a^t \frac{h_{n-2}(t,\sigma(s))}{r^{\frac{1}{p+q-1}}(s)}\Delta s\right)^{(p+q-1)}\Delta t\right)^{\frac{p}{p+q}}.$$

Proof. From Taylor's formula, since $y^{\Delta^i}(a) = 0,$ for $i = 0,1,2,\ldots,n-1,$ we have

$$y(t) = \int_a^t h_{n-1}(t,\sigma(s))y^{\Delta^n}(s)\Delta s, \quad \text{for } t \in [a,\tau]_{\mathbb{T}}. \tag{3.4.43}$$

This implies that

$$|y(t)| \le \int_a^t \frac{h_{n-1}(t,\sigma(s))}{(r(s))^{\frac{1}{p+q}}}(r(s))^{\frac{1}{p+q}}\left|y^{\Delta^n}(s)\right|\Delta s.$$

Applying the Hölder inequality with

$$f(s) = \frac{h_{n-1}(t,\sigma(s))}{(r(s))^{\frac{1}{p+q}}}, \quad g(s) = (r(s))^{\frac{1}{p+q}}\left|y^{\Delta^n}(s)\right|,$$

$$\gamma = \frac{p+q}{p+q-1} \quad \text{and } \nu = p+q,$$

we have

$$\int_a^t h_{n-1}(t,\sigma(s))\left|y^{\Delta^n}(s)\right|\Delta s \le \left(\int_a^t \frac{h_{n-1}(t,\sigma(s))}{(r(s))^{\frac{1}{p+q-1}}}\Delta s\right)^{\frac{p+q-1}{p+q}}$$

$$\times \left(\int_a^t r(s)\left|y^{\Delta^n}(s)\right|^{p+q}\Delta s\right)^{\frac{1}{p+q}}.$$

Then, for $a \le t \le \tau,$ we get that

$$|y(t)|^p \le \left(\int_a^t \frac{h_{n-1}(t,\sigma(s))}{(r(s))^{\frac{1}{p+q-1}}}\Delta s\right)^{p(\frac{p+q-1}{p+q})}\left(\int_a^t r(s)\left|y^{\Delta^n}(s)\right|^{p+q}\Delta s\right)^{\frac{p}{p+q}}.$$

$$\tag{3.4.44}$$

Since $y^\sigma = y + \mu y^\Delta$, we have

$$y(t) + y^\sigma(t) = 2y(t) + \mu y^\Delta(t).$$

Applying the inequality (3.3.2), we get (here $p > 1$) that

$$|y + y^\sigma|^p \le 2^{p-1}(2^p |y|^p + \mu^p |y^\Delta|^p) = 2^{2p-1} |y|^p + 2^{p-1} \mu^p |y^\Delta|^p. \quad (3.4.45)$$

From (3.4.44), we get that

$$|y(t)|^p \left|y^{\Delta^n}(t)\right|^q \le \left(\int_a^t \frac{h_{n-1}(t, \sigma(s))}{(r(s))^{\frac{1}{p+q-1}}} \Delta s\right)^{p(\frac{p+q-1}{p+q})}$$

$$\times \left|y^{\Delta^n}(t)\right|^q \left(\int_a^t r(s) \left|y^{\Delta^n}(s)\right|^{p+q} \Delta s\right)^{\frac{p}{p+q}} \quad (3.4.46)$$

Also, by using (3.4.43), we get that

$$\left|y^\Delta(t)\right|^p \left|y^{\Delta^n}(t)\right|^q \le \left(\int_a^t \frac{h_{n-2}(t, \sigma(s))}{(r(s))^{\frac{1}{p+q-1}}} \Delta s\right)^{p(\frac{p+q-1}{p+q})}$$

$$\times \left|y^{\Delta^n}(t)\right|^q \left(\int_a^t r(s) \left|y^{\Delta^n}(s)\right|^{p+q} \Delta s\right)^{\frac{p}{p+q}}. \quad (3.4.47)$$

Substituting (3.4.47) and (3.4.46) into (3.4.45), we have

$$s(t) |y(t) + y^\sigma(t)|^p \left|y^{\Delta^n}(t)\right|^q$$

$$\le 2^{2p-1} s(t) \left(\int_a^t \frac{h_{n-1}(t, \sigma(s))}{(r(s))^{\frac{1}{p+q-1}}} \Delta s\right)^{p(\frac{p+q-1}{p+q})} \left|y^{\Delta^n}(t)\right|^q$$

$$\left(\int_a^t r(s) \left|y^{\Delta^n}(s)\right|^{p+q} \Delta s\right)^{\frac{p}{p+q}}$$

$$+2^{p-1} \mu^p(t) s(t) \left(\int_a^t \frac{h_{n-2}(t, \sigma(s))}{(r(s))^{\frac{1}{p+q-1}}} \Delta s\right)^{p(\frac{p+q-1}{p+q})}$$

$$\times \left|y^{\Delta^n}(t)\right|^q \left(\int_a^t r(s) \left|y^{\Delta^n}(s)\right|^{p+q} \Delta s\right)^{\frac{p}{p+q}}. \quad (3.4.48)$$

Setting

$$z(t) := \int_a^t r(s) \left|y^{\Delta^n}(s)\right|^{p+q} \Delta s, \quad (3.4.49)$$

we see that $z(a) = 0$, and

$$z^\Delta(t) = r(t) \left|y^{\Delta^n}(t)\right|^{p+q} > 0. \quad (3.4.50)$$

From this, we get that

$$\left|y^{\Delta^n}(t)\right|^{p+q} = \frac{z^{\Delta}(t)}{r(t)}, \quad \text{and} \quad \left|y^{\Delta^n}(t)\right|^q = \left(\frac{z^{\Delta}(t)}{r(t)}\right)^{\frac{q}{p+q}}. \tag{3.4.51}$$

From (3.4.46) and (3.4.51), since s is nonnegative on (a, τ), we have that

$$2^{2p-1} s(t) |y(t)|^p \left|y^{\Delta^n}(t)\right|^q$$

$$\leq 2^{2p-1} s(t) \left(\frac{1}{r(t)}\right)^{\frac{q}{p+q}} \times \left(\int_a^t \frac{h_{n-1}(t, \sigma(s))}{r^{\frac{1}{p+q-1}}(s)} \Delta s\right)^{p\left(\frac{p+q-1}{p+q}\right)}$$

$$\times (z(t))^{\frac{p}{p+q}} \left(z^{\Delta}(t)\right)^{\frac{q}{p+q}}.$$

This implies that

$$2^{2p-1} \int_a^\tau s(t) |y(t)|^p \left|y^{\Delta^n}(t)\right|^q \Delta t$$

$$\leq 2^{2p-1} \int_a^\tau s(t) \left(\frac{1}{r(t)}\right)^{\frac{q}{p+q}} \times \left(\int_a^t \frac{h_{n-1}(t, \sigma(s))}{r^{\frac{1}{p+q-1}}(s)} \Delta s\right)^{p\left(\frac{p+q-1}{p+q}\right)}$$

$$\times (z(t))^{\frac{p}{p+q}} \left(z^{\Delta}(t)\right)^{\frac{q}{p+q}} \Delta t.$$

Applying the Hölder inequality with indices $(p + q)/p$ and $(p + q)/q$ on the right-hand side of the last inequality, we have

$$2^{2p-1} \int_a^\tau s(t) |y(t)|^p \left|y^{\Delta^n}(t)\right|^q \Delta t$$

$$\leq 2^{2p-1} \left(\int_a^\tau s^{\frac{p+q}{p}}(t) \left(\frac{1}{r(t)}\right)^{\frac{q}{p}} \left(\int_a^t \frac{h_{n-1}(t, \sigma(s))}{r^{\frac{1}{p+q-1}}(s)} \Delta s\right)^{(p+q-1)} \Delta t\right)^{\frac{p}{p+q}}$$

$$\times \left(\int_a^\tau z^{\frac{p}{q}}(t) z^{\Delta}(t) \Delta t\right)^{\frac{q}{p+q}}. \tag{3.4.52}$$

From (3.4.50), and the chain rule formula (1.1.7), we obtain

$$z^{\frac{p}{q}}(t) z^{\Delta}(t) \leq \frac{q}{p+q} \left(z^{\frac{p+q}{q}}(t)\right)^{\Delta}. \tag{3.4.53}$$

Substituting (3.4.53) into (3.4.52) and using the fact that $z(a) = 0$, we have that

$$2^{2p-1} \int_a^\tau s(t) \, |y(t)|^p \left| y^{\Delta^n}(t) \right|^q \Delta t$$

$$\leq 2^{2p-1} \left(\int_a^\tau s^{\frac{p+q}{p}}(t) \left(\frac{1}{r(t)} \right)^{\frac{q}{p}} \left(\int_a^t \frac{h_{n-1}(t,\sigma(s))}{r^{\frac{1}{p+q-1}}(s)} \Delta s \right)^{(p+q-1)} \Delta t \right)^{\frac{p}{p+q}}$$

$$\times \left(\frac{p}{p+q} \right)^{\frac{q}{p+q}} \left(\int_a^\tau \left(z^{\frac{p+q}{q}}(s) \right)^\Delta \Delta s \right)^{\frac{q}{p+q}}$$

$$= \left(\int_a^\tau s^{\frac{p+q}{p}}(t) \left(\frac{1}{r(t)} \right)^{\frac{q}{p}} \left(\int_a^t \frac{h_{n-1}(t,\sigma(s))}{r^{\frac{1}{p+q-1}}(s)} \Delta s \right)^{(p+q-1)} \Delta t \right)^{\frac{p}{p+q}}$$

$$\times 2^{2p-1} \left(\frac{q}{p+q} \right)^{\frac{q}{p+q}} z(\tau).$$

Using (3.4.49), we have from the last inequality that

$$2^{2p-1} \int_a^\tau s(t) \, |y(t)|^p \left| y^{\Delta^n}(t) \right|^q \Delta t \leq K_1(a,\tau,p,q) \int_a^\tau r(t) \left| y^{\Delta^n}(t) \right|^{p+q} \Delta t.$$
$$(3.4.54)$$

Proceeding as above, we can also show that

$$2^{p-1} \int_a^\tau \mu^p(t) s(t) \left(\int_a^t \frac{h_{n-2}(t,\sigma(s))}{(r(s))^{\frac{1}{p+q-1}}} \Delta s \right)^{p\left(\frac{p+q-1}{p+q}\right)}$$

$$\times \left| y^{\Delta^n}(t) \right|^q \left(\int_a^t r(s) \left| y^{\Delta^n}(s) \right|^{p+q} \Delta s \right)^{\frac{p}{p+q}} \Delta t$$

$$\leq K_2(a,\tau,p,q) \int_a^\tau r(t) \left| y^{\Delta^n}(t) \right|^{p+q} \Delta t.$$
$$(3.4.55)$$

Integrating (3.4.48) from a to τ and using (3.4.54) and (3.4.55), we get that

$$\int_a^\tau s(t) \, |y(t) + y^\sigma(t)|^p \left| y^{\Delta^n}(t) \right|^q \Delta t$$

$$\leq (K_1(a,\tau,p,q) + K_2(a,\tau,p,q)) \int_a^\tau r(t) \left| y^{\Delta^n}(t) \right|^{p+q} \Delta t,$$

which is the desired inequality (3.4.42). The proof is complete. ∎

Similar reasoning as in the proof of Theorem 3.4.18, with $[a,\tau]$ replaced by $[b,\tau]_{\mathbb{T}}$ and $y(t)$ in (3.4.43) is replaced by

$$y(t) = (-1)^n \int_t^b g_{n-1}(\sigma(s),t) y^{\Delta^n}(s) \Delta s, \quad \text{for} \quad t \in [b,\tau]_{\mathbb{T}}$$

yields the following result.

Theorem 3.4.19 *Let* $\mathbb{T}$ *be a time scale with* τ, $b \in \mathbb{T}$ *and* p, q *be positive real numbers such that* $p > 1$, $1/p + 1/q = 1$, *and let* r, s *be nonnegative rd-continuous functions on* $(\tau, b)_{\mathbb{T}}$. *If* $y \in C_{rd}^{(n)}([\tau, b] \cap \mathbb{T})$ *with* $y^{\Delta^i}(b) = 0$, *for* $0 \leq i \leq n - 1$, *then*

$$\int_\tau^b s(t) \left| y(t) + y^\sigma(t) \right|^p \left| y^{\Delta^n}(t) \right|^q \Delta t \leq K^*(\tau, b, p, q) \int_\tau^b r(t) \left| y^{\Delta^n}(t) \right|^{p+q} \Delta t,$$

(3.4.56)

where $K^*(\tau, b, p, q) = K_1^*(\tau, b, p, q) + K_2^*(\tau, b, p, q)$,

$$K_1^*(\tau, b, p, q) = 2^{2p-1} \left(\frac{q}{p+q} \right)^{\frac{q}{p+q}}$$
$$\times \left(\int_\tau^b \frac{(s(t))^{\frac{p+q}{p}}}{(r(t))^{\frac{q}{p}}} \left(\int_t^b \frac{g_{n-1}(\sigma(s), t)}{r^{\frac{1}{p+q-1}}(s)} \Delta s \right)^{(p+q-1)} \Delta t \right)^{\frac{p}{p+q}}.$$

and

$$K_2^*(\tau, b, p, q) = 2^{2p-1} \left(\frac{q}{p+q} \right)^{\frac{q}{p+q}}$$
$$\times \left(\int_\tau^b \frac{\mu^{p+q}(t)(s(t))^{\frac{p+q}{p}}}{(r(t))^{\frac{q}{p}}} \left(\int_t^b \frac{g_{n-2}(\sigma(s), t)}{r^{\frac{1}{p+q-1}}(s)} \Delta s \right)^{(p+q-1)} \Delta t \right)^{\frac{p}{p+q}}.$$

In the following, we assume that there exists $\tau \in (a, b)_{\mathbb{T}}$ such that

$$K_1(p, q) = K_1(a, \tau, p, q) = K_1^*(\tau, b, p, q) < \infty,$$
$$K_2(p, q) = K_2(a, \tau, p, q) = K_2^*(\tau, b, p, q) < \infty,$$

where $K_1(a, \tau, p, q)$, $K_2(\tau, b, p, q)$, $K_1^*(a, \tau, p, q)$, and $K_2^*(\tau, b, p, q)$ are defined before. Note that

$$\int_a^b s(t) \left| y(t) + y^\sigma(t) \right|^p \left| y^{\Delta^n}(t) \right|^q \Delta t$$
$$= \int_a^\tau s(t) \left| y(t) + y^\sigma(t) \right|^p \left| y^{\Delta^n}(t) \right|^q \Delta t + \int_\tau^b s(t) \left| y(t) + y^\sigma(t) \right|^p \left| y^{\Delta^n}(t) \right|^q \Delta t,$$

so combining Theorems 3.4.18 and 3.4.19 yields the following result.

Theorem 3.4.20 *Let* $\mathbb{T}$ *be a time scale with* a, $b \in \mathbb{T}$ *and* p, q *be positive real numbers such that* $p > 1$, $1/p + 1/q = 1$, *and let* r, s *be nonnegative rd-continuous functions on* $(a, b)_{\mathbb{T}}$. *If* $y \in C_{rd}^{(n)}([a, b] \cap \mathbb{T})$ *with* $y^{\Delta^i}(a) = y^{\Delta^i}(b) = 0$, *for* $i = 0, 1, 2, \ldots, n - 1$, *then*

$$\int_a^b s(t) \left| y(t) + y^\sigma(t) \right|^p \left| y^{\Delta^n}(t) \right|^q \Delta t \leq K(a, b) \int_a^b r(t) \left| y^{\Delta^n}(t) \right|^{p+q} \Delta t,$$

(3.4.57)

where $K(a, b) = K_1(p, q) + K_2(p, q)$.

For $r = s$ in Theorem 3.4.18, we obtain the following result.

Corollary 3.4.4 *Let* $\mathbb{T}$ *be a time scale with* $a, \tau \in \mathbb{T}$ *and* p, q *be positive real numbers such that* $p > 1$, $1/p + 1/q = 1$, *and let* r *be a nonnegative rd-continuous function on* $(a, \tau)_{\mathbb{T}}$. *If* $y \in C_{rd}^{(n)}([a, \tau] \cap \mathbb{T})$ *with* $y^{\Delta^i}(a) = 0$, *for* $i = 0, 1, 2, \ldots, n-1$, *then*

$$\int_a^\tau r(t) \left| y(t) + y^\sigma(t) \right|^p \left| y^{\Delta^n}(t) \right|^q \Delta t \leq K^*(a, \tau, p, q) \int_a^\tau r(t) \left| y^{\Delta^n}(t) \right|^{p+q} \Delta t,$$

$$(3.4.58)$$

where

$$K^*(a, \tau, p, q) = 2^{2p-1} \left(\frac{q}{p+q} \right)^{\frac{q}{p+q}}$$

$$\times \left(\int_a^\tau r(t) \left(\int_a^t \frac{h_{n-1}(t, \sigma(s))}{r^{\frac{1}{p+q-1}}(s)} \Delta s \right)^{(p+q-1)} \Delta t \right)^{\frac{p}{p+q}}$$

$$+ 2^{p-1} \left(\frac{q}{p+q} \right)^{\frac{q}{p+q}}$$

$$\times \left(\int_a^\tau \mu^{p+q}(t) r(t) \left(\int_a^t \frac{h_{n-2}(t, \sigma(s))}{r^{\frac{1}{p+q-1}}(s)} \Delta s \right)^{(p+q-1)} \Delta t \right)^{\frac{p}{p+q}}.$$

From Theorems 3.4.19 and 3.4.20 one can derive similar results by setting $r = s$. Setting $r = 1$ in (3.4.58), we have the following result.

Corollary 3.4.5 *Let* $\mathbb{T}$ *be a time scale with* $a, \tau \in \mathbb{T}$ *and* p, q *be positive real numbers such that* $p > 1$, $1/p + 1/q = 1$. *If* $y \in C_{rd}^{(n)}([a, \tau] \cap \mathbb{T})$ *is delta differentiable with* $y^{\Delta^i}(a) = 0$, *for* $i = 0, 1, 2, \ldots, n-1$, *then*

$$\int_a^\tau \left| y(t) + y^\sigma(t) \right|^p \left| y^{\Delta^n}(t) \right|^q \Delta t \leq L(a, b, p, q) \int_a^\tau \left| y^{\Delta^n}(t) \right|^{p+q} \Delta t, \quad (3.4.59)$$

where

$$L(a, \tau, p, q) = 2^{2p-1} \left(\frac{q}{p+q} \right)^{\frac{q}{p+q}} \left(\int_a^\tau \left(\int_a^t h_{n-1}(t, \sigma(s)) \Delta s \right)^{(p+q-1)} \Delta t \right)^{\frac{p}{p+q}}$$

$$+ 2^{p-1} \left(\frac{q}{p+q} \right)^{\frac{q}{p+q}} (G(t))^{\frac{p}{p+q}},$$

where $G(t) = \int_a^\tau \mu^{p+q}(t) \left(\int_a^t h_{n-2}(t, \sigma(s)) \Delta s \right)^{(p+q-1)} \Delta t$.

Next we present Opial inequalities involving several functions and their higher-order derivatives. The generalized Taylor's formula and generalized polynomials will be used.

Theorem 3.4.21 Let $\ell, n \in \mathbb{N}$, $a, b \in \mathbb{T}$ and $y_j \in \mathrm{C}^{\ell}_{\mathrm{rd}}(\mathbb{T}, \mathbb{R})$ with $y_j^{\Delta^i}(a) = 0$ for all $j \in [1, n+1]_{\mathbb{N}}$ and all $i \in [0, \ell)_{\mathbb{N}_0}$. Then, the following inequality holds

$$\int_a^b \sum_{j=1}^{n+1} \left| \left[\prod_{i=1}^{j-1} y_i^{\sigma}(\xi) \right] y_j^{\Delta^{\ell}}(\xi) \left[\prod_{i=j+1}^{n+1} y_i(\xi) \right] \right| \Delta\xi$$

$$\leq \left(\frac{1}{n+1} \int_a^b \left(\int_a^{\sigma(\xi)} \left[h_{\ell-1}(\sigma(\xi), \sigma(\zeta)) \right]^2 \Delta\zeta \right)^n \Delta\xi \right)^{\frac{1}{2}}$$

$$\times \sum_{j=1}^{n+1} \left(\int_a^b \left[y_j^{\Delta^{\ell}}(\xi) \right]^2 \Delta\xi \right)^{\frac{n+1}{2}}.$$

Proof. We have

$$y_j(t) = \int_a^t h_{\ell-1}(t, \sigma(\xi)) y_j^{\Delta^{\ell}}(\xi) \Delta\xi, \quad \text{for all } t \in [a, b]_{\mathbb{T}} \text{ and all } j \in [0, n+1]_{\mathbb{N}}. \tag{3.4.60}$$

Now, set

$$y_j(t) := \int_a^t \left[y_j^{\Delta^{\ell}}(\xi) \right]^2 \Delta\xi, \quad \text{for all } t \in [a, b]_{\mathbb{T}} \text{ and all } j \in [0, n+1]_{\mathbb{N}}. \tag{3.4.61}$$

Then

$$\left| \prod_{i=1}^{j-1} y_i^{\sigma}(t) y_j^{\Delta^{\ell}}(t) \prod_{i=j+1}^{n+1} y_i(t) \right| = \left| \left[\prod_{i=1}^{j-1} \int_a^{\sigma(t)} h_{\ell-1}(\sigma(t), \sigma(\xi)) y_i^{\Delta^{\ell}}(\xi) \Delta\xi \right] \left(y_j^{\Delta}(t) \right)^{\frac{1}{2}} \right.$$

$$\times \left. \prod_{i=j+1}^{n+1} \int_a^t h_{\ell-1}(t, \sigma(\xi)) y_i^{\Delta^{\ell}}(\xi) \Delta\xi \right|$$

$$\leq H(t) \left(\prod_{i=1}^{j-1} y_i^{\sigma}(t) y_j^{\Delta}(t) \prod_{i=j+1}^{n+1} y_i(t) \right)^{\frac{1}{2}} \tag{3.4.62}$$

for all $j \in [1, n+1]_{\mathbb{N}}$, where

$$H(t) := \left(\int_a^{\sigma(t)} \left[h_{\ell-1}(\sigma(t), \sigma(\xi)) \right]^2 \Delta\xi \right)^{\frac{n}{2}}, \quad \text{for all } t \in [a, b]_{\mathbb{T}}. \tag{3.4.63}$$

Integrating (3.4.62) from a to b and applying Hölder's inequality, we get

$$\int_a^b \left| \prod_{i=1}^{j-1} y_i^{\sigma}(\xi) y_j^{\Delta^{\ell}}(\xi) \prod_{i=j+1}^{n+1} y_i(\xi) \right| \Delta\xi$$

$$\leq \left(\int_a^b \left[H(\xi) \right]^2 \Delta\xi \right)^{\frac{1}{2}} \left(\int_a^b \prod_{i=1}^{j-1} y_i^{\sigma}(\xi) y_j^{\Delta}(\xi) \prod_{i=j+1}^{n+1} y_i(\xi) \Delta\xi \right)^{\frac{1}{2}}.$$

Then, summing the resulting inequality over $j \in [1, n+1]_{\mathbb{N}}$, we get

$$
\int_a^b \sum_{j=1}^{n+1} \left| \left[\prod_{i=1}^{j-1} y_i^\sigma(\xi) \right] y_j^{\Delta^\ell}(\xi) \left[\prod_{i=j+1}^{n+1} y_i(\xi) \right] \right| \Delta\xi
$$

$$
\leq \left(\int_a^b \left[H(\xi) \right]^2 \Delta\xi \right)^{\frac{1}{2}} \tag{3.4.64}
$$

$$
\times \sum_{j=1}^{n+1} \left(\int_a^b \left[\prod_{i=1}^{j-1} y_i^\sigma(\xi) \right] y_j^\Delta(\xi) \left[\prod_{i=j+1}^{n+1} y_i(\xi) \right] \Delta\xi \right)^{\frac{1}{2}}
$$

$$
\leq \left(\int_a^b \left[H(\xi) \right]^2 \Delta\xi \right)^{\frac{1}{2}}
$$

$$
\times \left((n+1) \sum_{j=1}^{n+1} \int_a^b \left[\prod_{i=1}^{j-1} y_i^\sigma(\xi) \right] y_j^\Delta(\xi) \left[\prod_{i=j+1}^{n+1} y_i(\xi) \right] \Delta\xi \right)^{\frac{1}{2}}
$$

$$
= \left((n+1) \int_a^b \left[H(\xi) \right]^2 \Delta\xi \right)^{\frac{1}{2}}
$$

$$
\times \left(\int_a^b \left[\prod_{j=1}^{n+1} y_j(\xi) \right]^\Delta \Delta\xi \right)^{\frac{1}{2}} \tag{3.4.65}
$$

$$
= \left((n+1) \int_a^b \left[H(\xi) \right]^2 \Delta\xi \right)^{\frac{1}{2}} \left(\prod_{j=1}^{n+1} y_j(b) \right)^{\frac{1}{2}}
$$

$$
\leq \left(\frac{1}{n+1} \int_a^b \left[H(\xi) \right]^2 \Delta\xi \right)^{\frac{1}{2}} \sum_{j=1}^{n+1} \left[y_j(b) \right]^{\frac{n+1}{2}}, \tag{3.4.66}
$$

where the elementary inequalities

$$
\sum_{j=1}^{n+1} \lambda_j^{\frac{1}{2}} \leq \left((n+1) \sum_{j=1}^{n+1} \lambda_j \right)^{\frac{1}{2}},
$$

$$
\left(\prod_{j=1}^{n+1} \lambda_j \right)^{\frac{1}{2}} \leq \frac{1}{n+1} \sum_{j=1}^{n+1} \lambda_j^{\frac{n+1}{2}},
$$

for all $\lambda_j \in \mathbb{R}_0^+$ and $j \in [1, n+1]_{\mathbb{N}}$ are employed. The proof is hence completed. ∎

In the rest of this section, we make use of the following notations and definitions. Let $n \in \mathbb{N}$ satisfy $n \geq 2$, $\mathbb{T}_i$ be a time scale for all $i \in [1, n]_{\mathbb{N}}$, and

$a_i, b_i \in \mathbb{T}_i$ satisfy $-\infty < a_i \leq b_i < \infty$ for all $i \in [1, n]_\mathbb{N}$. Set $\Omega_n := [a_1, b_1]_{\mathbb{T}_1} \times \cdots \times [a_n, b_n]_{\mathbb{T}_n}$, and for $y \in C_{\mathrm{rd}}(\Omega_n, \mathbb{R})$ (y belongs to $C_{\mathrm{rd}}([a_i, b_i]_{\mathbb{T}_i}, \mathbb{R})$ for each $i \in [1, n]_\mathbb{N}$) denote

$$
\begin{cases}
\int\limits_{\Omega_n} y(\xi)\Delta\xi = \int\limits_{a_1}^{b_1} \cdots \int\limits_{a_n}^{b_n} y(\xi_1, \ldots, \xi_n)\Delta\xi_1 \cdots \Delta\xi_n, \\
\mathrm{grad}_n^r y(t) := \left(\dfrac{\partial^r}{\Delta t_1^r} y(t), \ldots, \dfrac{\partial^r}{\Delta t_n^r} y(t) \right).
\end{cases}
\tag{3.4.67}
$$

and

$$
\|\mathrm{grad}_n^r y(t)\| := \left(\sum_{i=1}^n \left| \frac{\partial^r}{\Delta t_i^r} y(t) \right|^2 \right)^{\frac{1}{2}},
\tag{3.4.68}
$$

where $t = (t_1, t_2, \ldots, t_n) \in \Omega_n$ and $r \in \mathbb{N}$.

Theorem 3.4.22 *Let p and q be real constants such that $p \geq 1$ and $q \geq 1$, and $y_{\mathrm{rd}}^1(\Omega_n, \mathbb{R})$ (y belongs to $C_{\mathrm{rd}}^1([a_i, b_i]_{\mathbb{T}_i}, \mathbb{R})$ for each $i \in [1, n]_\mathbb{N}$) vanishes on the boundary $\partial\Omega_n$ of Ω_n. Then*

$$
\int\limits_{\Omega_n} |y(\xi)|^p \, \|\mathrm{grad}_n^1 y(\xi)\|^q \, \Delta\xi
$$

$$
\leq \frac{1}{2^p n} \left[\sum_{i=1}^n (b_i - a_i)^{\frac{p(p+q)}{q}} \right]^{\frac{q}{p+q}} \int\limits_{\Omega_n} \|\mathrm{grad}_n^1 y(\xi)\|^{p+q} \, \Delta\xi.
\tag{3.4.69}
$$

Proof. Clearly, because of the boundary condition on y, for all $t = (t_1, \ldots, t_n) \in \Omega_n$ and all $i \in [1, n]_\mathbb{N}$, we have

$$
\int\limits_{a_i}^{t_i} \frac{\partial}{\Delta\xi_i} y(t; \xi_i)\Delta\xi_i = y(t; t_i) - y(t; a_i) = y(t)
\tag{3.4.70}
$$

and similarly

$$
\int\limits_{t_i}^{b_i} \frac{\partial}{\Delta\xi_i} y(t; \xi_i)\Delta\xi_i = -y(t),
\tag{3.4.71}
$$

where $y(t; s_i) := y(t_1, \ldots, t_{i-1}, s_i, t_{i+1}, \ldots, t_n)$ for all $i \in [1, n]_\mathbb{N}$. Therefore, from (3.4.70) and (3.4.71), we have

$$
y(t) = \frac{1}{n} \sum_{i=1}^n \int\limits_{a_i}^{t_i} \frac{\partial}{\Delta\xi_i} y(t; \xi_i)\Delta\xi_i
\tag{3.4.72}
$$

and

$$
y(t) = -\frac{1}{n} \sum_{i=1}^n \int\limits_{t_i}^{b_i} \frac{\partial}{\Delta\xi_i} y(t; \xi_i)\Delta\xi_i
\tag{3.4.73}
$$

for all $t \in \Omega_n$. Taking (3.4.72) and (3.4.73) into account, we see that

$$
|y(t)| = \frac{1}{2} \left(|y(t)| + |y(t)| \right)
$$

$$
= \frac{1}{2n} \left(\left| \sum_{i=1}^{n} \int_{a_i}^{t_i} \frac{\partial}{\Delta \xi_i} y(t; \xi_i) \Delta \xi_i \right| + \left| \sum_{i=1}^{n} \int_{t_i}^{b_i} \frac{\partial}{\Delta \xi_i} y(t; \xi_i) \Delta \xi_i \right| \right)
$$

$$
\leq \frac{1}{2n} \left(\sum_{i=1}^{n} \int_{a_i}^{t_i} \left| \frac{\partial}{\Delta \xi_i} y(t; \xi_i) \right| \Delta \xi_i + \sum_{i=1}^{n} \int_{t_i}^{b_i} \left| \frac{\partial}{\Delta \xi_i} y(t; \xi_i) \right| \Delta \xi_i \right)
$$

$$
= \frac{1}{2n} \sum_{i=1}^{n} \int_{a_i}^{b_i} \left| \frac{\partial}{\Delta \xi_i} y(t; \xi_i) \right| \Delta \xi_i \qquad (3.4.74)
$$

holds for all $t \in \Omega_n$. Applying Hölder's inequality with the indices $(p+q)/(p+q-1)$ and $p+q$ to the right-hand side of (3.4.74), we get

$$
|y(t)| \leq \frac{1}{2n} \sum_{i=1}^{n} \left[(b_i - a_i)^{\frac{p+q-1}{p+q}} \left(\int_{a_i}^{b_i} \left| \frac{\partial}{\Delta \xi_i} y(t; \xi_i) \right|^{p+q} \Delta \xi_i \right)^{\frac{1}{p+q}} \right] \qquad (3.4.75)
$$

for all $t \in \Omega_n$. Raising both sides of (3.4.75) to the p-th power, and applying (3.4.1), we have

$$
|y(t)|^p \leq n^{p-1} \left(\frac{1}{2n} \right)^p
$$

$$
\times \sum_{i=1}^{n} \left[(b_i - a_i)^{\frac{p(p+q-1)}{p+q}} \left(\int_{a_i}^{b_i} \left| \frac{\partial}{\Delta \xi_i} y(t; \xi_i) \right|^{p+q} \Delta \xi_i \right)^{\frac{p}{p+q}} \right] \qquad (3.4.76)
$$

for all $t \in \Omega_n$. Multiplying both sides of (3.4.76) by $|\nabla_n^1 y|^q$, we have

$$
|y(t)|^p |\nabla_n^1 y(t)|^q \leq \frac{1}{2^p n} \sum_{i=1}^{n} \left[(b_i - a_i)^{\frac{p(p+q-1)}{p+q}} |\nabla_n^1 y(t)|^q \right. \qquad (3.4.77)
$$

$$
\left. \times \left(\int_{a_i}^{b_i} \left| \frac{\partial}{\Delta \xi_i} y(t; \xi_i) \right|^{p+q} \Delta \xi_i \right)^{\frac{p}{p+q}} \right],
$$

for all $t \in \Omega_n$. Here we note that

$$
\int_{a_i}^{b_i} \left| \frac{\partial}{\Delta \xi_i} y(t; \xi_i) \right|^{p+q} \Delta \xi_i, \qquad (3.4.78)
$$

does not depend on the i-th component t_i. Integrating both sides of (3.4.77) on Ω_n with respect to t, we obtain

$$\int_{\Omega_n} |y(\xi)|^p ||\mathrm{grad}_n^1 y(\xi)||^q \Delta\xi$$

$$\leq \frac{1}{2^p n} \sum_{i=1}^n [(b_i - a_i)^{\frac{p(p+q-1)}{p+q}} \int_{\Omega_n} ||\mathrm{grad}_n^1 y(\xi)||^q \left(\int_{a_i}^{b_i} \left| \frac{\partial}{\Delta\zeta_i} y(\xi; \zeta_i) \right|^{p+q} \Delta\zeta_i \right)^{\frac{p}{p+q}} \Delta\xi]$$

$$\leq \frac{1}{2^p n} \sum_{i=1}^n \left[(b_i - a_i)^{\frac{p(p+q-1)}{p+q}} \left(\int_{\Omega_n} ||\mathrm{grad}_n^1 y(\xi)||^{p+q} \Delta\xi \right)^{\frac{q}{p+q}} \right.$$

$$\times \left. \left(\int_{\Omega_n} \int_{a_i}^{b_i} \left| \frac{\partial}{\Delta\zeta_i} y(\xi; \zeta_i) \right|^{p+q} \Delta\zeta_i \Delta\xi \right)^{\frac{p}{p+q}} \right] \tag{3.4.79}$$

$$= \frac{1}{2^p n} \sum_{i=1}^n \left[(b_i - a_i)^{\frac{p(p+q-1)}{p+q}} (b_i - a_i)^{\frac{p}{p+q}} \left(\int_{\Omega_n} ||\mathrm{grad}_n^1 y(\xi)||^{p+q} \Delta\xi \right)^{\frac{q}{p+q}} \right.$$

$$\times \left. \left(\int_{\Omega_n} \left| \frac{\partial}{\Delta\xi_i} y(\xi) \right|^{p+q} \Delta\xi \right)^{\frac{p}{p+q}} \right] \tag{3.4.80}$$

$$= \frac{1}{2^p n} \sum_{i=1}^n \left[(b_i - a_i)^p \left(\int_{\Omega_n} ||\mathrm{grad}_n^1 y(\xi)||^{p+q} \Delta\xi \right)^{\frac{q}{p+q}} \right. \tag{3.4.81}$$

$$\times \left. \left(\int_{\Omega_n} \left| \frac{\partial}{\Delta\xi_i} y(\xi) \right|^{p+q} \Delta\xi \right)^{\frac{p}{p+q}} \right]. \tag{3.4.82}$$

Now using (3.4.1) we have

$$\int_{\Omega_n} \sum_{i=1}^n \left| \frac{\partial}{\Delta\xi_i} y(\xi) \right|^{p+q} \Delta\xi$$

$$= \int_{\Omega_n} \left(\left(\sum_{i=1}^n \left| \frac{\partial}{\Delta\xi_i} y(\xi) \right|^{p+q} \right)^{\frac{2}{p+q}} \right)^{\frac{p+q}{2}} \Delta\xi \tag{3.4.83}$$

$$\leq \int_{\Omega_n} ||\mathrm{grad}_n^1 y(\xi)||^{p+q} \Delta\xi. \tag{3.4.84}$$

Applying Hölder's inequality and (3.4.1) to the right-hand side of (3.4.81), and taking (3.4.84) into account, we have

$$\int_{\Omega_n} |y(\xi)|^p \big\|\mathrm{grad}_n^1 y(\xi)\big\|^q \Delta\xi$$

$$\leq \frac{1}{2^p n} \sum_{i=1}^n \left[(b_i - a_i)^{\frac{p(p+q)}{q}} \left(\int_{\Omega_n} \big\|\mathrm{grad}_n^1 y(\xi)\big\|^{p+q} \Delta\xi \right)^{\frac{q}{p+q}} \right.$$

$$\times \left. \left(\int_{\Omega_n} \Big| \frac{\partial}{\Delta\xi_i} y(\xi) \Big|^{p+q} \Delta\xi \right)^{\frac{p}{p+q}} \right] \tag{3.4.85}$$

$$= \frac{1}{2^p n} \left(\sum_{i=1}^n (b_i - a_i)^{\frac{p(p+q)}{p+q}} \int_{\Omega_n} \big\|\mathrm{grad}_n^1 y(\xi)\big\|^{p+q} \Delta\xi \right)^{\frac{q}{p+q}}$$

$$\times \left(\sum_{i=1}^n \int_{\Omega_n} \Big| \frac{\partial}{\Delta\xi_i} y(\xi) \Big|^{p+q} \Delta\xi \right)^{\frac{p}{p+q}} \tag{3.4.86}$$

$$\leq \frac{1}{2^p n} \left(\sum_{i=1}^n (b_i - a_i)^{\frac{p(p+q)}{p+q}} \right)^{\frac{q}{p+q}} \left(\int_{\Omega_n} \big\|\mathrm{grad}_n^1 y(\xi)\big\|^{p+q} \Delta\xi \right)^{\frac{q}{p+q}}$$

$$\times \left(\int_{\Omega_n} \big\|\mathrm{grad}_n^1 y(\xi)\big\|^{p+q} \Delta\xi \right)^{\frac{p}{p+q}} \tag{3.4.87}$$

$$= \frac{1}{2^p n} \left(\sum_{i=1}^n (b_i - a_i)^{\frac{p(p+q)}{p+q}} \right)^{\frac{q}{p+q}} \int_{\Omega_n} \big\|\mathrm{grad}_n^1 y(\xi)\big\|^{p+q} \Delta\xi. \tag{3.4.88}$$

The proof is complete. ∎

The following result considers several functions.

Theorem 3.4.23 *Let m be an integer satisfying $m \geq 2$, p_i and q_i be real constants such that $p_i \geq 1$ and $q_i \geq 1$ for all $i \in [1,m]_{\mathbb{N}}$, and $y_{i\mathrm{rd}}^1(\Omega_n, \mathbb{R})$ vanish on the boundary $\partial\Omega_n$ of Ω_n for all $i \in [1,k]_{\mathbb{N}}$. Then*

$$\int_{\Omega_n} \prod_{i=1}^k |y_i(\xi)|^{p_i} \big\|\mathrm{grad}_n^1 y_i(\xi)\big\|^{q_i} \Delta\xi \tag{3.4.89}$$

$$\leq \frac{1}{kn} \sum_{i=1}^k \left[\frac{1}{2^{kp_i}} \left(\sum_{j=1}^n (b_j - a_j)^{\frac{kp_i(p_i+q_i)}{q_i}} \right)^{\frac{q_i}{p_i+q_i}} \int_{\Omega_n} \big\|\mathrm{grad}_n^1 y_i(\xi)\big\|^{k(p_i+q_i)} \Delta\xi \right]$$

Proof. Applying the well-known inequality concerning the arithmetic mean and geometric mean , we get

$$\int_{\Omega_n} \prod_{i=1}^{k} |y_i(\xi)|^{p_i} ||\text{grad}_n^1 y_i(\xi)||^{q_i} \Delta\xi$$

$$= \int_{\Omega_n} \left(\prod_{i=1}^{k} |y_i(\xi)|^{kp_i} ||\text{grad}_n^1 y_i(\xi)||^{kq_i} \right)^{\frac{1}{k}} \Delta\xi \qquad (3.4.90)$$

$$\leq \frac{1}{k} \sum_{i=1}^{k} \int_{\Omega_n} |y_i(\xi)|^{kp_i} ||\text{grad}_n^1 y_i(\xi)||^{kq_i} \Delta\xi. \qquad (3.4.91)$$

Applying Theorem 3.4.22 on the right-hand side of (3.4.91), we see that (3.4.89) is true. ∎

Theorem 3.4.24 *Let p and q be real constants such that $p \geq 1$ and $q \geq 1$, r be a fixed integer satisfying $r \geq 1$, and $y_{\text{rd}}^r(\Omega_n, \mathbb{R})$ and each of its partial delta-derivatives up to the order $r - 1$ vanish on the boundary $\partial\Omega_n$ of Ω_n. Then*

$$\int_{\Omega_n} |y(\xi)|^p \, ||\text{grad}_n^r y(\xi)||^{q_i} \, \Delta\xi$$

$$\leq \frac{1}{2^{pn}} \left[\sum_{i=1}^{n} (b_i - a_i)^{\frac{p}{q+p}} \left(\int_{a_i}^{b_i} |h_{r-1}(t_i, \sigma(\xi_i))|^{\frac{p+q}{p+q-1}} \, \Delta\xi_i \right)^{\frac{p(p+q-1)}{p+q}} \right]^{\frac{q}{p+q}}$$

$$\times \int_{\Omega_n} ||\text{grad}_n^r y(\xi)||^{p+q} \, \Delta\xi \qquad (3.4.92)$$

Proof. Now

$$\int_{a_i}^{t_i} h_{r-1}(t_i, \sigma(\xi_i)) \frac{\partial^r}{\Delta\xi_i^r} y(t; \xi_i) \Delta\xi_i = y(t) \qquad (3.4.93)$$

and

$$\int_{t_i}^{b_i} h_{r-1}(t_i, \sigma(\xi_i)) \frac{\partial^r}{\Delta\xi_i^r} y(t; \xi_i) \Delta\xi_i = -y(t), \qquad (3.4.94)$$

for all $t = (t_1, \ldots, t_n) \in \Omega_n$ and all $i \in [1, n]_{\mathbb{N}}$. From (3.4.93) and (3.4.94), we get

$$
|y(t)| = \frac{1}{2n} \left(\left| \sum_{i=1}^{n} \int_{a_i}^{t_i} h_{r-1}(t_i, \sigma(\xi_i)) \frac{\partial^r}{\Delta \xi_i^r} y(t; \xi_i) \Delta \xi_i \right| \right.
$$

$$
\left. + \left| \sum_{i=1}^{n} \int_{t_i}^{b_i} h_{r-1}(t_i, \sigma(\xi_i)) \frac{\partial^r}{\Delta \xi_i^r} y(t; \xi_i) \Delta \xi_i \right| \right) \tag{3.4.95}
$$

$$
\leq \frac{1}{2n} \left(\sum_{i=1}^{n} \int_{a_i}^{t_i} |h_{r-1}(t_i, \sigma(\xi_i))| \left| \frac{\partial^r}{\Delta \xi_i^r} y(t; \xi_i) \right| \Delta \xi_i \right.
$$

$$
\left. + \sum_{i=1}^{n} \int_{t_i}^{b_i} |h_{r-1}(t_i, \sigma(\xi_i))| \left| \frac{\partial^r}{\Delta \xi_i^r} y(t; \xi_i) \right| \Delta \xi_i \right) \tag{3.4.96}
$$

$$
\leq \frac{1}{2n} \sum_{i=1}^{n} \int_{a_i}^{t_i} |h_{r-1}(t_i, \sigma(\xi_i))| \left| \frac{\partial^r}{\Delta \xi_i^r} y(t; \xi_i) \right| \Delta \xi_i \tag{3.4.97}
$$

holds for all $t \in \Omega_n$. Following similar reasoning as in Theorem 3.4.22 yields (3.4.92). ∎

Finally we combine Theorems 3.4.23 and 3.4.24.

Theorem 3.4.25 *Let m be a fixed integer satisfying $m \geq 2$, p_i and q_i be real constants such that $p_i \geq 1$ and $q_i \geq 1$ for all $i \in [1, m]_{\mathbb{N}}$, r be a fixed integer satisfying $r \geq 1$, and $y_{i \, \mathrm{rd}}^{r}(\Omega_n, \mathbb{R})$ and each of their partial derivatives up to the order $r - 1$ vanish on the boundary $\partial \Omega_n$ of Ω_n for all $i \in [1, k]_{\mathbb{N}}$. Then*

$$
\int_{\Omega_n} \prod_{i=1}^{k} |y_i(\xi)|^{p_i} \|\mathrm{grad}_n^r y_i(\xi)\|^{q_i} \, \Delta \xi
$$

$$
\leq \frac{1}{kn} \sum_{i=1}^{n} \frac{1}{2^{kp_i}} \left[\left(\sum_{j=1}^{k} (b_j - a_j)^{\frac{p_j}{p_j + q_j}} H^{\frac{p(p+q-1)}{p+q}}(a_j, b_j) \right)^{\frac{q_i}{p_i + q_i}} \right.
$$

$$
\left. \times \int_{\Omega_n} \|\mathrm{grad}_n^r y_i(\xi)\|^{p_i + q_i} \, \Delta \xi \right]. \tag{3.4.98}
$$

where $H(a_j, b_j) = \left(\int_{a_j}^{b_j} |h_{r-1}(t_i, \sigma(\xi_i))|^{\frac{k(p_j + q_j)}{k(p_j + q_j - 1)}} \, \Delta \xi_i \right)$.

3.5 Diamond-α Opial Inequalities

In this section, we present a sequence of Opial type inequalities for first and higher order diamond alpha derivatives on time scales. The results are adapted from [32, 47, 50]. Throughout this section, we say that a function

$y : [0, \tau]_\mathbb{T} \to \mathbb{R}$ is in the class $C^1_{\Diamond_\alpha}$ if y is $\Diamond_\alpha$-differentiable such that αy^Δ is rd-continuous and $(1 - \alpha)y^\nabla$ is ld-continuous and $\alpha(1 - \alpha)y^{\Diamond_\alpha}$ is continuous. Note that $C^1_{\Diamond_\alpha}$ is, for $\alpha \in (0, 1)$, equal to the class of functions that are Δ-differentiable and ∇-differentiable such that y^Δ is rd-continuous and y^∇ is ld-continuous, and $y^{\Diamond_\alpha}$ is continuous. Moreover, $C^1_{\Diamond_0}$ is equal to the class of functions that are ∇-differentiable such that y^∇ is ld-continuous and $C^1_{\Diamond_1}$ is equal to the class of functions that are Δ-differentiable such that y^∇ is ld-continuous.

Theorem 3.5.1 *Let* $\alpha \in [0, 1]$ *and* $h \in \mathbb{T}$ *with* $h > 0$. *For any* $y \in C^1_{\Diamond_\alpha}$ *with* $y(0) = 0$ *and* $\alpha(1 - \alpha)y^\nabla y^\Delta \geq 0$, *we have*

$$\alpha^3 \int_0^h \left|(y + y^\sigma)y^\Delta\right|(t)\Delta t + (1 - \alpha)^3 \int_0^h \left|(y + y^\rho)y^\nabla\right|(t)\nabla t \leq h \int_0^h \left(y^{\Diamond_\alpha}\right)^2 \Diamond_\alpha t.$$

Proof. By Theorem 3.1.1 we have

$$\alpha^3 \int_0^h \left|(y + y^\sigma)y^\Delta\right|(t)\Delta t \leq \alpha^3 h \int_0^h \left|y^\Delta\right|^2 (t)\Delta t,$$

and by Theorem 3.1.2

$$(1 - \alpha)^3 \int_0^h \left|(y + y^\rho)y^\nabla\right|(t)\nabla t \leq (1 - \alpha)^3 h \int_0^h \left|y^\nabla\right|^2 (t)\nabla t.$$

Also from the definition of $y^{\Diamond_\alpha}$, we see that

$$\left(y^{\Diamond_\alpha}\right)^2 = \alpha^2 \left(y^\Delta\right)^2 + 2\alpha(1 - \alpha)y^\Delta y^\nabla + (1 - \alpha)^2 \left(y^\nabla\right)^2.$$

Now, since $y^\Delta y^\nabla \geq 0$, and we get that

$$\alpha^2 \left(y^\Delta\right)^2 \leq \left(y^{\Diamond_\alpha}\right)^2, \quad \text{and} \ (1 - \alpha)^2 \left(y^\nabla\right)^2 \leq \left(y^{\Diamond_\alpha}\right)^2. \tag{3.5.1}$$

This implies that

$$\alpha^3 \int_0^h \left|(y + y^\sigma)y^\Delta\right|(t)\Delta t + (1 - \alpha)^3 \int_0^h \left|(y + y^\rho)y^\nabla\right|(t)\nabla t$$

$$\leq h \left[\alpha \int_0^h \left(y^{\Diamond_\alpha}(t)\right)^2 \Delta t + (1 - \alpha) \int_0^h \left(y^{\Diamond_\alpha}(t)\right)^2 \nabla t\right] = h \int_0^h \left(y^{\Diamond_\alpha}(t)\right)^2 \Diamond_\alpha t,$$

which completes the proof. ■

Notice that if we take $\alpha = 1$ and $\alpha = 0$, then Theorem 3.5.1 reduces to Theorem 3.1.1 and Theorem 3.1.2, respectively.

Theorem 3.5.2 *Let $\alpha \in [0,1]$ and $h \in \mathbb{T}$ with $h > 0$. For any $y \in C_{\Diamond_\alpha}^1$ with $y(0) = 0$ and $\alpha y^\Delta \geq 0$ and $(1-\alpha)y^\nabla \geq 0$, we have*

$$\alpha^3 \int_0^h |(y + y^\sigma)y^\Delta|\,(t)\Delta t + (1-\alpha)^3 \int_0^h |(y + y^\rho)y^\nabla|\,(t)\nabla t$$

$$\leq \beta \int_0^h (y^{\Diamond_\alpha})^2 \Diamond_\alpha t + 2\gamma(1 - 3\alpha + 3\alpha^2)(y(h) - y(0)),$$

where

$$\beta = \min_{u \in [0,h]_\mathbb{T}} \{u, h - u\} \quad and \quad \gamma = \max\{|y(0)|, |y(h)|\}.$$

Proof. Using the fact that $\alpha y^\Delta \geq 0$, it follows from Theorem 3.1.3 that

$$\alpha^3 \int_0^h |(y + y^\sigma)y^\Delta|\,(t)\Delta t \leq \alpha^3 \beta \int_0^h |y^\Delta|^2\,(t)\Delta t + 2\gamma\alpha^3 \int_0^h y^\Delta(t)\Delta t$$

$$= \alpha^3 \beta \int_0^h |y^\Delta|^2\,(t)\Delta t + 2\gamma\alpha^3(y(h) - y(0)).$$

Similarly, with help of $(1-\alpha)y^\nabla \geq 0$, one can get that

$$(1-\alpha)^3 \int_0^h |(y + y^\rho)y^\nabla|\,(t)\nabla t \leq \beta(1-\alpha)^3 \int_0^h |y^\nabla|^2\,(t)\nabla t$$

$$+ 2\gamma(1-\alpha)^3(y(h) - y(0)).$$

Combining these inequalities and also using (3.5.1), we obtain that

$$\alpha^3 \int_0^h |(y + y^\sigma)y^\Delta|\,(t)\Delta t + (1-\alpha)^3 \int_0^h |(y + y^\rho)y^\nabla|\,(t)\nabla t$$

$$\leq \beta \left\{ \alpha^3 \int_0^h |y^\Delta|^2\,(t)\Delta t + (1-\alpha)^3 \int_0^h |y^\nabla|^2\,(t)\nabla t \right\}$$

$$+ 2\gamma \left[\alpha^3 + (1-\alpha)^3\right](y(h) - y(0))$$

$$\leq \beta \left\{ \alpha \int_0^h (y^{\Diamond_\alpha})_\alpha^2 \Delta t + (1-\alpha) \int_0^h (y^{\Diamond_\alpha})^2 \nabla t \right\}$$

$$+ 2\gamma \left[\alpha^3 + (1-\alpha)^3\right](y(h) - y(0))$$

$$= \beta \int_0^h (y^{\Diamond_\alpha})^2 \Diamond_\alpha t + 2\gamma \left[1 - 3\alpha + 3\alpha^2\right](y(h) - y(0)),$$

which completes the proof. ∎

Corollary 3.5.1 *Let $\alpha \in [0,1]$ and $h \in \mathbb{T}$ with $h > 0$. For any $y \in C_{\Diamond_\alpha}^1$ with $y(0) = y(h) = 0$ and $\alpha y^\Delta \geq 0$ and $(1-\alpha)y^\nabla \geq 0$, we have*

$$\alpha^3 \int_0^h |(y + y^\sigma)y^\Delta|\,(t)\Delta t + (1-\alpha)^3 \int_0^h |(y + y^\rho)y^\nabla|\,(t)\nabla t \leq \beta \int_0^h (y^{\Diamond_\alpha})^2 \Diamond_\alpha t,$$

where β is defined as in Theorem 3.5.2.

Theorem 3.5.3 *Let $\alpha \in [0,1]$ and $h \in \mathbb{T}$ with $h > 0$.   Assume that g : $[0,h]_\mathbb{T} \to \mathbb{R}$ is a nonincreasing continuous function. For any $y \in C^1_{\Diamond_\alpha}$ with $y(0) = 0$ and $\alpha(1-\alpha)y^\Delta y^\nabla \geq 0$, we have*

$$\alpha^3 \int_0^h \left[g^\sigma \left|(y + y^\sigma)y^\Delta\right|\right](t)\Delta t + (1-\alpha)^3 \int_0^h \left[g^\rho \left|(y + y^\rho)y^\nabla\right|\right](t)\nabla t$$

$$\leq h \int_0^h g(t) \left(y^{\Diamond_\alpha}\right)^2 \Diamond_\alpha t.$$

Proof. By Theorem 3.1.10, we get that

$$\alpha^3 \int_0^h \left[g^\sigma \left|(y + y^\sigma)y^\Delta\right|\right](t)\Delta t \leq \alpha^3 h \int_0^h g(t)(y^\Delta(t))^2 \Delta t.$$

In a similar manner, we can obtain that

$$(1-\alpha)^3 \int_0^h \left[g^\rho \left|(y + y^\rho)y^\nabla\right|\right](t)\nabla t \leq (1-\alpha)^3 h \int_0^h g(t)(y^\nabla(t))^2 \nabla t.$$

Combining these inequalities and using (3.5.1), we get that

$$\alpha^3 \int_0^h \left[g^\sigma \left|(y + y^\sigma)y^\Delta\right|\right](t)\Delta t + (1-\alpha)^3 \int_0^h \left[g^\rho \left|(y + y^\rho)y^\nabla\right|\right](t)\nabla t$$

$$\leq h \left\{\alpha^3 \int_0^h g(t)(y^\Delta(t))^2 \Delta t + (1-\alpha)^3 \int_0^h g(t)(y^\nabla(t))^2 \nabla t\right\}$$

$$\leq h \left\{\alpha \int_0^h g(t)(y^{\Diamond_\alpha})^2 \Delta t + (1-\alpha) \int_0^h g(t)(y^{\Diamond_\alpha})^2 \nabla t\right\}$$

$$= h \int_0^h g(t)(y^{\Diamond_\alpha})^2 \Diamond_\alpha t,$$

which completes the proof. ∎

In the following, we deal with two weight functions and generalize some delta Opial inequalities obtained in Sect. 3.1 to the diamond-alpha case. For our purposes, we offer the following slight but essential improvement of Theorem 3.1.11.

Theorem 3.5.4 *Assume that $a \in \mathbb{T}$, $b \in (a,\infty)_\mathbb{T}$, r, $s \in C_{rd}([a,b]_\mathbb{T},(0,\infty))$ and $f \in C^1_{rd}([a,b]_\mathbb{T},\mathbb{R})$.*
   If $f(a) = 0$, then

$$\int_a^b s(t)\left|\left(f^2\right)^\Delta(t)\right|\Delta t \leq K \int_a^b r(t)|f^\Delta(t)|^2 \Delta t,$$

where

$$K = \sqrt{\int_a^b s^2(t)\left(R^2\right)^\Delta(t)\Delta t} \quad with \quad R(t) = \int_a^t \frac{\Delta s}{r(s)}.$$

Proof. Define $g(t) := \int_a^t r(s)|f^\Delta(s)|^2 \Delta s$. Then $g(a) = 0$, $g^\Delta(t) := r(t)|f^\Delta(t)|^2$ so that $|f^\Delta(t)| = \sqrt{\frac{g^\Delta(t)}{r(t)}} = \sqrt{R^\Delta(t)g^\Delta(t)}$, and

$$
\begin{aligned}
|f(t)| &= |f(t) - f(a)| = \left| \int_a^t f^\Delta(s) \Delta s \right| \le \int_a^t |f^\Delta(s)| \, \Delta s \\
&= \int_a^t \frac{1}{\sqrt{r(s)}} \left(\sqrt{r(s)} \, |f^\Delta(s)| \right) \Delta s \\
&\le \sqrt{\frac{\Delta s}{r(s)}} \sqrt{r(s) \, |f^\Delta(s)|^2 \, \Delta s} = \sqrt{R(t)g(t)},
\end{aligned}
$$

where we have used the time scales Cauchy–Schwarz inequality. Thus

$$
\begin{aligned}
\left| \left(f^2 \right)^\Delta (t) \right| &= \left| (f(t) + f(\sigma(t))) f^\Delta(t) \right| \\
&\le (|f(t)| + |f(\sigma(t))|) \left| f^\Delta(t) \right| \\
&\le \left(\sqrt{R(t)g(t)} + \sqrt{R(\sigma(t))g(\sigma(t))} \right) \sqrt{R^\Delta(t)g^\Delta(t)} \\
&\le \sqrt{R(t) + R(\sigma(t))} \sqrt{g(t) + g(\sigma(t))} \sqrt{R^\Delta(t)g^\Delta(t)} \\
&= \sqrt{(R^2)^\Delta (t)} \sqrt{(g^2)^\Delta (t)},
\end{aligned}
$$

where we have used the classical Cauchy–Schwarz inequality, and hence

$$
\begin{aligned}
\int_a^b s(t) \left| \left(f^2 \right)^\Delta (t) \right| \Delta t &\le \int_a^b s(t) \sqrt{(R^2)^\Delta (t)} \sqrt{(g^2)^\Delta (t)} \Delta t \\
&\le \sqrt{\int_a^b s^2(t) (R^2)^\Delta (t) \Delta t} \sqrt{\int_a^b (g^2)^\Delta (t) \Delta t} \\
&= K \sqrt{g^2(b)} = Kg(b) = K \int_a^b r(t)|f^\Delta(t)|^2 \Delta t,
\end{aligned}
$$

where we have used the time scales Cauchy–Schwarz inequality one last time. ∎

Similarly, we may prove the following nabla result.

Theorem 3.5.5 *Assume that* $a \in \mathbb{T}$, $b \in (a, \infty)_\mathbb{T}$, $r, s \in C_{ld}([a, b]_\mathbb{T}, (0, \infty))$ *and* $f \in C_{ld}^1([a, b]_\mathbb{T}, \mathbb{R})$.
If $f(a) = 0$, *then*

$$
\int_a^b s(t) \left| \left(f^2 \right)^\nabla (t) \right| \nabla t \le L \int_a^b r(t)|f^\nabla(t)|^2 \nabla t,
$$

where

$$
L = \sqrt{\int_a^b s^2(t) (S^2)^\nabla (t) \nabla t} \quad with \quad S(t) = \int_a^t \frac{\nabla s}{r(s)}.
$$

Now we are ready to prove the corresponding diamond-alpha inequality.

Theorem 3.5.6 *Assume that* $\alpha \in [0,1]$, $a \in \mathbb{T}$, $b \in (a,\infty)_\mathbb{T}$ $r,s \in$ $C([a,b]_\mathbb{T},(0,\infty))$ *and* $f \in C^1_{\Diamond_\alpha}([a,b]_\mathbb{T},\mathbb{R})$. *If* $\alpha(1-\alpha)f^\Delta f^\nabla \geq 0$ *and* $f(a) = 0$, *then*

$$\alpha^5 \int_a^b s(t)\left|\left(f^2\right)^\Delta (t)\right|\Delta t + (1-\alpha)^5 \int_a^b s(t)\left|\left(f^2\right)^\nabla (t)\right|\nabla t$$

$$\leq \Lambda \int_a^b r(t)|f^{\Diamond_\alpha}(t)|^2 \Diamond_\alpha(t),$$

where

$$\Lambda = \sqrt{\int_a^b s^2(t)\left(T^2\right)^{\Diamond_\alpha}(t)\Diamond_\alpha t} \quad with \quad T(t) = \int_a^t \frac{\Diamond_\alpha s}{r(s)}.$$

Proof. From Theorems 3.5.4 and 3.5.5, we have

$$\alpha^5 \int_a^b s(t)\left|\left(f^2\right)^\Delta (t)\right|\Delta t + (1-\alpha)^5 \int_a^b s(t)\left|\left(f^2\right)^\nabla (t)\right|\nabla t$$

$$\leq \alpha^5 K \int_a^b r(t)|f^\Delta(t)|^2\Delta t + (1-\alpha)^5 L \int_a^b r(t)|f^\nabla(t)|^2\nabla t$$

$$= \alpha^3 K \int_a^b r(t)|\alpha f^\Delta(t)|^2\Delta t + (1-\alpha)^3 L \int_a^b r(t)|(1-\alpha)f^\nabla(t)|^2\nabla t$$

$$\leq \alpha^3 K \int_a^b r(t)|f^{\Diamond_\alpha}(t)|^2\Delta t + (1-\alpha)^3 L \int_a^b r(t)|f^{\Diamond_\alpha}(t)|^2\nabla t$$

$$= (\alpha^2 K)\left(\alpha \int_a^b r(t)|f^{\Diamond_\alpha}(t)|^2\Delta t\right) + ((1-\alpha)^2 L)\left((1-\alpha)\int_a^b r(t)|f^{\Diamond_\alpha}(t)|^2\nabla t\right)$$

$$\leq \tilde{\Lambda}\sqrt{\left(\alpha \int_a^b r(t)|f^{\Diamond_\alpha}(t)|^2\Delta t\right)^2 + \left((1-\alpha)\int_a^b r(t)|f^{\Diamond_\alpha}(t)|^2\nabla t\right)^2}$$

$$\leq \tilde{\Lambda}\sqrt{\left(\int_a^b r(t)|f^{\Diamond_\alpha}(t)|^2\Diamond_\alpha t\right)^2} = \tilde{\Lambda}\int_a^b r(t)|f^{\Diamond_\alpha}(t)|^2\Diamond_\alpha t,$$

where we have used the classical Cauchy–Schwarz inequality and

$$\tilde{\Lambda} = \sqrt{\alpha^4 K^2 + (1-\alpha)^4 L^2}$$

$$= \sqrt{\alpha^4 \int_a^b s^2(t)\left(R^2\right)^\Delta (t)\Delta t + (1-\alpha)^4 \int_a^b s^2(t)\left(S^2\right)^\nabla (t)\nabla t}$$

$$\leq \sqrt{\alpha \int_a^b s^2(t)\left(T^2\right)^{\Diamond_\alpha}(t)\Delta t + (1-\alpha)\int_a^b s^2(t)\left(T^2\right)^{\Diamond_\alpha}(t)\nabla t}$$

$$= \sqrt{\int_a^b s^2(t)\left(T^2\right)^{\Diamond_\alpha}(t)\Diamond_\alpha t} = \Lambda,$$

where we have used the inequalities

$$\alpha^3 \left(R^2\right)^{\Delta} \le \left(T^2\right)^{\Diamond_\alpha} \quad \text{and} \quad (1-\alpha)^3 \left(S^2\right)^{\Delta} \le \left(T^2\right)^{\Diamond_\alpha}. \tag{3.5.2}$$

Now we show (3.5.2) in order to complete the proof. Note first that by [52, Theorem 5.37], we have

$$R^{\Delta} = S^{\nabla} = \frac{1}{r}, \quad R^{\nabla} = \frac{1}{r^\rho} \quad \text{and } S^{\Delta} = \frac{1}{r^\sigma},$$

and all of these derivatives are positive. Using these relations and the time scales product rules, we have

$$\left(R^2\right)^{\nabla} = \frac{R+R^\rho}{r^\rho}, \quad \left(S^2\right)^{\Delta} = \frac{S+S^\sigma}{r^\sigma}, \quad \left(R^2\right)^{\Delta} = \frac{R+R^\sigma}{r},$$

$$\left(S^2\right)^{\nabla} = \frac{S+S^\rho}{r}, \quad (RS)^{\Delta} = \frac{S}{r} + \frac{R^\sigma}{r^\sigma}, \quad (RS)^{\nabla} = \frac{S}{r^\rho} + \frac{R^\rho}{r},$$

and again all of these derivatives are positive. Since $T = \alpha R + (1-\alpha)S$, the calculation

$$
\begin{aligned}
\left(T^2\right)^{\Diamond_\alpha} &= \alpha \left(T^2\right)^{\Delta} + (1-\alpha) \left(T^2\right)^{\nabla} \\
&= \alpha \left(\alpha^2 \left(R^2\right)^{\Delta} + 2\alpha(1-\alpha)(RS)^{\Delta} + (1-\alpha)^2 \left(S^2\right)^{\Delta}\right) \\
&\quad + (1-\alpha)\left(\alpha^2 \left(R^2\right)^{\nabla} + 2\alpha(1-\alpha)(RS)^{\nabla} + (1-\alpha)^2 \left(S^2\right)^{\nabla}\right) \\
&= \alpha^3 \left(R^2\right)^{\Delta} + (1-\alpha)^3 \left(S^2\right)^{\nabla} + 2\alpha^2(1-\alpha)(RS)^{\Delta} \\
&\quad + 2\alpha(1-\alpha)^2 (RS)^{\nabla} + \alpha(1-\alpha)^2 \left(S^2\right)^{\Delta} + \alpha^2(1-\alpha)\left(R^2\right)^{\nabla}
\end{aligned}
$$

confirms the validity of the inequalities (3.5.2). ∎

Following the same steps as in the proofs of all previous results in this section, we can establish the following result.

Theorem 3.5.7 *Assume that* $\alpha \in [0,1]$, $a \in \mathbb{T}$, $b \in (a, \infty)_{\mathbb{T}}$ $r, s \in C([a,b]_{\mathbb{T}}, (0,\infty))$ *and* $f \in C^1_{\Diamond_\alpha}([a,b]_{\mathbb{T}}, \mathbb{R})$. *If* $\alpha(1-\alpha) f^{\Delta} f^{\nabla} \ge 0$ *and* $f(b) = 0$, *then*

$$\alpha^5 \int_a^b s(t)| \left(f^2\right)^{\Delta}(t)|\Delta t + (1-\alpha)^5 \int_a^b s(t)| \left(f^2\right)^{\nabla}(t)|\nabla t$$

$$\le \Omega \int_a^b r(t)|f^{\Diamond_\alpha}(t)|^2 \Diamond_\alpha(t),$$

where

$$\Omega = \sqrt{\int_a^b s^2(t) \left(T^2\right)^{\Diamond_\alpha}(t)\Diamond_\alpha t} \quad \text{with} \quad T(t) = \int_t^b \frac{\Diamond_\alpha s}{r(s)}.$$

The last result combines Theorems 3.5.6 and 3.5.7.

Theorem 3.5.8 *Assume that* $\alpha \in [0,1]$, $a \in \mathbb{T}$, $b \in (a,\infty)_\mathbb{T}$ $r, s \in$
$C([a,b]_\mathbb{T},(0,\infty))$ *and* $f \in C^1_{\Diamond_\alpha}([a,b]_\mathbb{T},\mathbb{R})$. *If* $\alpha (1 - \alpha) f^\Delta f^\nabla \geq 0$ *and* $f(a) =$
$f(b) = 0$, *then*

$$\alpha^5 \int_a^b s(t)|\left(f^2\right)^\Delta (t)|\Delta t + (1-\alpha)^5 \int_a^b s(t)|\left(f^2\right)^\nabla (t)|\nabla t$$

$$\leq \ \beta \int_a^b r(t)|f^{\Diamond_\alpha}(t)|^2 \Diamond_\alpha (t),$$

where

$$\beta := \min v(u)$$

with

$$v(u) := \max \left\{ \sqrt{\int_a^u s^2(t) \left(T_a^2\right)^{\Diamond_\alpha} (t) \Diamond_\alpha t}, \sqrt{\int_u^b s^2(t) \left(T_b^2\right)^{\Diamond_\alpha} (t) \Diamond_\alpha t} \right\},$$

and T_c *for* $c \in \mathbb{T}$ *is defined by* $T_c(t) = \int_c^t \frac{\Diamond_\alpha s}{r(s)}.$

In the following, we obtain a sequence of Opial inequalities for first order diamond alpha derivatives on time scales and establish some higher order inequalities. Throughout, we say that a function $f : [0,h]_\mathbb{T} \to \mathbb{R}$ is in the class $C^n_{\Diamond_\alpha}$ if f is n $\Diamond_\alpha$-differentiable such that αf^{Δ^n} is rd-continuous, $(1-\alpha)f^{\nabla^n}$ is ld-continuous, and $\alpha(1 - \alpha)f^{\Diamond_\alpha^n}$ is continuous. To prove the results we need the following Lemmas.

Lemma 3.5.1 *Let* $i \in \mathbb{N}$ *and* $j \in \mathbb{N}_0$. *Assume* f *is* $(i + j)$ *times* $\Diamond_\alpha$-differentiable. *If* $(1-\alpha)f^{\Diamond_\alpha^{i-1}\Delta^j\nabla} \geq 0$, *then*

$$f^{\Diamond_\alpha^i \Delta^j} \geq \alpha f^{\Diamond_\alpha^{i-1}\Delta^{j+1}},$$

and if $\alpha f^{\Diamond_\alpha^{i-1}\Delta\nabla^j} \geq 0$, *then*

$$f^{\Diamond_\alpha^i \nabla^j} \geq (1 - \alpha) f^{\Diamond_\alpha^{i-1}\nabla^{j+1}}.$$

Proof. We have

$$f^{\Diamond_\alpha^i} = \left(f^{\Diamond_\alpha^{i-1}}\right)^{\Diamond_\alpha} = \alpha \left(f^{\Diamond_\alpha^{i-1}}\right)^\Delta + (1-\alpha) \left(f^{\Diamond_\alpha^{i-1}}\right)^\nabla,$$

so by using the assumption we have that

$$f^{\Diamond_\alpha^i \Delta^j} = \alpha \left(f^{\Diamond_\alpha^{i-1}}\right)^{\Delta^{j+1}} + (1-\alpha) \left(f^{\Diamond_\alpha^{i-1}}\right)^{\nabla\Delta^j} \geq \alpha \left(f^{\Diamond_\alpha^{i-1}}\right)^{\Delta^j},$$

and

$$f^{\Diamond_\alpha^i \nabla^j} = \alpha \left(f^{\Diamond_\alpha^{i-1}}\right)^{\Delta\nabla^j} + (1-\alpha) \left(f^{\Diamond_\alpha^{i-1}}\right)^{\nabla^{j+1}} \geq (1-\alpha) \left(f^{\Diamond_\alpha^{i-1}}\right)^{\nabla^{j+1}},$$

which completes the proof. ■

Lemma 3.5.2 *Let $n \in \mathbb{N}$. Assume f is n times $\Diamond_\alpha$-differentiable. If $(1-\alpha)f^{\Diamond_\alpha^{n-j-1}\Delta^j\nabla} \geq 0$ for all $j \in \{0,1,\ldots,n-1\}$, then*

$$f^{\Diamond_\alpha^n} \geq \alpha^n f^{\Delta^n},$$

and if $\alpha f^{\Diamond_\alpha^{n-j-1}\Delta\nabla^j} \geq 0$ for all $j \in \{0,1,\ldots,n-1\}$, then

$$f^{\Diamond_\alpha^n} \geq (1-\alpha)^n f^{\nabla^n}.$$

Proof. We use Lemma 3.5.1 for $i = n - j$, where $j \in \{0,1,\ldots,n-1\}$, to obtain

$$f^{\Diamond_\alpha^n} = f^{\Diamond_\alpha^n\Delta^0} \geq \alpha\left(f^{\Diamond_\alpha^{n-1}}\right)^{\Delta^1} \geq \alpha^2\left(f^{\Diamond_\alpha^{n-2}}\right)^{\Delta^2} \geq \ldots \geq \alpha^n\left(f^{\Diamond_\alpha^0}\right)^{\Delta^n} = \alpha^n f^{\Delta^n},$$

and similarly

$$
\begin{aligned}
f^{\Diamond_\alpha^n} &= \left(f^{\Diamond_\alpha^n}\right)^{\nabla^0} \geq (1-\alpha)\left(f^{\Diamond_\alpha^{n-1}}\right)^{\nabla^1} \geq (1-\alpha)^2\left(f^{\Diamond_\alpha^{n-2}}\right)^{\nabla^2} \\
&\geq \ldots \geq (1-\alpha)^n\left(f^{\Diamond_\alpha^0}\right)^{\nabla^n} = (1-\alpha)^n f^{\nabla^n},
\end{aligned}
$$

which completes the proof. ∎

Theorem 3.5.9 *Let $\mathbb{T}$ be a time scale, m, $n \in \mathbb{N}$, $\alpha \in [0,1]$ and $h \in \mathbb{T}$ with $h > 0$. Assume that f is n times $\Diamond_\alpha$-differentiable with $(1-\alpha)f^{\Diamond_\alpha^{n-j-1}\Delta^j\nabla} \geq 0$ and $\alpha f^{\Diamond_\alpha^{n-j-1}\Delta^j\nabla} \geq 0$, $\alpha f^{\Delta^n} \geq 0$ and $(1-\alpha)f^{\nabla^n} \geq 0$ for all $j \in \{0,1,\ldots, n-1\}$. If*

$$\alpha f^{\Delta^j}(0) = (1-\alpha)f^{\nabla^j}(0) = 0, \text{ for } j \in \{0,1,\ldots,n-1\},$$

then

$$
\begin{aligned}
&\alpha^{n(m+1)+1}\int_0^h \left|\left(\sum_{k=0}^m f^k(f^\sigma)^{m-k}\right)f^{\Delta^n}\right|(t)\Delta t \\
&+ (1-\alpha)^{n(m+1)+1}\int_0^h \left|\left(\sum_{k=0}^m f^k(f^\rho)^{m-k}\right)f^{\nabla^n}\right|(t)\nabla t \\
&\leq h^{mn}\int_0^h \left(f^{\Diamond_\alpha^n}\right)^{m+1}(t)\Diamond_\alpha t.
\end{aligned}
$$

Proof. It follows from Theorem 3.4.16 that

$$\alpha^{n(m+1)+1}\int_0^h \left|\left(\sum_{k=0}^m f^k(f^\sigma)^{m-k}\right)f^{\Delta^n}\right|(t)\Delta t \leq \alpha^{n(m+1)+1}h^{mn}\int_0^h \left(f^{\Delta^n}(t)\right)^{m+1}\Delta t.$$

$$(3.5.3)$$

Similarly, we get

$$(1-\alpha)^{n(m+1)+1} \int_0^h \left| \left(\sum_{k=0}^m f^k (f^\rho)^{m-k} \right) f^{\nabla^n} \right| (t) \nabla t$$

$$\leq (1-\alpha)^{n(m+1)+1} \int_0^h (f^{\nabla^n}(t))^{m+1} \nabla t. \qquad (3.5.4)$$

Also

$$\alpha^{n(m+1)} \left(f^{\Delta^n} \right)^{m+1} \leq \left(f^{\Diamond_\alpha^n} \right)^{m+1}, \text{ and } (1-\alpha)^{n(m+1)} (f^{\nabla^n}(t))^{m+1} \leq \left(f^{\Diamond_\alpha^n} \right)^{m+1}. \qquad (3.5.5)$$

Hence, combining (3.5.3)–(3.5.5), we obtain that

$$\alpha^{n(m+1)+1} \int_0^h \left| \left(\sum_{k=0}^m f^k (f^\sigma)^{m-k} \right) f^{\Delta^n} \right| (t) \Delta t$$

$$+ (1-\alpha)^{n(m+1)+1} \int_0^h \left| \left(\sum_{k=0}^m f^k (f^\rho)^{m-k} \right) f^{\nabla^n} \right| (t) \nabla t$$

$$\leq h^{mn} \left\{ \alpha \int_0^h \left(f^{\Diamond_\alpha^n} \right)^{m+1} (t) \Diamond_\alpha t + (1-\alpha) \int_0^h \left(f^{\Diamond_\alpha^n} \right)^{m+1} (t) \Diamond_\alpha t \right\}$$

$$= h^{mn} \int_0^h \left(f^{\Diamond_\alpha^n} \right)^{m+1} (t) \Diamond_\alpha t,$$

which completes the proof. ∎

Chapter 4

Lyapunov Inequalities

You know that I write slowly. This is chiefly because I am never satisfied until I have said as much as possible in a few words, and writing briefly takes far more time than writing at length.

Gauss (1777–1855).

In 1906 Lyapunov [105] proved an inequality giving the distance between two consecutive zeros of solutions of second order differential equations. It is proved that, if the differential equation

$$y^{''}(t) + p(t)y(t) = 0, \tag{4.0.1}$$

has a nontrivial solution $y(t)$ with $y(a) = y(b) = 0$ $(a < b)$ and $y(t) \neq 0$ for $t \in (a, b)$, then

$$\int_a^b p(t)dt > \frac{4}{b-a}, \tag{4.0.2}$$

where p is a positive real valued function defined on $[a, b]$. If the difference equation

$$\Delta^2 y(n) + p(n)y(n+1) = 0, \tag{4.0.3}$$

has a nontrivial solution $y(n)$ satisfying $y(0) = y(N) = 0$, where $p(n)$ is a positive sequence, then the Lyapunov inequality is given by

$$\sum_{k=0}^{N-1} p(n) \geq \begin{cases} \frac{2}{m+1}, & \text{if } N = 2m+2, \\ \frac{2m+1}{m(m+1)}, & \text{if } N = 2m+1. \end{cases}$$

The chapter is organized as follows. In Sect. 4.1 we present some Lyapunov type inequalities for second order linear dynamic equations and in Sect. 4.2

© Springer International Publishing Switzerland 2014
R. Agarwal et al., *Dynamic Inequalities On Time Scales*,
DOI 10.1007/978-3-319-11002-8_4

we present results for half-linear dynamic equations. Section 4.3 considers dynamic equations with damping terms and in Sect. 4.4 we consider Hamiltonian systems on time scales.

Throughout this chapter (usually without mentioning) the integrals in the statements of the theorems are assumed to exist.

4.1 Second Order Linear Equations

In this section, we establish some Lyapunov type inequalities for Sturm–Liouville linear dynamic equations on time scales and then establish some sufficient conditions for disconjugacy of solutions. The results in this section are adapted from [48, 90, 123, 125, 128]. First, we consider the Sturm–Liouville dynamic equation

$$y^{\Delta\Delta}(t) + p(t)y^{\sigma}(t) = 0, \tag{4.1.1}$$

together with the quadratic functional

$$\mathcal{F}(y) = \int_a^b \left\{ (y^{\Delta}(t))^2 - p(y^{\sigma})^2(t) \right\} \Delta t,$$

where $p(t)$ is a positive rd-continuous function defined on $\mathbb{T}$.

By a solution of (4.1.1), we mean a continuous function $y : [a, \sigma^2(b)]_{\mathbb{T}} \to \mathbb{R}$, which is twice differentiable on $[a, b]_{\mathbb{T}}$ with y^{Δ^2} rd-continuous. It is known that (4.1.1) admits a unique solution when $y(a)$ and $y^{\Delta}(a)$ are prescribed. We say y has a generalized zero at some $c \in [a, \sigma(b)]_{\mathbb{T}}$ provided that $y(c)y^{\sigma}(c) \leq 0$ holds, and (4.1.1) is called disconjugate on $[a, b]_{\mathbb{T}}$ if there is no nontrivial solution of (4.1.1) with at least two generalized zeros in $[a, b]_{\mathbb{T}}$. Finally, (4.1.1) is said to be disfocal on $[a, \sigma^2(b)]_{\mathbb{T}}$ provided there is no nontrivial solution y of (4.1.1) with a generalized zero in $[a, \sigma^2(b)]_{\mathbb{T}}$ followed by a generalized zero of y^{Δ} in $[a, \sigma(b)]_{\mathbb{T}}$.

Lemma 4.1.1 *If x solves (4.1.1) and if $\mathcal{F}(y)$ is defined, then*

$$\mathcal{F}(y) - \mathcal{F}(x) = \mathcal{F}(y - x) + 2(y - x)(b)x^{\Delta}(b) - 2(y - x)x^{\Delta}(a).$$

Proof. *Under the above assumptions we find*

$$\mathcal{F}(y) - \mathcal{F}(x) - \mathcal{F}(y - x)$$

$$= \int_a^b \left\{ (y^{\Delta})^2 - p(y^{\sigma})^2 - (x^{\Delta})^2 + p(x^{\sigma})^2 \right.$$

$$\left. - (y^{\Delta} - x^{\Delta})^2 + p(y^{\sigma} - x^{\sigma})^2 \right\} (t)\Delta t$$

$$
= \int_a^b \{(y^\Delta)^2 - p(y^\sigma)^2 - (x^\Delta)^2 + p(x^\sigma)^2 - (y^\Delta)^2 + 2y^\Delta x^\Delta - (x^\Delta)^2
$$

$$
+ p(y^\sigma)^2 - 2py^\sigma x^\sigma + p(x^\sigma)^2\} (t)\Delta t
$$

$$
= 2\int_a^b \{y^\Delta x^\Delta - py^\sigma x^\sigma + p(x^\sigma)^2 - (x^\Delta)^2\} (t)\Delta t
$$

$$
= 2\int_a^b \left\{ y^\Delta x^\Delta + y^\sigma x^{\Delta^2} - x^\sigma x^{\Delta^2} - (x^\Delta)^2 \right\} (t)\Delta t
$$

$$
= 2\int_a^b \{yx^\Delta - xx^\Delta\}^\Delta \Delta(t) = 2\int_a^b \{(y-x)x^\Delta\}^\Delta \Delta t
$$

$$
= 2(y(b) - x(b))x^\Delta(b) - 2(y(a) - x(a))x^\Delta(a),
$$

where we have used the product rule. ∎

Lemma 4.1.2 *If $\mathcal{F}(y)$ is defined, then for any $r, s \in \mathbb{T}$ with $a \le r < s \le b$*

$$
\int_r^s (y^\Delta(t))^2 \Delta t \ge \frac{(y(s) - y(r))^2}{s - r}.
$$

Proof. Let

$$
x(t) = \frac{y(s) - y(r)}{s - r} t + \frac{sy(r) - ry(s)}{s - r}.
$$

Then x solves the Sturm–Liouville equation (4.1.1) with $p = 0$ and therefore we may apply Lemma 4.1.1 to $\mathcal{F}_0$ defined by

$$
\mathcal{F}_0(x) = \int_r^s (x^\Delta(t))^2 \Delta t,
$$

to find

$$
\begin{aligned}
\mathcal{F}_0(y) &= \mathcal{F}_0(x) + \mathcal{F}_0(y - x) + (y - x)(s)x^\Delta(s) - (y - x)(r)x^\Delta(r) \\
&= \mathcal{F}_0(x) + \mathcal{F}_0(y - x) \\
&\ge \mathcal{F}_0(x) = \int_r^s \left\{ \frac{y(s) - y(r)}{s - r} \right\}^2 \Delta t = \frac{(y(s) - y(r))^2}{s - r},
\end{aligned}
$$

and this completes the proof. ∎

The following lemma will be used later (see [51]).

Lemma 4.1.3 *Equation (4.1.1) is disconjugate on $[a, b]_{\mathbb{T}}$ if and only if*

$$\mathcal{F}(y) = \int_a^b \left\{ (y^\Delta(t))^2 - p(y^\sigma)^2(t) \right\} \Delta t > 0,$$

for all nontrivial solutions y with $y(a) = y(b) = 0$.

The following theorem gives the Lyapunov type inequality for the second order dynamic equation (4.1.1).

Theorem 4.1.1 *If $y(t)$ is a nontrivial solution of (4.1.1) with $y(a) = y(b) = 0$ $(a < b)$, then*

$$\int_a^b p(t)\Delta t > \frac{b - a}{f(d)}, \tag{4.1.2}$$

where $f(d) = \max\{f(t) : t \in [a, b]\}$ and $f(t) = (t - a)(b - t)$.

Proof. From Lemma 4.1.1, since y is a nontrivial solution of (4.1.1) with $y(a) = y(b) = 0$, we have that

$$\mathcal{F}(y) = \int_0^b \left\{ (y^\Delta(t))^2 - p(y^\sigma)^2(t) \right\} \Delta t = 0.$$

Also, since y is nontrivial, we see that

$$M := \max\{y^2(t) : t \in [a, b] \cap \mathbb{T}\}, \tag{4.1.3}$$

is defined and positive. Now let $c \in [a, b]$ be such that $y^2(c) = M$. Applying the above and Lemma 4.1.2, twice (once with $r = a$ and $s = c$ and a second time with $r = c$ and $s = b$), we find

$$
\begin{aligned}
M \int_0^b p(t)\Delta t \ &\geq\ \int_0^b \left\{ p(y^\sigma)^2(t) \right\} \Delta t \\
&=\ \int_0^b (y^\Delta(t))^2 \Delta t = \int_0^b (y^\Delta(t))^2 \Delta t + \int_0^b (y^\Delta(t))^2 \Delta t \\
&\geq\ \frac{(y(c) - y(a))^2}{c - a} + \frac{(y(b) - y(c))^2}{b - c} \\
&=\ y^2(c) \left\{ \frac{1}{c - a} + \frac{1}{b - c} \right\} = M \frac{b - a}{f(c)} \geq M \frac{b - a}{f(d)},
\end{aligned}
$$

where the last inequality holds since $f(d) = \max\{f(t) : t \in [a, b] \cap \mathbb{T}\}$. Hence, dividing by $M > 0$ yields the desired inequality. The proof is complete. ∎

Example 4.1.1 *We use the notation from the proof of Theorem 4.1.1.*

(i). If $\mathbb{T} = \mathbb{R}$ *, then*

$$\min\left\{\left|\frac{a+b}{2} - s\right| : s \in [a,b]\right\} = 0, \text{ and so that } d = \frac{a+b}{2}.$$

Hence $f(d) = ((b-a)^2/4)$ *and the Lyapunov inequality from Theorem 4.1.1 reads*

$$\int_0^b p(t)dt \geq \frac{4}{b-a}.$$

(ii). If $\mathbb{T} = \mathbb{Z}$, *then we consider two cases. First, if* $a + b$ *is even, then*

$$\min\left\{\left|\frac{a+b}{2} - s\right| : s \in [a,b] \cap \mathbb{Z}\right\} = 0, \text{ and so that } d = \frac{a+b}{2}.$$

Hence $f(d) = ((b-a)^2/4)$ *and the Lyapunov inequality reads*

$$\sum_{t=a}^{b-1} p(t) \geq \frac{4}{b-a}.$$

If $a + b$ *is odd, then*

$$\min\left\{\left|\frac{a+b}{2} - s\right| : s \in [a,b] \cap \mathbb{Z}\right\} = \frac{1}{2}, \text{ and so that } d = \frac{a+b-1}{2}.$$

Then, we have $f(d) = ((b-a)^2 - 1/4)$ *and the Lyapunov inequality reads*

$$\sum_{t=a}^{b-1} p(t) \geq \frac{4}{b-a}\left\{\frac{1}{1 - (1/(b-a)^2)}\right\}.$$

As an application of Theorem 4.1.1, we now prove a sufficient condition for the disconjugacy of (4.1.1).

Theorem 4.1.2 *If* p *satisfies*

$$\int_a^b p(t)\Delta(t) < \frac{b-a}{f(d)}, \tag{4.1.4}$$

then (4.1.1) is disconjugate on $[a,b]_{\mathbb{T}}$.

Proof. Suppose that (4.1.4) holds. For the sake of contradiction we assume that (4.1.1) is not disconjugate. But then, by Lemma 4.1.3, there exists a nontrivial y with $y(a) = y(b) = 0$ such that $\mathcal{F}(y) \leq 0$. Using this y, we now define M by (4.1.3) and we find

$$M \int_a^b p(t)\Delta t \geq \int_a^b p(t)(y^\sigma(t))^2 \Delta t \geq \int_a^b (y^\Delta(t))^2 \Delta t \geq \frac{M(b-a)}{f(d)},$$

where the last inequality follows as in the proof of Theorem 4.1.1. Hence, after dividing by $M > 0$, we arrive at

$$\int_a^b p(t)\Delta t \geq \frac{b-a}{f(d)},$$

which contradicts (4.1.4) and hence completes the proof. ∎

Remark 4.1.1 *Note that in both condition (4.1.2) and (4.1.4) we could replace $(b-a)/f(d)$ by $4/(b-a)$, and Theorems 4.1.1 and 4.1.2 would remain true. This is because for $a \leq c \leq b$, we have*

$$\frac{1}{c-a} + \frac{1}{b-c} = \frac{(a+b-2c)^2}{(b-a)(c-a)(b-c)} + \frac{4}{b-a} \geq \frac{4}{b-a}.$$

In the following, we apply Opial type inequalities on time scales to prove some Lyapunov type inequalities for the second-order dynamic equation (4.1.1).

Theorem 4.1.3 *Assume that y is a nontrivial solution of the second-order dynamic equation (4.1.1) with $y(a) = y^{\Delta\sigma}(b) = 0$. Then, we have*

$$K_P(\sigma(b), a) = \left(2 \int_a^{\sigma(b)} |P(t)|^2 \left[\sigma(t) - a\right] \Delta t \right)^{1/2} \geq 1, \tag{4.1.5}$$

where

$$P(t) := \int_t^{\sigma(b)} p(s)\Delta s, \quad \text{for} \quad t \in [a, \sigma(b)]_{\mathbb{T}}. \tag{4.1.6}$$

Proof. Now

$$\int_a^{\sigma(b)} y^\sigma(t) y^{\Delta^2}(t)\Delta t = y^\sigma(b) y^{\Delta\sigma}(b) - y(a)y^\Delta(a) - \int_a^{\sigma(b)} \left[y^\Delta(t)\right]^2 \Delta t$$

$$= - \int_a^{\sigma(b)} \left[y^\Delta(t)\right]^2 \Delta t, \tag{4.1.7}$$

and using (4.1.6), we get that

$$\int_a^{\sigma(b)} p(t)\left[y^\sigma(t)\right]^2 \Delta t = -\int_a^{\sigma(b)} P^\Delta(t)\left[y^\sigma(t)\right]^2 \Delta t$$

$$= P(a)\left[y(a)\right]^2 + \int_a^{\sigma(b)} P(t)\left[\left[y(t)\right]^2\right]^\Delta \Delta t$$

$$= \int_a^{\sigma(b)} P(t)\left[\left[y(t)\right]^2\right]^\Delta \Delta t$$

$$= \int_a^{\sigma(b)} P(t)\left(\left[y(t) + y^\sigma(t)\right]y^\Delta(t)\right)\Delta t$$

$$\leq \int_a^{\sigma(b)} \left|P(t)\right|\left(\left[y(t) + y^\sigma(t)\right]y^\Delta(t)\right)\Delta t. \tag{4.1.8}$$

Multiplying (4.1.1) by y^σ and integrating from a to $\sigma(b)$ and using Theorem 3.1.7, (4.1.7) and (4.1.8), we get

$$\int_a^{\sigma(b)} \left(y^\Delta(t)\right)^2 \Delta t \leq \int_a^{\sigma(b)} \left|P(t)\right|\left(\left[y(t) + y^\sigma(t)\right]y^\Delta(t)\right)\Delta t$$

$$\leq K_P(\sigma(b), a)\int_a^{\sigma(b)} \left[y^\Delta(t)\right]^2, \tag{4.1.9}$$

Clearly, (4.1.5) follows from (4.1.9) by dividing by

$$\int_a^{\sigma(b)} \left[y^\Delta(t)\right]^2 \Delta t,$$

on both sides. The proof is complete. ∎

Remark 4.1.2 *The conclusion of Theorem 4.1.3 also holds for the second order dynamic inequality*

$$y^{\Delta^2}(t) + p(t)y^\sigma(t) \geq 0, \qquad for \quad t \in [a, b]_{\mathbb{T}}, \tag{4.1.10}$$

with $y(a) = 0$ and $y(b)y^{\Delta\sigma}(b) \leq 0$.

Similar reasoning by considering Theorem 3.1.8 instead of Theorem 3.1.7 yields the following result.

Theorem 4.1.4 *Assume that x is a nontrivial solution of (4.1.1) with $x^\Delta(a) = x^{\sigma^2}(b) = 0$. Then, we have*

$$L_P(\sigma^2(b), a) = \left(2\int_a^{\sigma^2(b)} (P(t))^2 \left[\sigma^2(b) - t\right]\Delta t\right)^{1/2} \geq 1,$$

where

$$P(t) := \int_a^t p(s)\Delta s, \qquad for \quad t \in [a, \sigma(b)]_{\mathbb{T}}.$$

Remark 4.1.3 *The conclusion of Theorem 4.1.4 also holds for (4.1.10) with* $x(a)x^\Delta(a) \geq 0$ *and* $x^{\sigma^2}(b) = 0$.

In the following, we establish a disconjugacy result for solutions of (4.1.1).

Theorem 4.1.5 *Assume that y is a nontrivial solution of (4.1.1) with $y(a) = y^{\sigma^2}(b) = 0$, and let $P \in C^1_{rd}([a, b]_\mathbb{T}, \mathbb{R})$ be a function satisfying $P^\Delta \equiv p$ on $[a, b]_\mathbb{T}$. Then, we have*

$$\min_{c \in [a, \sigma^2(b)]_\mathbb{T}} \left\{ \max \left\{ K_P(\sigma^2(b), c), L_P(c, a) \right\} \right\} \geq 1. \qquad (4.1.11)$$

Proof. Similar reasoning as in the proof of Theorem 4.1.3 yields the desired inequality (4.1.11) by applying Corollary 3.1.2 instead of Theorem 3.1.7. ∎

Corollary 4.1.1 *Assume that y is a nontrivial solution of (4.1.1) with $y(a) = 0$, and let $P \in C^1_{rd}([a, b]_\mathbb{T}, \mathbb{R})$ be a function as in Theorem 4.1.5. If*

$$\min_{c \in [a, \sigma^2(b)]_\mathbb{T}} \left\{ \max \left\{ K_P(\sigma^2(b), c), L_P(c, a) \right\} \right\} < 1,$$

then $y^{\sigma^2}(b) \neq 0$.

Next we consider the second order dynamic equation on $[a, b]$

$$[r(t)y^\Delta(t)]^\Delta + q(t)y^\sigma(t) = 0, \qquad t \in [a, b], \qquad (4.1.12)$$

on an arbitrary time scale $\mathbb{T}$, where r is a positive rd-continuous function and q is rd-continuous function and

$$\int_\alpha^\beta 1/r(t)\Delta t < \infty, \quad \text{and} \quad \int_\alpha^\beta |q(t)|\,\Delta t < \infty. \qquad (4.1.13)$$

We obtain lower bounds for the spacing $\beta - \alpha$ where y is a solution of (4.1.12) satisfying some conditions at α and β.

By a solution of (4.1.12) on an interval $\mathbb{T}$, we mean a nontrivial real-valued function $y \in C_{rd}(\mathbb{T})$, which has the property that $r(t)y^\Delta(t) \in C^1_{rd}(\mathbb{T})$ and satisfies Eq. (4.1.12) on $\mathbb{T}$. We say that (4.1.12) is right disfocal (left disfocal) on $[\alpha, \beta]_\mathbb{T}$ if the solutions of (4.1.12) such that $y^\Delta(\alpha) = 0$ $(y^\Delta(\beta) = 0)$ have no generalized zeros in $[\alpha, \beta]_\mathbb{T}$.

Theorem 4.1.6 *Suppose y is a nontrivial solution of (4.1.12). If $y(\alpha) = y^\Delta(\beta) = 0$, then*

$$\left[\sqrt{2} \left(\int_\alpha^\beta \frac{Q^2(t)}{r(t)} \left(\int_\alpha^t \frac{\Delta u}{r(u)} \right) \Delta t \right)^{\frac{1}{2}} + \sup_{\alpha \leq t \leq \beta} \left| \mu(t) \frac{Q(t)}{r(t)} \right| \right] \geq 1, \qquad (4.1.14)$$

where $Q(t) = \int_t^\beta q(s)ds$. If instead $y^\Delta(\alpha) = y(\beta) = 0$, then

$$\left[\sqrt{2}\left(\int_\alpha^\beta \frac{Q^2(t)}{r(t)}\left(\int_t^\beta \frac{\Delta u}{r(u)}\right)\Delta t\right)^{\frac{1}{2}} + \sup_{\alpha \leq t \leq \beta}\left|\mu(t)\frac{Q(t)}{r(t)}\right|\right] \geq 1, \quad (4.1.15)$$

where $Q(t) = \int_\alpha^t q(s)ds$.

Proof. We prove (4.1.14). Multiplying (4.1.12) by y^σ and integrating by parts, we have

$$\int_\alpha^\beta y^\sigma(t)\left(r(t)y^\Delta(t)\right)^\Delta \Delta t = y(t)r(t)y^\Delta(t)\big|_\alpha^\beta - \int_\alpha^\beta r(t)(y^\Delta(t))^2\Delta t$$

$$= -\int_\alpha^\beta q(t)\left(y^\sigma(t)\right)^2 \Delta t.$$

Using the assumptions that $y(\alpha) = y^\Delta(\beta) = 0$ and $Q(t) = \int_t^\beta q(s)\Delta s$, we get that

$$\int_\alpha^\beta r(t)\left(y^\Delta(t)\right)^2 \Delta t = \int_\alpha^\beta q(t)\left(y^\sigma(t)\right)^2 \Delta t = -\int_\alpha^\beta Q^\Delta(t)\left(y^\sigma(t)\right)^2 \Delta t.$$

Integrating by parts the right-hand side and using the fact that $y(\alpha) = 0 = Q(\beta)$, we see that

$$\int_\alpha^\beta r(t)\left(y^\Delta(t)\right)^2 \Delta t = \int_\alpha^\beta Q(t)\left(y(t) + y^\sigma(t)\right)y^\Delta(t)\Delta t$$

$$\leq \int_\alpha^\beta |Q(t)|\,|y(t) + y^\sigma(t)|\,\left|y^\Delta(t)\right| \Delta t.$$

Applying the inequality (3.1.23) with $s = Q$, we have

$$\int_\alpha^\beta r(t)\left(y^\Delta(t)\right)^2 \Delta t \leq \left[\sqrt{2}\left(\int_\alpha^\beta \frac{Q^2(t)}{r(t)}\left(\int_\alpha^t \frac{\Delta u}{r(u)}\right)\Delta t\right)^{\frac{1}{2}} + \sup_{\alpha \leq t \leq \beta}\left|\mu(t)\frac{Q(t)}{r(t)}\right|\right]$$

$$\times \int_\alpha^\beta r(t)\left|y^\Delta(t)\right|^2 \Delta t.$$

This implies that

$$\left[\sqrt{2}\left(\int_\alpha^\beta \frac{Q^2(t)}{r(t)}\left(\int_\alpha^t \frac{\Delta u}{r(u)}\right)\Delta t\right)^{\frac{1}{2}} + \sup_{\alpha \leq t \leq \beta}\left|\mu(t)\frac{Q(t)}{r(t)}\right|\right] \geq 1,$$

which is the desired inequality (4.1.14). The proof of (4.1.15) is similar to the proof of (4.1.14) by using integration by parts and Theorem 3.1.12 instead of Theorem 3.1.11. The proof is complete. ∎

As a special case of Theorem 4.1.6, when $r(t) = 1$, we have the following results for Eq. (4.1.1).

Corollary 4.1.2 *Suppose y is a nontrivial solution of (4.1.1). If $y(\alpha) = y^\Delta(\beta) = 0$, then*

$$\left[\sqrt{2} \left(\int_\alpha^\beta Q^2(t)\,(t - \alpha)\,\Delta t \right)^{\frac{1}{2}} + \sup_{\alpha \le t \le \beta} (\mu(t)\,|Q(t)|) \right] \ge 1,$$

where $Q(t) = \int_t^\beta q(s)ds$. If instead $y^\Delta(\alpha) = y(\beta) = 0$, then

$$\left[\sqrt{2} \left(\int_\alpha^\beta Q^2(t)\,(\beta - t)\,\Delta t \right)^{\frac{1}{2}} + \sup_{\alpha \le t \le \beta} (\mu(t)\,|Q(t)|) \right] \ge 1,$$

where $Q(t) = \int_\alpha^t q(s)ds$.

Remark 4.1.4 *Note that if $\mathbb{T} = \mathbb{R}$ then $\mu(t) = 0$ and Eq. (4.1.12) (when $r(t) = 1$) becomes*

$$y''(t) + q(t)y(t) = 0. \tag{4.1.16}$$

In this case the result in Corollary 4.1.2 reduces to a result obtained by Brown and Hinton [57].

Corollary 4.1.3 ([57]). *Suppose y is a solution of Eq. (4.1.16). If $y(\alpha) = y'(\beta) = 0$, then*

$$2 \int_\alpha^\beta Q^2(s)(s - \alpha)ds > 1, \tag{4.1.17}$$

where $Q(t) = \int_t^\beta q(s)ds$. If instead $y'(\alpha) = y(\beta) = 0$, then

$$2 \int_\alpha^\beta Q^2(s)(\beta - s)ds > 1, \tag{4.1.18}$$

where $Q(t) = \int_\alpha^t q(s)ds$.

Remark 4.1.5 *Note that if $\mathbb{T} = \mathbb{N}$, then $\mu(t) = 1$ and Eq. (4.1.12) (when $r(t) = 1$) becomes*

$$\Delta^2 y(n) + q(n)y(n + 1) = 0, \tag{4.1.19}$$

and the result in Corollary 4.1.2 reduces to the following result.

Corollary 4.1.4 *Suppose y is a solution of Eq. (4.1.19). If $y(\alpha) = \Delta y(\beta) = 0$, then*

$$\sqrt{2} \left(\sum_{n=\alpha}^{\beta-1} (Q(n))^2\,(n - \alpha) \right)^{\frac{1}{2}} + \sup_{\alpha \le n \le \beta} |Q(n)| > 1,$$

where $Q(n) = \sum_{s=n}^{\beta-1} q(s)$. If instead $\Delta y(\alpha) = y(\beta) = 0$, then

$$\sqrt{2} \left(\sum_{n=\alpha}^{\beta-1} (Q(n))^2 (\beta - n) \right)^{\frac{1}{2}} + \sup_{\alpha \leq n \leq \beta} |Q(n)| > 1,$$

where $Q(n) = \sum_{s=\alpha}^{n-1} q(s)$.

Theorem 4.1.7 *Suppose that y is a nontrivial solution of (4.1.12). If $y(\alpha) = y^{\Delta}(\beta) = 0$, then*

$$\sqrt{2} \left(\sup_{\alpha \leq t \leq \beta} \frac{Q^2(t)}{r(t)} \int_{\alpha}^{\beta} \frac{1}{r(t)} \left(\int_{\alpha}^{t} \frac{\Delta u}{r(u)} \right) \Delta t \right)^{\frac{1}{2}} + \sup_{\alpha \leq t \leq \beta} \left| \frac{Q(t)}{r(t)} \right| \mu(t) \geq 1,$$

$$(4.1.20)$$

where $Q(t) = \int_{t}^{\beta} q(s)ds$. If instead $y^{\Delta}(\alpha) = y(\beta) = 0$, then

$$\sqrt{2} \left(\sup_{\alpha \leq t \leq \beta} \frac{Q^2(t)}{r(t)} \int_{\alpha}^{\beta} \frac{1}{r(t)} \left(\int_{t}^{\beta} \frac{\Delta u}{r(u)} \right) \Delta t \right)^{\frac{1}{2}} + \sup_{\alpha \leq t \leq \beta} \left| \frac{Q(t)}{r(t)} \right| \mu(t) \geq 1,$$

$$(4.1.21)$$

where $Q(t) = \int_{\alpha}^{t} q(s)ds$.

Proof. We prove (4.1.20). Multiplying (4.1.12) by y^{σ} and integrating by parts and following the proof of Theorem 4.1.6, we have

$$\int_{\alpha}^{\beta} r(t) \left(y^{\Delta}(t) \right)^2 \Delta t = \int_{\alpha}^{\beta} q(t) \left(y^{\sigma}(t) \right)^2 \Delta t = - \int_{\alpha}^{\beta} Q^{\Delta}(t) \left(y^{\sigma}(t) \right)^2 \Delta t.$$

Integrating by parts the right-hand side and using the fact that $y(\alpha) = 0 = Q(\beta)$, we see that

$$\int_{\alpha}^{\beta} r(t) \left(y^{\Delta}(t) \right)^2 \Delta t \leq \int_{\alpha}^{\beta} |Q(t)| \, |y(t) + y^{\sigma}(t)| \, \left| y^{\Delta}(t) \right| \Delta t$$

$$\leq \sup_{\alpha \leq t \leq \beta} \left| \frac{Q(t)}{r(t)} \right| \int_{\alpha}^{\beta} r(t) \, |y(t) + y^{\sigma}(t)| \, \left| y^{\Delta}(t) \right| \Delta t.$$

Applying the inequality (3.1.37) with (3.1.38) and cancelling the term $\int_{\alpha}^{\beta} r(t) \left(y^{\Delta}(t) \right)^2 \Delta t$, we get the desired inequality (4.1.20). The proof of (4.1.21) is similar to the proof of (4.1.20) by using integration by parts and Corollary 3.1.4 instead of Corollary 3.1.3. The proof is complete. ∎

As a special case of Theorem 4.1.7, when $r(t) = 1$, we have the following result.

Corollary 4.1.5 *Suppose that y is a nontrivial solution of (4.1.1). If $y(\alpha) = y^\Delta(\beta) = 0$, then*

$$\sup_{\alpha \leq t \leq \beta} |Q(t)| (\beta - \alpha) + \sup_{\alpha \leq t \leq \beta} |Q(t)| \mu(t) \geq 1,$$

where $Q(t) = \int_t^\beta q(s)ds$. If instead $y^\Delta(\alpha) = y(\beta) = 0$, then

$$\sup_{\alpha \leq t \leq \beta} |Q(t)| (\beta - \alpha) + \sup_{\alpha \leq t \leq \beta} |Q(t)| \mu(t) \geq 1,$$

where $Q(t) = \int_\alpha^t q(s)ds$.

As special case of Corollary 4.1.5, when $\mathbb{T} = \mathbb{R}$, (note that in this case $\mu(t) = 0$), we have the following result due to Harris and Kong [73] for the second order differential equation (4.1.16).

Corollary 4.1.6 *Suppose that y is a nontrivial solution of (4.1.16). If $y(\alpha) = y'(\beta) = 0$, then*

$$(\beta - \alpha) \max_{\alpha \leq t \leq \beta} \left| \left(\int_t^\beta q(s)ds \right) \right| \geq 1. \qquad (4.1.22)$$

If instead $y'(\alpha) = y(\beta) = 0$, then

$$(\beta - \alpha) \max_{\alpha \leq t \leq \beta} \left| \int_\alpha^t q(s)ds \right| \geq 1. \qquad (4.1.23)$$

As a special case of Corollary 4.1.5, when $\mathbb{T} = \mathbb{N}$ (note that in this case $\mu(t) = 1$), we have the following result for the second order difference equation (4.1.19).

Corollary 4.1.7 *Suppose y is a solution of Eq. (4.1.19). If $\Delta y(\alpha) = y(\beta) = 0$, then*

$$\sup_{\alpha \leq n \leq \beta} |Q(n)| (\beta + 1 - \alpha) > 1,$$

where $Q(n) = \sum_{s=n}^{\beta-1} q(s)$. If instead $y(\alpha) = \Delta y(\beta) = 0$, then

$$\sup_{\alpha \leq n \leq \beta} |Q(n)| (\beta + 1 - \alpha) > 1 > 1,$$

where $Q(n) = \sum_{s=\alpha}^{n-1} q(s)$.

Remark 4.1.6 *The above results yield sufficient conditions for the disfocality of (4.1.12), i.e., sufficient conditions so that there does not exist a nontrivial solution y satisfying either $y(\alpha) = y^\Delta(\beta) = 0$ or $y^\Delta(\alpha) = y(\beta) = 0$.*

Now, we assume that there exists a unique $h \in [\alpha, \beta]_{\mathbb{T}}$ such that

$$\int_\alpha^h \frac{\Delta t}{r(t)} = \int_h^\beta \frac{\Delta t}{r(t)}. \tag{4.1.24}$$

Note that when $r(t) = 1$, we see that $(h - \alpha) = (\beta - h)$, so that the unique solution of (4.1.24) is given by $h = (\alpha + \beta)/2$.

Theorem 4.1.8 *Assume that (4.1.24) holds and $Q^\Delta(t) = q(t)$. Suppose that y is a nontrivial solution of (4.1.12). If $y(\alpha) = y(\beta) = 0$, then*

$$\left[\sqrt{2} \left(\int_\alpha^\beta \frac{Q^2(t)}{r(t)} \left(\int_\alpha^h \frac{\Delta u}{r(u)} \right) \Delta t \right)^{\frac{1}{2}} + \sup_{\alpha \leq t \leq \beta} \mu(t) \left| \frac{Q(t)}{r(t)} \right| \right] \geq 1. \tag{4.1.25}$$

Proof. As in the proof of Theorem 4.1.6 by multiplying (4.1.12) by $y^\sigma(t)$, integrating by parts and using $y(\alpha) = y(\beta) = 0$, we have that

$$\int_\alpha^\beta r(t) \left| y^\Delta(t) \right|^2 dt \leq \int_\alpha^\beta |Q(t)| \, |y(t) + y^\sigma(t)|^\gamma \, |y^\Delta(t)| \, dt. \tag{4.1.26}$$

Then

$$\int_\alpha^\beta r(t) \left| y^\Delta(t) \right|^2 dt \leq K(\alpha, \beta) \int_\alpha^\beta r(t) \left| y^\Delta(t) \right|^2 dt,$$

where $K(\alpha, \beta)$ is defined as in (4.2.10). From the last inequality, after cancelling the term $\int_\alpha^\beta r(t) \left| y^\Delta(t) \right|^2 \Delta t$, we get the desired inequality (4.1.25). This completes the proof. ∎

When $r(t) = 1$, (note that in this case $h = (\alpha + \beta)/2$), we have the following result for Eq. (4.1.1).

Theorem 4.1.9 *Assume that $Q^\Delta(t) = q(t)$. Suppose that y is a nontrivial solution of (4.1.1). If $y(\alpha) = y(\beta) = 0$, then*

$$\left[\sqrt{\beta - \alpha} \left(\int_\alpha^\beta Q^2(t) \Delta t \right)^{\frac{1}{2}} + \sup_{\alpha \leq t \leq \beta} (\mu(t) \, |Q(t)|) \right] \geq 1.$$

Remark 4.1.7 *The results in Theorems 4.1.8 and 4.1.9 yield sufficient conditions for the disconjugacy of (4.3.1), i.e., sufficient conditions so that there does not exist a nontrivial solution y satisfying $y(\alpha) = y(\beta) = 0$.*

As a special case of Theorem 4.1.9, when $\mathbb{T} = \mathbb{R}$ and $\mathbb{T} = \mathbb{N}$, we have the following results for the second order differential equation (4.1.16) and the second order difference equation (4.1.19).

Corollary 4.1.8 *Assume that $Q'(t) = q(t)$. Suppose that y is a nontrivial solution of (4.1.16). If $y(\alpha) = y(\beta) = 0$, then*

$$\int_\alpha^\beta \left(\int_\alpha^t q(u) du \right)^2 dt \geq \frac{1}{\beta - \alpha}. \tag{4.1.27}$$

Corollary 4.1.9. *Assume that $\Delta Q(n) = q(n)$. Suppose that y is a nontrivial solution of (4.1.19). If $y(\alpha) = y(\beta) = 0$, then*

$$\left[\sqrt{\beta - \alpha} \left(\sum_{n=\alpha}^{n-1} Q^2(n) \right)^{\frac{1}{2}} + \sup_{\alpha \le n \le \beta} |Q(n)| \right] \ge 1.$$

4.2   Second Order Half-Linear Equation

In this section, we consider some second order half-linear dynamic equations on time scales and establish Lyapunov inequalities. First we consider the second-order half-linear dynamic equation of the form

$$\left(r(t)\varphi(x^{\Delta}) \right)^{\Delta} + p(t)\varphi(x^{\sigma}(t)) = 0, \tag{4.2.1}$$

on an arbitrary time scale $\mathbb{T}$, where $\varphi(u) = |u|^{\gamma-1} u$, $\gamma > 0$ is a positive constant, r and p are real rd-continuous positive functions defined on $\mathbb{T}$ with $r(t) \ne 0$. The results for (4.2.1) are adapted from [130].

Theorem 4.2.1 *Let $x(t)$ be a positive solution of (4.2.1) on $\mathbb{T}$ satisfying $x(a) = x(b) = 0$, $x(t) \ne 0$ for $t \in (a,b)$ and $x(t)$ has a maximum at a point $c \in (a, b)$. Then*

$$\left(\int_a^b r^{\frac{-1}{\gamma}}(t)\Delta t \right)^{\gamma} \int_a^b p(t)\Delta t \ge 2^{\gamma+1}. \tag{4.2.2}$$

Proof. Let

$$M = |x(c)| = \left| \int_a^c x^{\Delta}(t)\Delta t \right| = \left| \int_c^b x^{\Delta}(t)\Delta t \right|. \tag{4.2.3}$$

From (4.2.3), we observe that

$$2M = \left| \int_a^c x^{\Delta}(t)\Delta t \right| + \left| \int_c^b x^{\Delta}(t)\Delta t \right| \le \int_a^c \left| x^{\Delta}(t) \right| \Delta t + \int_c^b \left| x^{\Delta}(t) \right| \Delta t.$$

This implies that

$$2M \le \int_a^b \left| x^{\Delta}(t) \right| \Delta t = \int_a^b r^{\frac{-1}{\gamma+1}}(t)(r^{\frac{1}{\gamma+1}}(t) \left| x^{\Delta}(t) \right|)\Delta t. \tag{4.2.4}$$

From this we get

$$(2M)^{\gamma+1} \le \left(\int_a^b r^{\frac{-1}{\gamma+1}}(t)(r^{\frac{1}{\gamma+1}}(t) \left| x^{\Delta}(t) \right|)\Delta t \right)^{\gamma+1}. \tag{4.2.5}$$

Applying the Hölder inequality with $f(t) = r^{\frac{-1}{\gamma+1}}(t)$, $g(t) = r^{\frac{1}{\gamma+1}}(t)\left|x^{\Delta}(t)\right|$, $p = \gamma + 1$ and $q = \frac{\gamma+1}{\gamma}$, we obtain

$$\int_a^b r^{\frac{-1}{\gamma+1}}(t)(r^{\frac{1}{\gamma+1}}(t)\left|x^{\Delta}(t)\right|)\Delta t$$

$$\leq \left(\int_a^b \left(r^{\frac{-1}{\gamma+1}}(t)\right)^{\frac{\gamma+1}{\gamma}}\Delta t\right)^{\frac{\gamma}{\gamma+1}} \left(\int_a^b \left(r^{\frac{1}{\gamma+1}}(t)\left|x^{\Delta}(t)\right|\right)^{\gamma+1}\Delta t\right)^{\frac{1}{\gamma+1}}$$

$$= \left(\int_a^b r^{\frac{-1}{\gamma}}(t)\Delta t\right)^{\frac{\gamma}{\gamma+1}} \left(\int_a^b \left(r^{\frac{1}{\gamma+1}}(t)\left|x^{\Delta}(t)\right|\right)^{\gamma+1}\Delta t\right)^{\frac{1}{\gamma+1}}$$

$$= \left(\int_a^b r^{\frac{-1}{\gamma}}(t)\Delta t\right)^{\frac{\gamma}{\gamma+1}} \left(\int_a^b r(t)(\left|x^{\Delta}(t)\right|)^{\gamma+1}\Delta t\right)^{\frac{1}{\gamma+1}}.$$

Thus

$$\left(\int_a^b r^{\frac{-1}{\gamma+1}}(t)(r^{\frac{1}{\gamma+1}}(t)\left|x^{\Delta}(t)\right|)\Delta t\right)^{\gamma+1} \leq \left(\int_a^b r^{\frac{-1}{\gamma}}(t)\Delta t\right)^{\gamma} \left(\int_a^b r(t)(\left|x^{\Delta}(t)\right|)^{\gamma+1}\Delta t\right).$$
(4.2.6)

Substituting (4.2.6) in (4.2.5), we have

$$(2M)^{\gamma+1} \leq \left(\int_a^b r^{\frac{-1}{\gamma}}(t)\Delta t\right)^{\gamma} \left(\int_a^b r(t)(\left|x^{\Delta}(t)\right|)^{\gamma+1}\Delta t\right). \qquad (4.2.7)$$

Using integration by parts we see that (note $x(a) = x(b) = 0$)

$$\int_a^b r(t)(\left|x^{\Delta}(t)\right|)^{\gamma+1}\Delta t = \int_a^b x^{\Delta}(t)\left(r(t)(\left|x^{\Delta}(t)\right|)^{\gamma-1}x^{\Delta}(t)\right)\Delta t$$

$$= -\int_a^b \left[r(t)(\left|x^{\Delta}(t)\right|)^{\gamma-1}x^{\Delta}(t)\right]^{\Delta} x^{\sigma}(t)\Delta t. \quad (4.2.8)$$

Now (4.2.1) implies that

$$\int_a^b r(t)(\left|x^{\Delta}(t)\right|)^{\gamma+1}\Delta t = \int_a^b p(t)\left(x^{\sigma}(t)\right)^{\gamma+1}\Delta t.$$

This and (4.2.7) imply that

$$(2M)^{\gamma+1} \leq \left(\int_a^b r^{\frac{-1}{\gamma}}(t)\Delta t\right)^{\gamma} \left(\int_a^b p(t)\left(x^{\sigma}(t)\right)^{\gamma+1}\Delta t\right)$$

$$\leq M^{\gamma+1} \left(\int_a^b r^{\frac{-1}{\gamma}}(t)\Delta t\right)^{\gamma} \left(\int_a^b p(t)\Delta t\right).$$

Now, dividing by $M^{\gamma+1}$, we have

$$\left(\int_a^b r^{\frac{-1}{\gamma}}(t)\Delta t\right)^\gamma \left(\int_a^b p(t)\Delta t\right) \geq 2^{\gamma+1},$$

which is the desired inequality (4.2.2). The proof is complete. ∎

Remark 4.2.1 *Note the inequality with $\gamma = 1$ and $r(t) = 1$, reduces to the inequality*

$$\int_a^b p(t)\Delta t > \frac{4}{b-a}. \tag{4.2.9}$$

Now, we consider the half-linear delay dynamic equation

$$(r(t)(\varphi(x^\Delta(t)))^\Delta + p(t)(\varphi(x(\tau(t))) = 0, \tag{4.2.10}$$

on an arbitrary time scale $\mathbb{T}$, where $\gamma > 0$ is a positive constant, r and p are real rd-continuous positive functions defined on $\mathbb{T}$ with $r(t) \neq 0$, $\tau : \mathbb{T} \to \mathbb{T}$, $\tau(t) \leq t$ for all $t \in \mathbb{T}$, $\lim_{t\to\infty} \tau(t) = \infty$, and

$$\int_{t_0}^\infty \left(\frac{1}{r(t)}\right)^{\frac{1}{\gamma}} \Delta t = \infty. \tag{4.2.11}$$

Note that when the condition (4.2.11) holds, then the positive solution $x(t)$ of (4.2.10) satisfies $x^\Delta(t) > 0$. Under this condition, we see, since $\tau(t) \leq t$, that $x(\tau(t))/x^\sigma(t) \leq 1$. Using this claim we have the following result for (4.2.10).

Corollary 4.2.1 *Assume that (4.2.11) holds and let $x(t)$ be a positive solution of (4.2.10) on $\mathbb{T}$ satisfying $x(a) = x(b) = 0$, $x(t) \neq 0$ for $t \in (a, b)$ and $x(t)$ has a maximum at a point $c \in (a, b)$. Then*

$$\left(\int_a^b r^{\frac{-1}{\gamma}}(t)\Delta t\right)^\gamma \int_a^b p(t)\Delta t \geq 2^{\gamma+1}.$$

Proof. We proceed as in the proof of Theorem 4.2.1, to get

$$(2M)^{\gamma+1} \leq \left(\int_a^b r^{\frac{-1}{\gamma}}(t)\Delta t\right)^\gamma \left(\int_a^b r(t)(|x^\Delta(t)|)^{\gamma+1}\Delta t\right).$$

Using integration by parts we see that (note $x(a) = x(b) = 0$)

$$\int_a^b r(t)(|x^\Delta(t)|)^{\gamma+1}\Delta t = -\int_a^b \left[r(t)(|x^\Delta(t)|)^{\gamma-1}x^\Delta(t)\right]^\Delta x^\sigma(t)\Delta t.$$

Now (4.2.10) implies that

$$\int_a^b r(t)(|x^\Delta(t)|)^{\gamma+1}\Delta t = \int_a^b p(t)\left(\frac{x(\tau(t))}{x^\sigma(t)}\right)^\gamma (x^\sigma(t))^{\gamma+1}\,\Delta t.$$

Using the above claim, since $x(\tau(t))/x^\sigma(t) \leq 1$, we have

$$\int_a^b r(t)(|x^\Delta(t)|)^{\gamma+1}\Delta t \leq \int_a^b p(t)\,(x^\sigma(t))^{\gamma+1}\,\Delta t.$$

The remainder of the proof is similar to the proof in Theorem 4.2.1 and hence is omitted. ∎

In the following, we establish some sufficient conditions for the disconjugacy of (4.2.1).

Theorem 4.2.2 *Let r and p satisfy*

$$\int_a^b p(t)\Delta t < \begin{cases} \dfrac{r^{\gamma+1}(a)}{r^\gamma(b)}\dfrac{(b-c)^\gamma+(c-a)^\gamma}{(c-a)^\gamma(b-c)^\gamma}, & \text{if } r(t) \text{ is increasing,} \\[2mm] \dfrac{r^{\gamma+1}(b)}{r^\gamma(a)}\dfrac{(b-c)^\gamma+(c-a)^\gamma}{(c-a)^\gamma(b-c)^\gamma}, & \text{if } r(t) \text{ is decreasing.} \end{cases}$$

$$(4.2.12)$$

Then (4.2.1) is disconjugate in $\mathbb{T}$.

Proof. Suppose that (4.2.12) holds and assume for the sake of contradiction that (4.2.1) is not disconjugate. Then there exists a nontrivial solution x with $x(a) = x(b) = 0$. Using this x, and integrate by parts to see that (note $x(a) = x(b) = 0$)

$$\int_a^b r(t)(|x^\Delta(t)|)^{\gamma+1}\Delta t = \int_a^b x^\Delta(t)\left(r(t)(|x^\Delta(t)|)^{\gamma-1}x^\Delta(t)\right)\Delta t$$

$$= -\int_a^b \left[r(t)(|x^\Delta(t)|)^{\gamma-1}x^\Delta(t)\right]^\Delta x(t)\Delta t.$$

Now (4.2.1) implies that

$$\int_a^b r(t)(|x^\Delta(t)|)^{\gamma+1}\Delta t = \int_a^b p(t)\,(x(t))^{\gamma+1}\,\Delta t$$

Then, we have

$$M^{\gamma+1}\int_a^b p(t)\Delta t \geq \int_a^b p(t)\,|x(t)|^{\gamma+1}\,\Delta t \geq \int_a^b r(t)\,|x^\Delta(t)|^{\gamma+1}\,\Delta t$$

$$= \int_a^c r(t)\,|x^\Delta(t)|^{\gamma+1}\,\Delta t + \int_c^b r(t)\,|x^\Delta(t)|^{\gamma+1}\,\Delta t, \quad (4.2.13)$$

where M is defined as in Theorem 4.2.1. Now, since

$$\int_a^c r(t) \left|x^\Delta(t)\right| \Delta t = \int_a^c r^{\frac{\gamma}{\gamma+1}}(t) \left(r^{\frac{1}{\gamma+1}}(t) \left|x^\Delta(t)\right|\right) \Delta t,$$

we have after applying the Hölder inequality with $f(t) = r^{\frac{\gamma}{\gamma+1}}(t)$, $g(t) = r^{\frac{1}{\gamma+1}}(t) \left|x^\Delta(t)\right|$, $p = \gamma + 1$ and $q = \frac{\gamma+1}{\gamma}$, that

$$\int_a^c r^{\frac{\gamma}{\gamma+1}}(t) \left(r^{\frac{1}{\gamma+1}}(t) \left|x^\Delta(t)\right|\right) \Delta t$$

$$\leq \left(\int_a^c \left(r^{\frac{\gamma}{\gamma+1}}(t)\right)^{\frac{\gamma+1}{\gamma}} \Delta t\right)^{\frac{\gamma}{\gamma+1}} \left(\int_a^c \left(r^{\frac{1}{\gamma+1}}(t) \left|x^\Delta(t)\right|\right)^{\gamma+1} \Delta t\right)^{\frac{1}{\gamma+1}}$$

$$= \left(\int_a^c r(t) \Delta t\right)^{\frac{\gamma}{\gamma+1}} \left(\int_a^c r(t) \left|x^\Delta(t)\right|^{\gamma+1} \Delta t\right)^{\frac{1}{\gamma+1}}.$$

Then

$$\left(\int_a^c r(t) \Delta t\right)^{\gamma} \left(\int_a^c r(t) \left|x^\Delta(t)\right|^{\gamma+1} \Delta t\right)$$

$$\geq \left(\int_a^c r^{\frac{\gamma}{\gamma+1}}(t) \left(r^{\frac{1}{\gamma+1}}(t) x^\Delta(t)\right) \Delta t\right)^{\gamma+1} = \left(\int_a^c r(t) \left|x^\Delta(t)\right| \Delta t\right)^{\gamma+1}.$$

This implies that

$$\left(\int_a^c r(t) \left|x^\Delta(t)\right|^{\gamma+1} \Delta t\right) \geq \frac{\left(\int_a^c r(t) \left|x^\Delta(t)\right| \Delta t\right)^{\gamma+1}}{\left(\int_a^c r(t) \Delta t\right)^{\gamma}}. \qquad (4.2.14)$$

Also we see that

$$\left(\int_c^b r(t) \left|x^\Delta(t)\right|^{\gamma+1} \Delta t\right) \geq \frac{\left(\int_c^b r(t) \left|x^\Delta(t)\right| \Delta t\right)^{\gamma+1}}{\left(\int_c^b r(t) \Delta t\right)^{\gamma}}. \qquad (4.2.15)$$

Substituting (4.2.14) and (4.2.15) into (4.2.13), we have

$$M^{\gamma+1} \int_a^b p(t) \Delta t$$

$$\geq \frac{\left(\int_a^c r(t) \left|x^\Delta(t)\right| \Delta t\right)^{\gamma+1}}{\left(\int_a^c r(t) \Delta t\right)^{\gamma}} + \frac{\left(\int_c^b r(t) \left|x^\Delta(t)\right| \Delta t\right)^{\gamma+1}}{\left(\int_c^b r(t) \Delta t\right)^{\gamma}}$$

$$
\geq
\begin{cases}
\dfrac{\left(r(a)\int_a^c \left|x^\Delta(t)\right|\Delta t\right)^{\gamma+1}}{\left(\int_a^c r(t)\Delta t\right)^\gamma} + \dfrac{\left(r(a)\int_c^b \left|x^\Delta(t)\right|\Delta t\right)^{\gamma+1}}{\left(\int_c^b r(t)\Delta t\right)^\gamma}, & \text{if } r(t) \text{ is increasing,} \\[4mm]
\dfrac{\left(r(b)\int_a^c \left|x^\Delta(t)\right|\Delta t\right)^{\gamma+1}}{\left(\int_a^c r(t)\Delta t\right)^\gamma} + \dfrac{\left(r(b)\int_c^b \left|x^\Delta(t)\right|\Delta t\right)^{\gamma+1}}{\left(\int_c^b r(t)\Delta t\right)^\gamma}, & \text{if } r(t) \text{ is decreasing,}
\end{cases}
$$

$$
\geq
\begin{cases}
\dfrac{r^{\gamma+1}(a)M^{\gamma+1}}{r^\gamma(b)(c-a)^\gamma} + \dfrac{r^{\gamma+1}(a)M^{\gamma+1}}{r^\gamma(b)(b-c)^\gamma}, & \text{if } r(t) \text{ is increasing,} \\[4mm]
\dfrac{r^{\gamma+1}(b)M^{\gamma+1}}{r^\gamma(a)(c-a)^\gamma} + \dfrac{r^{\gamma+1}(b)M^{\gamma+1}}{r^\gamma(a)(b-c)^\gamma}, & \text{if } r(t) \text{ is decreasing.}
\end{cases}
$$

Dividing by $M^{\gamma+1}$, we have

$$
\int_a^b p(t)\Delta t \geq
\begin{cases}
\dfrac{r^{\gamma+1}(a)}{r^\gamma(b)}\dfrac{(b-c)^\gamma + (c-a)^\gamma}{(c-a)^\gamma(b-c)^\gamma}, & \text{if } r(t) \text{ is increasing,} \\[4mm]
\dfrac{r^{\gamma+1}(b)}{r^\gamma(a)}\dfrac{(b-c)^\gamma + (c-a)^\gamma}{(c-a)^\gamma(b-c)^\gamma}, & \text{if } r(t) \text{ is decreasing,}
\end{cases}
$$

which is a contradiction with (4.2.12) and hence completes the proof. ∎

As a consequence from Theorem 4.2.2, by using the fact that

$$
\left(\frac{x_1^\gamma + x_2^\gamma}{2}\right)^{\frac{1}{\gamma}} \geq \frac{2x_1 x_2}{x_1 + x_2}, \quad \text{for } x_1 = c-a \text{ and } x_2 = b-c,
$$

we have the following result.

Theorem 4.2.3 . *If r and p satisfy*

$$
\int_a^b p(t)\Delta t <
\begin{cases}
\dfrac{r^{\gamma+1}(a)}{r^\gamma(b)}\dfrac{2^{\gamma+1}}{(b-a)^\gamma}, & \text{if } r(t) \text{ is increasing,} \\[4mm]
\dfrac{r^{\gamma+1}(b)}{r^\gamma(a)}\dfrac{2^{\gamma+1}}{(b-a)^\gamma}, & \text{if } r(t) \text{ is decreasing.}
\end{cases}
\tag{4.2.16}
$$

Then (4.2.1) is disconjugate in $\mathbb{T}$.

We end this section by applying Opial type inequalities to establish some Lyapunov type inequalities for the second order half-linear dynamic equation

$$
(r(t)(y^\Delta(t))^\gamma)^\Delta + q(t)\,(y^\sigma(t))^\gamma = 0, \quad \text{on } [a,b]_\mathbb{T},
\tag{4.2.17}
$$

where $\mathbb{T}$ is an arbitrary time scale. The results are adapted from [133]. For Eq. (4.2.17), we assume that $0 < \gamma \leq 1$ is a quotient of odd positive integers, r and q are real rd-continuous functions defined on $\mathbb{T}$ with $r(t) > 0$. We obtain lower bounds for the spacing $\beta - \alpha$ where y is a solution of (4.2.17) satisfying some conditions at α and β.

To simplify the presentation of the results, we define

$$M(\beta) \quad : \quad = \sup_{\alpha \le t \le \beta} \mu^\gamma(t) \frac{|Q(t)|}{r(t)}, \quad \text{where} \quad Q(t) = \int_t^\beta q(s)\Delta s,$$

$$M(\alpha) \quad : \quad = \sup_{\alpha \le t \le \beta} \mu^\gamma(t) \frac{|Q(t)|}{r(t)}, \quad \text{where} \quad Q(t) = \int_\alpha^t q(s)\Delta s.$$

Note that when $\mathbb{T} = \mathbb{R}$, we have $M(\alpha) = 0 = M(\beta)$, and when $\mathbb{T} = \mathbb{Z}$, we have

$$M(\beta) = \sup_{\alpha \le t \le \beta} \frac{\left|\sum_{s=t}^{\beta-1} q(s)\right|}{r(t)}, \quad \text{and } M(\alpha) = \sup_{\alpha \le t \le \beta} \frac{\left|\sum_{s=\alpha}^{t-1} q(s)\right|}{r(t)}. \quad (4.2.18)$$

Theorem 4.2.4 *Suppose that y is a nontrivial solution of (4.2.17) and y^Δ does not change sign in $(\alpha, \beta)_\mathbb{T}$. If $y(\alpha) = y^\Delta(\beta) = 0$, then*

$$\frac{2}{(\gamma+1)^{\frac{\gamma}{\gamma+1}}} \times \left(\int_\alpha^\beta \frac{|Q(x)|^{\frac{\gamma+1}{\gamma}}}{r^{\frac{1}{\gamma}}(x)} \left(\int_\alpha^x \frac{\Delta t}{r^{\frac{1}{\gamma}}(t)} \right)^\gamma \Delta x \right)^{\frac{\gamma}{\gamma+1}} + 2^{1-\gamma} M(\beta) \ge 1,$$

$$(4.2.19)$$

where $Q(t) = \int_t^\beta q(s)\Delta s$. If $y^\Delta(\alpha) = y(\beta) = 0$, then

$$\frac{2}{(\gamma+1)^{\frac{\gamma}{\gamma+1}}} \left(\int_\alpha^\beta \frac{|Q(x)|^{\frac{\gamma+1}{\gamma}}}{r^{\frac{1}{\gamma}}(x)} \left(\int_x^\beta \frac{\Delta t}{r^{\frac{1}{\gamma}}(t)} \right)^\gamma \Delta x \right)^{\frac{\gamma}{\gamma+1}} + 2^{1-\gamma} M(\alpha) \ge 1,$$

$$(4.2.20)$$

where $Q(t) = \int_\alpha^t q(s)\Delta s$.

Proof. We prove (4.2.19). Without loss of generality we may assume that $y(t) > 0$ in $[\alpha, \beta]_\mathbb{T}$. Multiplying (4.2.17) by y^σ and integrating by parts, we have

$$\int_\alpha^\beta \left(r(t) \left(y^\Delta(t)\right)^\gamma \right)^\Delta y^\sigma(t)\Delta t \quad = \quad r(t) \left(y^\Delta(t)\right)^\gamma y(t) \Big|_\alpha^\beta$$

$$- \int_\alpha^\beta r(t) \left(y^\Delta(t)\right)^{\gamma+1} \Delta t$$

$$= \quad - \int_\alpha^\beta q(t) \left(y^\sigma(t)\right)^{\gamma+1} \Delta t.$$

Using the assumptions that $y(\alpha) = y^\Delta(\beta) = 0$ and $Q(t) = \int_t^\beta q(s)\Delta s$, we have

$$\int_\alpha^\beta r(t) \left(y^\Delta(t)\right)^{\gamma+1} \Delta t = \int_\alpha^\beta q(t) \left(y^\sigma(t)\right)^{\gamma+1} \Delta t = - \int_\alpha^\beta Q^\Delta(t) \left(y^\sigma(t)\right)^{\gamma+1} \Delta t.$$

$$(4.2.21)$$

Integrating by parts the right-hand side we see that

$$\int_\alpha^\beta r(t)\left(y^\Delta(t)\right)^{\gamma+1}\Delta t = -\left.Q(t)(y(t))^{\gamma+1}\right|_\alpha^\beta + \int_\alpha^\beta Q(t)\left(y^{\gamma+1}(t)\right)^\Delta \Delta t.$$

Again using the facts that $y(\alpha) = 0 = Q(\beta)$, we obtain

$$\int_\alpha^\beta r(t)\left(y^\Delta(t)\right)^{\gamma+1}dt = \int_\alpha^\beta Q(t)\left(y^{\gamma+1}(t)\right)^\Delta dt. \tag{4.2.22}$$

Applying the chain rule formula and the inequality (3.3.2), we see that

$$\left|\left(y^{\gamma+1}(t)\right)^\Delta\right| \leq (\gamma+1)\int_0^1 |hy^\sigma(t)+(1-h)y(t)|^\gamma\,dh\,|y^\Delta(t)|$$

$$\leq (\gamma+1)|y^\Delta(t)|\int_0^1 |hy^\sigma(t)|^\gamma\,dh$$

$$+(\gamma+1)|y^\Delta(t)|\int_0^1 |(1-h)y(t)|^\gamma\,dh$$

$$= |y^\Delta(t)|\,|y^\sigma(t)|^\gamma + |y^\Delta(t)|\,|y(t)|^\gamma$$

$$\leq 2^{1-\gamma}|y^\sigma(t)+y(t)|^\gamma\,|y^\Delta(t)|. \tag{4.2.23}$$

This and (4.2.22) imply that

$$\int_\alpha^\beta r(t)\left|y^\Delta(t)\right|^{\gamma+1}\Delta t \leq 2^{1-\gamma}\int_\alpha^\beta |Q(t)|\,|y(t)+y^\sigma(t)|^\gamma\,|y^\Delta(t)|\,\Delta t.$$

Applying the inequality (3.3.22) with $s(t) = |Q(t)|$, $p = \gamma$ and $q = 1$, we have

$$\int_\alpha^\beta r(t)\left|y^\Delta(t)\right|^{\gamma+1}\Delta t \leq 2^{1-\gamma}K_1(\alpha,\beta,\gamma,1)\int_\alpha^\beta r(t)\left|y^\Delta(t)\right|^{\gamma+1}\Delta t, \tag{4.2.24}$$

where

$$K_1(\alpha,\beta,\gamma,1) = M(\beta) + 2^\gamma\left(\frac{1}{\gamma+1}\right)^{\frac{1}{\gamma+1}}$$

$$\times\left(\int_\alpha^\beta |Q(x)|^{\frac{\gamma+1}{\gamma}}r^{-\frac{1}{\gamma}}(x)\left(\int_\alpha^x r^{\frac{-1}{\gamma}}(t)\Delta t\right)^\gamma\Delta x\right)^{\frac{\gamma}{\gamma+1}}.$$

Then, we have from (4.2.24) after cancelling the term $\int_\alpha^\beta r(t)\left|y^\Delta(t)\right|^{\gamma+1}\Delta t$, that

$$2^{1-\gamma}M(\beta) + \frac{2}{(\gamma+1)^{\frac{1}{\gamma+1}}}\times\left(\int_\alpha^\beta\frac{|Q(x)|^{\frac{\gamma+1}{\gamma}}}{r^{\frac{1}{\gamma}}(x)}\left(\int_\alpha^x\frac{\Delta t}{r^{\frac{1}{\gamma}}(t)}\right)^\gamma\Delta x\right)^{\frac{\gamma}{\gamma+1}} \geq 1,$$

which is the desired inequality (4.2.19). The proof of (4.2.20) is similar
to (4.2.19) by using integration by parts and (3.3.29) of Theorem 3.3.5 and
(3.3.30) instead of (3.3.23). The proof is complete. ∎

As a special case of Theorem 4.2.4, when $r(t) = 1$, we have the following
result.

Corollary 4.2.2 *Suppose that y is a nontrivial solution of*

$$\left(\left(y^{\Delta}(t) \right)^{\gamma} \right)^{\Delta} + q(t) \left(y^{\sigma}(t) \right)^{\gamma} = 0, \quad t \in [\alpha, \beta]_{\mathbb{T}}, \tag{4.2.25}$$

and y^{Δ} does not change sign in $(\alpha, \beta)_{\mathbb{T}}$. If $y(\alpha) = y^{\Delta}(\beta) = 0$, then

$$\frac{2}{(\gamma+1)^{\frac{1}{\gamma+1}}} \times \left[\int_{\alpha}^{\beta} |Q(t)|^{\frac{1+\gamma}{\gamma}} (t-\alpha)^{\gamma} \Delta t \right]^{\frac{\gamma}{\gamma+1}} + 2^{1-\gamma} \sup_{\alpha \le t \le \beta} (\mu^{\gamma}(t) |Q(t)|) \ge 1, \tag{4.2.26}$$

where $Q(t) = \int_{t}^{\beta} q(s)\Delta s$. If $y^{\Delta}(\alpha) = y(\beta) = 0$, then

$$\frac{2}{(\gamma+1)^{\frac{1}{\gamma+1}}} \left[\int_{\alpha}^{\beta} |Q(t)|^{\frac{1+\gamma}{\gamma}} (\beta-t)^{\gamma} \Delta t \right]^{\frac{\gamma}{\gamma+1}} + 2^{1-\gamma} \sup_{\alpha \le t \le \beta} (\mu^{\gamma}(t) |Q(t)|) \ge 1, \tag{4.2.27}$$

where $Q(t) = \int_{\alpha}^{t} q(s)\Delta s$.

Corollary 4.2.3 *Suppose that y is a nontrivial solution of (4.2.25) and y^{Δ}
does not change sign in $(\alpha, \beta)_{\mathbb{T}}$, and $\gamma \le 1$ is a quotient of odd positive
integers. If $y(\alpha) = y^{\Delta}(\beta) = 0$, then*

$$\frac{2(\beta-\alpha)^{\gamma}}{(\gamma+1)} \max_{\alpha \le t \le \beta} \left| \int_{t}^{\beta} q(s)\Delta s \right| + 2^{1-\gamma} \sup_{\alpha \le t \le \beta} \left(\mu^{\gamma}(t) \left| \int_{t}^{\beta} q(s)\Delta s \right| \right) \ge 1, \tag{4.2.28}$$

whereas if $y^{\Delta}(\alpha) = y(\beta) = 0$, then

$$\frac{2(\beta-\alpha)^{\gamma}}{(\gamma+1)} \max_{\alpha \le t \le \beta} \left| \int_{\alpha}^{t} q(s)\Delta s \right| + 2^{1-\gamma} \sup_{\alpha \le t \le \beta} \left(\mu^{\gamma}(t) \left| \int_{\alpha}^{t} q(s)\Delta s \right| \right) \ge 1. \tag{4.2.29}$$

As a special when $\mathbb{T} = \mathbb{R}$, we have $M(\alpha) = M(\beta) = 0$ and we consider
the second order half-linear differential equation

$$\left((y^{'}(t))^{\gamma} \right)^{'} + q(t)(y(t))^{\gamma} = 0, \quad \alpha \le t \le \beta, \tag{4.2.30}$$

where $\gamma \le 1$ is a quotient of odd positive integers.

Corollary 4.2.4 *Assume that $\gamma \leq 1$ is a quotient of odd positive integers. Suppose that y is a nontrivial solution of (4.2.30) and y' does not change sign in (α, β). If $y(\alpha) = y'(\beta) = 0$, then*

$$\frac{2}{(\gamma + 1)} (\beta - \alpha)^\gamma \sup_{\alpha \leq t \leq \beta} \left| \int_t^\beta q(s)ds \right| \geq 1. \tag{4.2.31}$$

If instead $y'(\alpha) = y(\beta) = 0$, then

$$\frac{2}{(\gamma + 1)} (\beta - \alpha)^\gamma \sup_{\alpha \leq t \leq \beta} \left| \int_\alpha^t q(s)ds \right| \geq 1. \tag{4.2.32}$$

As a special when $\mathbb{T} = \mathbb{Z}$, we see that $M(\alpha)$ and $M(\beta)$ are defined as in (4.2.18) and we consider the second order half-linear difference equation

$$\Delta((\Delta y(n))^\gamma) + q(n)(y(n + 1))^\gamma = 0, \quad \alpha \leq n \leq \beta, \tag{4.2.33}$$

where $\gamma \leq 1$ is a quotient of odd positive integers.

Corollary 4.2.5 *Suppose that y is a nontrivial solution of (4.3.17) and $\Delta y(n)$ does not change sign in $(\alpha, \beta)_\mathbb{T}$, and $\gamma \leq 1$ is a quotient of odd positive integers. If $y(\alpha) = \Delta y(\beta) = 0$, then*

$$\frac{2(\beta - \alpha)^\gamma}{(\gamma + 1)} \max_{\alpha \leq n \leq \beta} \left| \sum_{s=n}^{\beta-1} q(s) \right| + 2^{1-\gamma} \sup_{\alpha \leq n \leq \beta} \left(\left| \sum_{s=n}^{\beta-1} q(s) \right| \right) \geq 1,$$

whereas if $\Delta y(\alpha) = y(\beta) = 0$, then

$$\frac{2(\beta - \alpha)^\gamma}{(\gamma + 1)} \max_{\alpha \leq n \leq \beta} \left| \sum_{s=\alpha}^{n-1} q(s) \right| + 2^{1-\gamma} \sup_{\alpha \leq n \leq \beta} \left(\left| \sum_{s=\alpha}^{n-1} q(s) \right| \right) \geq 1.$$

Remark 4.2.2 *The above results yield sufficient conditions for the disfocality of (4.3.1), i.e., sufficient conditions so that there does not exist a nontrivial solution y satisfying either $y(\alpha) = y^\Delta(\beta) = 0$, or $y^\Delta(\alpha) = y(\beta) = 0$.*

Next we employ Theorem 3.3.6 to determine a lower bound for the distance between consecutive zeros of solutions of (4.2.17). Note that the applications of the above results allow the use of arbitrary anti-derivative Q in the above arguments. In the following, we assume that $Q^\Delta(t) = q(t)$ and there exists $h \in (\alpha, \beta)$ which is the unique solution of the equation

$$K_1(\alpha, \beta) = K_1(\alpha, \beta, h) = K_1(\alpha, h, \beta) < \infty, \tag{4.2.34}$$

where

$$K_1(\alpha, \beta, h) = \frac{2^\gamma}{(\gamma + 1)^{\frac{1}{\gamma+1}}} \times \left(\int_\alpha^\beta \frac{|Q(x)|^{\frac{\gamma+1}{\gamma}}}{r^{\frac{1}{\gamma}}(x)} \left(\int_\alpha^h \frac{\Delta t}{r^{\frac{1}{\gamma}}(t)} \right)^\gamma \Delta x \right)^{\frac{\gamma}{\gamma+1}},$$

and

$$K_1(\alpha, h, \beta) = \frac{2^\gamma}{(\gamma+1)^{\frac{1}{\gamma+1}}} \left(\int_\alpha^\beta \frac{|Q(x)|^{\frac{\gamma+1}{\gamma}}}{r^{\frac{1}{\gamma}}(x)} \left(\int_h^\beta \frac{\Delta t}{r^{\frac{1}{\gamma}}(t)} \right)^\gamma \Delta x \right)^{\frac{\gamma}{\gamma+1}}.$$

Theorem 4.2.5 *Assume that $Q^\Delta(t) = q(t)$. Suppose y is a nontrivial solution of (4.2.17) and $y^\Delta(t)$ does not change sign in (α, β). If $y(\alpha) = y(\beta) = 0$, then*

$$K_1(\alpha, \beta) \geq 1, \tag{4.2.35}$$

where $K_1(\alpha, \beta)$ is defined as in (4.2.34).

Proof. Multiply (4.2.17) by $y^\sigma(t)$, and proceed as in Theorem 4.2.4 and use $y(\alpha) = y(\beta) = 0$, to get

$$\int_\alpha^\beta r(t) \left(y^\Delta(t)\right)^{\gamma+1} \Delta t = \int_\alpha^\beta q(t) \left(y(t)\right)^{\gamma+1} \Delta t = \int_\alpha^\beta Q^\Delta(t) \left(y^\sigma(t)\right)^{\gamma+1} \Delta t.$$

Integrating by parts the right-hand side, we see that

$$\int_\alpha^\beta r(t) \left(y^\Delta(t)\right)^{\gamma+1} \Delta t = Q(t)(y(t))^{\gamma+1}\big|_\alpha^\beta + \int_\alpha^\beta (-Q(t)) \left(y^{\gamma+1}(t)\right)^\Delta \Delta t.$$

Again using the facts that $y(\alpha) = 0 = y(\beta)$, we obtain

$$\int_\alpha^\beta r(t) \left|y^\Delta(t)\right|^{\gamma+1} \Delta t \leq \int_\alpha^\beta |Q(t)| \, |y(t) + y^\sigma(t)|^\gamma \left|y^\Delta(t)\right| \Delta t.$$

Applying the inequality (3.3.31) with $s(t) = |Q(t)|$, $p = \gamma$ and $q = 1$, we have

$$\int_\alpha^\beta r(t) \left|y^\Delta(t)\right|^{\gamma+1} dt \leq 2^{1-\gamma} K_1(\alpha, \beta) \int_\alpha^\beta r(t) \left|y^\Delta(t)\right|^{\gamma+1} \Delta t.$$

From this inequality, after cancelling $\int_\alpha^\beta \left|y^\Delta(t)\right|^{\gamma+1} \Delta t$, we get the desired inequality (4.2.35). This completes the proof. ∎

4.3 Second Order Equations with Damping Terms

In this section we consider the second-order half-linear dynamic equation with a damping term

$$\left(r(t) \left(x^\Delta(t)\right)^\gamma\right)^\Delta + p(t) \left(x^\Delta(t)\right)^\gamma + q(t) \left(x^\sigma(t)\right)^\gamma = 0, \ \ t \in [\alpha, \beta]_\mathbb{T}, \tag{4.3.1}$$

where $\mathbb{T}$ is an arbitrary time scale and $\sigma(t)$ is the forward jump operator on $\mathbb{T}$ which is defined by $\sigma(t) := \inf\{s \in \mathbb{T} : s > t\}$.

We say that a solution x of (4.3.1) has a generalized zero at t if $x(t) = 0$, and has a generalized zero in $(t, \sigma(t))$ in the case $x(t)x^\sigma(t) < 0$ and $\mu(t) > 0$.

Equation (4.3.1) is disconjugate on the interval $[t_0, b]_\mathbb{T}$, if there is no nontrivial solution of (4.3.1) with two (or more) generalized zeros in $[t_0, b]_\mathbb{T}$. We say that (4.3.1) is right disfocal (left disfocal) on $[\alpha, \beta]_\mathbb{T}$ if the solutions of (4.3.1) such that $x^\Delta(\alpha) = 0$ $(x^\Delta(\beta) = 0)$ have no generalized zeros in $[\alpha, \beta]_\mathbb{T}$. For Eq. (4.3.1) the point $\beta > \alpha$ is called a right focal point of α if the solution of (4.3.1) with initial conditions $x(\alpha) \neq 0$, $x^\Delta(\alpha) = 0$ satisfies $x(\beta) = 0$. The left focal point is defined similarly.

We will assume that $\gamma \geq 1$ is a quotient of odd positive integers, r, p and q are real rd-continuous functions defined on $\mathbb{T}$ with $r(t) > 0$ and $\mu(t)\,|p(t)| \leq r(t)/c$ where c is a positive constant such that $c \geq 1$. We also assume that $\sup \mathbb{T} = \infty$, and define the time scale interval $[a, b]_\mathbb{T}$ by $[a, b]_\mathbb{T} := [a, b] \cap \mathbb{T}$. To simplify the presentation of the results, we define

$$\Lambda(\beta) \quad : \quad = \sup_{\alpha \leq t \leq \beta} \mu^\gamma(t) \frac{|Q(t)|}{r(t)}, \text{ where } Q(t) = \int_t^\beta q(s)\Delta s,$$

$$\Lambda(\alpha) \quad : \quad = \sup_{\alpha \leq t \leq \beta} \mu^\gamma(t) \frac{|Q(t)|}{r(t)}, \text{ where } Q(t) = \int_\alpha^t q(s)\Delta s,$$

$$R_\alpha(t) \quad : \quad = \int_\alpha^t \frac{\Delta s}{r^{\frac{1}{\gamma}}(s)}, \quad \text{and} \quad R_\beta(t) := \int_t^\beta \frac{\Delta s}{r^{\frac{1}{\gamma}}(s)}.$$

Note that when $\mathbb{T} = \mathbb{R}$, we have $\Lambda(\alpha) = 0 = \Lambda(\beta)$ and when $\mathbb{T} = \mathbb{Z}$, we have

$$\Lambda(\beta) = \sup_{\alpha \leq t \leq \beta} \frac{\left| \sum_{s=t}^{\beta-1} q(s) \right|}{r(t)}, \text{ and } \Lambda(\alpha) = \sup_{\alpha \leq t \leq \beta} \frac{\left| \sum_{s=\alpha}^{t-1} q(s) \right|}{r(t)}. \quad (4.3.2)$$

Now, we are ready to state and prove the main results.

Theorem 4.3.1 *Suppose that x is a nontrivial solution of (4.3.1) and x^Δ does not change sign on $(\alpha, \beta)_\mathbb{T}$. If $x(\alpha) = x^\Delta(\beta) = 0$, then*

$$2^{2\gamma-2}\Lambda(\beta) + \frac{2^{3\gamma-2}}{(\gamma+1)^{\frac{1}{\gamma+1}}} \times \left(\int_\alpha^\beta \frac{|Q(t)|^{\frac{\gamma+1}{\gamma}}}{r^{\frac{1}{\gamma}}(t)} (R_\alpha(t))^\gamma \Delta t \right)^{\frac{\gamma}{\gamma+1}}$$

$$+ \left(\frac{\gamma}{1+\gamma} \right)^{\frac{\gamma}{\gamma+1}} \times \left(\int_\alpha^\beta \frac{|p(t)|^{\gamma+1}}{r^\gamma(t)} (R_\alpha(t))^\gamma \Delta t \right)^{\frac{1}{\gamma+1}} \geq 1 - \frac{1}{c}, \quad (4.3.3)$$

where $Q(t) = \int_t^\beta q(s)\Delta s$. If instead $x^\Delta(\alpha) = x(\beta) = 0$, then

$$2^{2\gamma-2}\Lambda(\alpha) + \frac{2^{3\gamma-2}}{(\gamma+1)^{\frac{1}{\gamma+1}}} \times \left(\int_\alpha^\beta \frac{|Q(t)|^{\frac{\gamma+1}{\gamma}}}{r^{\frac{1}{\gamma}}(t)} (R_\beta(t))^\gamma \Delta t \right)^{\frac{\gamma}{\gamma+1}}$$

$$+ \left(\frac{\gamma}{1+\gamma} \right)^{\frac{\gamma}{\gamma+1}} \times \left(\int_\alpha^\beta \frac{|p(t)|^{\gamma+1}}{r^\gamma(t)} (R_\beta(t))^\gamma \Delta t \right)^{\frac{1}{\gamma+1}} \geq 1 - \frac{1}{c}, \quad (4.3.4)$$

where $Q(t) = \int_\alpha^t q(s)\Delta s$.

Proof. We prove (4.3.3). Without loss of generality we may assume that $x(t) \geq 0$ in $[\alpha, \beta]_{\mathbb{T}}$. Multiplying (4.3.1) by x^σ and integrating by parts, we have

$$\int_\alpha^\beta \left(r(t) \left(x^\Delta(t) \right)^\gamma \right)^\Delta x^\sigma(t) \Delta t + \int_\alpha^\beta p(t) x^\sigma(t) \left(x^\Delta(t) \right)^\gamma \Delta t$$

$$= \; r(t) \left(x^\Delta(t) \right)^\gamma x(t) \Big|_\alpha^\beta - \int_\alpha^\beta r(t) \left(x^\Delta(t) \right)^{\gamma+1} \Delta t$$

$$+ \int_\alpha^\beta p(t) x^\sigma(t) \left(x^\Delta(t) \right)^\gamma \Delta t = - \int_\alpha^\beta q(t) \left(x^\sigma(t) \right)^{\gamma+1} \Delta t.$$

Using the assumption $x(\alpha) = x^\Delta(\beta) = 0$ we have

$$- \int_\alpha^\beta r(t) \left(x^\Delta(t) \right)^{\gamma+1} \Delta t + \int_\alpha^\beta p(t) x^\sigma(t) \left(x^\Delta(t) \right)^\gamma \Delta t = - \int_\alpha^\beta q(t) \left(x^\sigma(t) \right)^{\gamma+1} \Delta t.$$

This implies (note that $Q(t) = \int_t^\beta q(s) \Delta s$) that

$$\int_\alpha^\beta r(t) \left(x^\Delta(t) \right)^{\gamma+1} \Delta t = \int_\alpha^\beta p(t) x^\sigma(t) \left(x^\Delta(t) \right)^\gamma \Delta t - \int_\alpha^\beta Q^\Delta(t) \left(x^\sigma(t) \right)^{\gamma+1} \Delta t.$$
$$(4.3.5)$$

Integrating by parts the right-hand side, we see that

$$\int_\alpha^\beta r(t) \left(x^\Delta(t) \right)^{\gamma+1} \Delta t \; = \; \int_\alpha^\beta p(t) x^\sigma(t) \left(x^\Delta(t) \right)^\gamma \Delta t$$

$$- Q(t)(x(t))^{\gamma+1} \Big|_\alpha^\beta + \int_\alpha^\beta Q(t) \left(x^{\gamma+1}(t) \right)^\Delta \Delta t.$$

Again using the assumptions $x(\alpha) = 0$ and $Q(\beta) = 0$, we obtain

$$\int_\alpha^\beta r(t) \left(x^\Delta(t) \right)^{\gamma+1} dt = \int_\alpha^\beta p(t) x^\sigma(t) \left(x^\Delta(t) \right)^\gamma \Delta t + \int_\alpha^\beta Q(t) \left(x^{\gamma+1}(t) \right)^\Delta \Delta t.$$
$$(4.3.6)$$

Applying the chain rule formula

$$\left(x^\lambda(t) \right)^\Delta = \lambda \int_0^1 \left[h x^\sigma(t) + (1-h) x(t) \right]^{\lambda-1} dh \, x^\Delta(t), \quad \text{for } \lambda > 0, \quad (4.3.7)$$

and the inequality

$$a^\lambda + b^\lambda \leq (a+b)^\lambda \leq 2^{\lambda-1}(a^\lambda + b^\lambda), \quad \text{if } a, \, b \geq 0, \, \lambda \geq 1, \quad (4.3.8)$$

we see that

$$
\begin{aligned}
\left|\left(x^{\gamma+1}(t)\right)^{\Delta}\right| &\leq (\gamma+1) \int_0^1 |hx^\sigma(t) + (1-h)x(t)|^\gamma \, dh \, |x^\Delta(t)| \\
&\leq 2^{\gamma-1}(\gamma+1) |x^\Delta(t)| \int_0^1 |hx^\sigma(t)|^\gamma \, dh \\
&\quad + 2^{\gamma-1}(\gamma+1) |x^\Delta(t)| \int_0^1 |(1-h)x(t)|^\gamma \, dh \\
&= 2^{\gamma-1} |x^\Delta(t)| \, |x^\sigma(t)|^\gamma + 2^{\gamma-1} |x^\Delta(t)| \, |x(t)|^\gamma \\
&\leq 2^{\gamma-1} |x^\sigma(t) + x(t)|^\gamma \, |x^\Delta(t)| .
\end{aligned}
\tag{4.3.9}
$$

This and (4.3.6) imply that

$$
\begin{aligned}
\int_\alpha^\beta r(t) \left|x^\Delta(t)\right|^{\gamma+1} \Delta t &\leq \int_\alpha^\beta |p(t)| \, |x^\sigma(t)| \, |x^\Delta(t)|^\gamma \, \Delta t \\
&\quad + 2^{\gamma-1} \int_\alpha^\beta |Q(t)| \, |x(t) + x^\sigma(t)|^\gamma \, |x^\Delta(t)| \, \Delta t
\end{aligned}
\tag{4.3.10}
$$

Applying the inequality (3.3.3) on the integral $\int_\alpha^\beta |Q(t)| \, |x(t) + x^\sigma(t)|^\gamma$ $|x^\Delta(t)| \, \Delta t$, with $s(t) = |Q(t)|$, $p = \gamma$, $q = 1$, we have

$$
\int_\alpha^\beta |Q(t)| \, |x(t) + x^\sigma(t)|^\gamma \, |x^\Delta(t)| \, \Delta t \leq K_1(\alpha, \beta, \gamma, 1) \int_\alpha^\beta r(t) \left|x^\Delta(t)\right|^{\gamma+1} \Delta t,
\tag{4.3.11}
$$

where

$$
K_1(\alpha, \beta, \gamma, 1) = 2^{2\gamma-2} \Lambda(\beta) + 2^{3\gamma-2} \frac{1}{(\gamma+1)^{\frac{1}{\gamma+1}}} \left(\int_\alpha^\beta \frac{|Q(x)|^{\frac{\gamma+1}{\gamma}}}{r^{\frac{1}{\gamma}}(x)} \left(R_\alpha(x)\right)^\gamma \Delta x \right)^{\frac{\gamma}{\gamma+1}} .
$$

Using that fact that $x^\sigma = x(t) + \mu(t)x^\Delta(t)$, we see that

$$
\begin{aligned}
\int_\alpha^\beta |p(t)| \, |x^\sigma(t)| \left|\left(x^\Delta(t)\right)^\gamma\right| \Delta t &= \int_\alpha^\beta |p(t)| \, |x(t) + \mu(t)x^\Delta(t)| \, |x^\Delta(t)|^\gamma \, \Delta t \\
&\leq \int_\alpha^\beta |p(t)| \, |x(t)| \, |x^\Delta(t)|^\gamma \, \Delta t \\
&\quad + \int_\alpha^\beta \mu(t) \, |p(t)| \, |x^\Delta(t)|^{\gamma+1} \, \Delta t .
\end{aligned}
$$

Applying the inequality (3.2.33) on the integral $\int_\alpha^\beta |p(t)| \, |x(t)| \, |x^\Delta(t)|^\gamma \, \Delta t$ with $s(t) = |p(t)|$, $p = 1$ and $q = \gamma$, we see that

$$
\int_\alpha^\beta |p(t)| \, |x(t)| \, |x^\Delta(t)|^\gamma \, \Delta t \leq G_1(\alpha, \beta, 1, \gamma) \int_\alpha^\beta r(t) \left|x^\Delta(t)\right|^{\gamma+1} \Delta t, \tag{4.3.12}
$$

where

$$G_1(\alpha, \beta, 1, \gamma) = \left(\frac{\gamma}{1+\gamma}\right)^{\frac{\gamma}{\gamma+1}} \times \left(\int_\alpha^\beta \frac{|p(t)|^{\gamma+1}}{(r(t))^\gamma} (R_\alpha(t))^\gamma \, \Delta t\right)^{\frac{1}{\gamma+1}}.$$

Using the assumption that $0 \le p(t)\mu(t) \le r(t)/c$, we see that

$$\int_\alpha^\beta p(t) |x^\sigma(t)| |x^\Delta(t)|^\gamma \, \Delta t \le G_1(\alpha, \beta, 1, \gamma) \int_\alpha^\beta r(t) |x^\Delta(t)|^{\gamma+1} \, \Delta t$$
$$+ \frac{1}{c} \int_\alpha^\beta r(t) |x^\Delta(t)|^{\gamma+1} \, \Delta t. \qquad (4.3.13)$$

Substituting (4.3.11) and (4.3.13) into (4.3.10), we have

$$\left(1 - \frac{1}{c}\right) \int_\alpha^\beta r(t) |x^\Delta(t)|^{\gamma+1} \, \Delta t \le K_1(\alpha, \beta, \gamma, 1) \int_\alpha^\beta r(t) |x^\Delta(t)|^{\gamma+1} \, \Delta t$$
$$+ G_1(\alpha, \beta, 1, \gamma) \int_\alpha^\beta r(t) |x^\Delta(t)|^{\gamma+1} \, \Delta t. \qquad (4.3.14)$$

Then, we have from (4.3.14) that

$$1 - \frac{1}{c} \le K_1(\alpha, \beta, \gamma, 1) + G_1(\alpha, \beta, 1, \gamma)$$
$$= 2^{2\gamma-2}\Lambda(\beta) + \frac{2^{3\gamma-2}}{(\gamma+1)^{\frac{1}{\gamma+1}}} \left(\int_\alpha^\beta \frac{|Q(t)|^{\frac{\gamma+1}{\gamma}}}{r^{\frac{1}{\gamma}}(t)} (R_\alpha(t))^\gamma \, \Delta t\right)^{\frac{\gamma}{\gamma+1}}$$
$$+ \left(\frac{\gamma}{1+\gamma}\right)^{\frac{\gamma}{\gamma+1}} \left(\int_\alpha^\beta \frac{|p(t)|^{\gamma+1}}{r^\gamma(t)} (R_\alpha(t))^\gamma \, \Delta t\right)^{\frac{1}{\gamma+1}},$$

which is the desired inequality (4.3.3). The proof of (4.3.4) is similar to (4.3.3) using Theorems 3.2.9 and 3.3.2. The proof is complete. ∎

In Theorem 4.3.1 if $r(t) = 1$, then we have the following result.

Corollary 4.3.1 *Suppose that x is a nontrivial solution of (4.3.1) and x^Δ does not change sign in $(\alpha, \beta)_\mathbb{T}$. If $x(\alpha) = x^\Delta(\beta) = 0$, then*

$$2^{2\gamma-2}\Lambda(\beta) + \frac{2^{3\gamma-2}}{(\gamma+1)^{\frac{1}{\gamma+1}}} \times \left(\int_\alpha^\beta |Q(t)|^{\frac{\gamma+1}{\gamma}} (t-\alpha)^\gamma \, \Delta t\right)^{\frac{\gamma}{\gamma+1}}$$
$$+ \left(\frac{\gamma}{1+\gamma}\right)^{\frac{\gamma}{\gamma+1}} \times \left(\int_\alpha^\beta |p(t)|^{\gamma+1} (t-\alpha)^\gamma \, \Delta t\right)^{\frac{1}{\gamma+1}} \ge 1 - \frac{1}{c},$$

where $Q(t) = \int_t^\beta q(s)\Delta s$. If instead $x^\Delta(\alpha) = x(\beta) = 0$, then

$$2^{2\gamma-2}\Lambda(\alpha) + \frac{2^{3\gamma-2}}{(\gamma+1)^{\frac{1}{\gamma+1}}} \times \left(\int_\alpha^\beta |Q(t)|^{\frac{\gamma+1}{\gamma}} (\beta-t)^\gamma \,\Delta t \right)^{\frac{\gamma}{\gamma+1}}$$

$$+ \left(\frac{\gamma}{1+\gamma} \right)^{\frac{\gamma}{\gamma+1}} \times \left(\int_\alpha^\beta |p(t)|^{\gamma+1} (\beta-t)^\gamma \,\Delta t \right)^{\frac{1}{\gamma+1}} \geq 1 - \frac{1}{c},$$

where $Q(t) = \int_\alpha^t q(s)\Delta s$.

As a special case of Theorem 4.3.1, when $\gamma = 1$, we have the following result.

Corollary 4.3.2 *Suppose that x is a nontrivial solution of (4.3.1) and x^Δ does not change sign in $(\alpha, \beta)_{\mathbb{T}}$. If $x(\alpha) = x^\Delta(\beta) = 0$, then*

$$\Lambda(\beta) + \sqrt{2}\left(\int_\alpha^\beta \frac{|Q(t)|^2}{r(t)} r_\alpha(t)\Delta t \right)^{\frac{1}{2}} + \frac{1}{\sqrt{2}}\left(\int_\alpha^\beta \frac{p^2(t)}{r(t)} R_\alpha(t)\Delta t \right)^{\frac{1}{2}} \geq 1 - \frac{1}{c},$$

where $R_\alpha(t) = \int_\alpha^t \frac{\Delta s}{r(s)}$ and $Q(t) = \int_t^\beta q(s)\Delta s$. If instead $x^\Delta(\alpha) = x(\beta) = 0$, then

$$\Lambda(\alpha) + \sqrt{2}\left(\int_\alpha^\beta \frac{|Q(t)|^2}{r(t)} r_\beta(t)\Delta t \right)^{\frac{1}{2}} + \frac{1}{\sqrt{2}}\left(\int_\alpha^\beta \frac{p^2(t)}{r(t)} R_\beta(t)\Delta t \right)^{\frac{1}{2}} \geq 1 - \frac{1}{c},$$

where $R_\beta(t) = \int_t^\beta \frac{\Delta s}{r(s)}$ and $Q(t) = \int_\alpha^t q(s)\Delta s$.

As a special case of Corollary 4.3.2, when $p(t) = 0$, we have the following result.

Corollary 4.3.3 *Suppose that x is a nontrivial solution of*

$$\left(r(t)x^\Delta(t) \right)^\Delta + q(t)x^\sigma(t) = 0, \quad t \in [\alpha, \beta]_{\mathbb{T}}, \tag{4.3.15}$$

and x^Δ does not change sign in $(\alpha, \beta)_{\mathbb{T}}$. If $x(\alpha) = x^\Delta(\beta) = 0$, then

$$\sqrt{2}\left(\int_\alpha^\beta \frac{|Q(t)|^2}{r(t)} \left(\int_\alpha^t \frac{\Delta t}{r(t)} \right) \Delta t \right)^{\frac{1}{2}} + \Lambda(\beta) \geq 1,$$

where $Q(t) = \int_t^\beta q(s)\Delta s$. If instead $x^\Delta(\alpha) = x(\beta) = 0$, then

$$\sqrt{2}\left(\int_\alpha^\beta \frac{|Q(t)|^2}{r(t)} \left(\int_t^\beta \frac{\Delta t}{r(t)} \right) \Delta t \right)^{\frac{1}{2}} + \Lambda(\alpha) \geq 1,$$

where $Q(t) = \int_\alpha^t q(s)\Delta s$.

Remark 4.3.1 *Theorem 4.3.1 yield sufficient conditions for the disfocality of (4.3.1), i.e., sufficient conditions so that there does not exist a nontrivial solution x satisfying $x(\alpha) = x^\Delta(\beta) = 0$ or $x^\Delta(\alpha) = x(\beta) = 0$.*

On a time scale $\mathbb{T}$, we note from the chain rule (4.3.7) that

$$
\begin{aligned}
\left((t-a)^{\lambda+\delta}\right)^\Delta &= (\lambda+\delta)\int_0^1 [h(\sigma(t)-a) + (1-h)(t-a)]^{\lambda+\delta-1}\,dh \\
&\geq (\lambda+\delta)\int_0^1 [h(t-a) + (1-h)(t-a)]^{\lambda+\delta-1}\,dh \\
&= (\lambda+\delta)(t-a)^{\lambda+\delta-1}.
\end{aligned}
$$

This implies that

$$
\int_a^\tau (t-a)^{(\lambda+\delta-1)}\Delta t \leq \int_a^\tau \frac{1}{(\lambda+\delta)}\left((t-a)^{\lambda+\delta}\right)^\Delta \Delta t = \frac{(\tau-a)^{\lambda+\delta}}{(\lambda+\delta)}.
$$
(4.3.16)

Now using the maximum of $|Q|$ and $|p|$ on $[\alpha, \beta]_\mathbb{T}$ and substituting (4.3.16) into the results of Corollary 4.3.1, we have the following result.

Corollary 4.3.4 *Suppose that x is a nontrivial solution of (4.3.1) and x^Δ does not change sign in $(\alpha, \beta)_\mathbb{T}$. If $x(\alpha) = x^\Delta(\beta) = 0$, then*

$$
\frac{2^{3\gamma-2}(\beta-\alpha)^\gamma}{(\gamma+1)} \max_{\alpha \leq t \leq \beta}\left|\int_t^\beta q(s)\Delta s\right| + \frac{\gamma^{\frac{\gamma}{\gamma+1}}}{\gamma+1}(\beta-\alpha)\max_{\alpha\leq t\leq\beta}|p(t)|
$$

$$
+2^{2\gamma-2}\sup_{\alpha\leq t\leq\beta}\mu^\gamma(t)\left|\int_t^\beta q(s)\Delta s\right| \geq 1 - \frac{1}{c}.
$$

If instead $x^\Delta(\alpha) = x(\beta) = 0$, then

$$
\frac{2^{3\gamma-2}(\beta-\alpha)^\gamma}{(\gamma+1)} \max_{\alpha \leq t \leq \beta}\left|\int_\alpha^t q(s)\Delta s\right| + \frac{\gamma^{\frac{\gamma}{\gamma+1}}}{\gamma+1}(\beta-\alpha)\max_{\alpha\leq t\leq\beta}|p(t)|
$$

$$
+2^{2\gamma-2}\sup_{\alpha\leq t\leq\beta}\mu^\gamma(t)\left|\int_\alpha^t q(s)\Delta s\right| \geq 1 - \frac{1}{c},
$$

As a special when $\mathbb{T} = \mathbb{Z}$, we see that $\Lambda(\alpha)$ and $\Lambda(\beta)$ are defined as in (4.3.2) and we consider the second order half-linear difference equation

$$
\Delta(\Delta x(n))^\gamma + p(n)(\Delta x(n))^\gamma + q(n)(x(n+1))^\gamma = 0, \quad \alpha \leq n \leq \beta, \quad (4.3.17)
$$

where $\gamma \geq 1$ is a quotient of odd positive integers and $p(n) \leq 1/c$.

Corollary 4.3.5 *Suppose that x is a nontrivial solution of (4.3.17) and $\Delta x(n)$ does not change sign in $(\alpha, \beta)_{\mathrm{T}}$. If $x(\alpha) = \Delta x(\beta) = 0$, then*

$$1 - \frac{1}{c} \leq \frac{2^{3\gamma-2}(\beta-\alpha)^\gamma}{(\gamma+1)} \max_{\alpha \leq n \leq \beta} \left| \sum_{s=n}^{\beta-1} q(s) \right| + 2^{2\gamma-2} \sup_{\alpha \leq n \leq \beta} \left| \sum_{s=n}^{\beta-1} q(s) \right|$$

$$+ \frac{\gamma^{\frac{\gamma}{\gamma+1}}}{\gamma+1}(\beta-\alpha) \max_{\alpha \leq n \leq \beta} |p(n)|.$$

If instead $\Delta x(\alpha) = x(\beta) = 0$, then

$$1 - \frac{1}{c} \leq \frac{2^{3\gamma-2}(\beta-\alpha)^\gamma}{(\gamma+1)} \max_{\alpha \leq n \leq \beta} \left| \sum_{s=\alpha}^{n-1} q(s) \right| + 2^{2\gamma-2} \sup_{\alpha \leq n \leq \beta} \left| \sum_{s=\alpha}^{n-1} q(s) \right|$$

$$+ \frac{\gamma^{\frac{\gamma}{\gamma+1}}}{\gamma+1}(\beta-\alpha) \max_{\alpha \leq n \leq \beta} |p(n)|.$$

If we apply the inequality

$$|a + b|^\lambda \leq 2^{\lambda-1}\left(|a|^\lambda + |b|^\lambda\right), \text{ where } a, b \text{ are real numbers and } \lambda \geq 1,$$

with $a = x(t)$ and $b = \mu(t)hx^\Delta(t)$, then we have from (4.3.7) that

$$\left|\left(x^{\gamma+1}(t)\right)^\Delta\right| \leq (\gamma+1)\left|x^\Delta(t)\right| \int_0^1 \left|x(t) + \mu(t)hx^\Delta(t)\right|^\gamma dh$$

$$\leq 2^{\gamma-1}(\gamma+1)\left|x^\Delta(t)\right| \int_0^1 |x(t)|^\gamma dh$$

$$+ 2^{\gamma-1}(\gamma+1)\left|x^\Delta(t)\right| \int_0^1 \left|\mu(t)hx^\Delta(t)\right|^\gamma dh$$

$$= 2^{\gamma-1}(\gamma+1)\left|x^\Delta(t)\right| |x(t)|^\gamma + 2^{\gamma-1}\mu(t)\left|x^\Delta(t)\right|^{\gamma+1}. \quad (4.3.18)$$

Substituting (4.3.18) into (4.3.6), we have that

$$\int_\alpha^\beta r(t)\left|x^\Delta(t)\right|^{\gamma+1} dt \leq \int_\alpha^\beta |p(t)|\left|x^\sigma(t)\right|\left|x^\Delta(t)\right|^\gamma \Delta t$$

$$+ 2^{\gamma-1}(\gamma+1)\int_\alpha^\beta |Q(t)|\left|x^\Delta(t)\right| |x(t)|^\gamma \Delta t$$

$$+ 2^{\gamma-1}\int_\alpha^\beta \mu(t)|Q(t)|\left|x^\Delta(t)\right|^{\gamma+1} \Delta t. \quad (4.3.19)$$

Using the inequality

$$\int_\alpha^\beta |p(t)|\left|x^\sigma(t)\right|\left|x^\Delta(t)\right|^\gamma \Delta t \leq \int_\alpha^\beta |p(t)||x(t)|\left|x^\Delta(t)\right|^\gamma \Delta t$$

$$+ \int_\alpha^\beta \mu(t)|p(t)|\left|x^\Delta(t)\right|^{\gamma+1} \Delta t,$$

we have from (4.3.19) that

$$\int_\alpha^\beta r(t)\left|x^\Delta(t)\right|^{\gamma+1} dt \;\le\; \int_\alpha^\beta |p(t)|\,|x(t)|\left|x^\Delta(t)\right|^\gamma \Delta t$$

$$+2^{\gamma-1}(\gamma+1)\int_\alpha^\beta |Q(t)|\left|x^\Delta(t)\right||x(t)|^\gamma \Delta t$$

$$+\int_\alpha^\beta \mu(t)(|p(t)|+2^{\gamma-1}|Q(t)|\left|x^\Delta(t)\right|^{\gamma+1}\Delta t$$

$$(4.3.20)$$

We now apply Opial inequalities to obtain results when the condition $\mu(t)$ $|p(t)| \le r(t)/c$ is replaced by the new condition $\mu(t)(|p(t)| + 2^{\gamma-1}|Q(t)|) \le r(t)/c$.

Now, applying the inequality (3.2.33) on the term

$$\int_\alpha^\beta |Q(t)|\left|x^\Delta(t)\right||x(t)|^\gamma \Delta t,\ \text{with}\ s(t)=|Q(t)|,\ p=\gamma\ \text{and}\ q=1,$$

we have

$$\int_\alpha^\beta |Q(t|)\,|x(t)|^\gamma \left|x^\Delta(t)\right|\Delta t \le K_1^*(\alpha,\beta,\gamma,1)\int_\alpha^\beta r(t)\left|x^\Delta(t)\right|^{\gamma+1}\Delta t,$$

where

$$K_1^*(\alpha,\beta,\gamma,1)=\left(\frac{1}{\gamma+1}\right)^{\frac{1}{\gamma+1}}\left(\int_\alpha^\beta \frac{|Q(t)|^{\frac{\gamma+1}{\gamma}}}{(r(t))^{\frac{1}{\gamma}}}R_\alpha^\gamma(t)\Delta t\right)^{\frac{\gamma}{\gamma+1}}.$$

Using the inequality

$$\int_\alpha^\beta |p(t)|\,|x(t)|\left|x^\Delta(t)\right|^\gamma \Delta t \le G_1(\alpha,\beta,1,\gamma)\int_\alpha^\beta r(t)\left|x^\Delta(t)\right|^{\gamma+1}\Delta t,$$

where

$$G_1(\alpha,\beta,1,\gamma)=\left(\frac{\gamma}{1+\gamma}\right)^{\frac{\gamma}{\gamma+1}}\times\left(\int_\alpha^\beta \frac{|p(t)|^{\gamma+1}}{(r(t))^\gamma}R_\alpha^\gamma(t)\Delta t\right)^{\frac{1}{\gamma+1}}.$$

and proceeding as in the proof of Theorem 4.3.1, we obtain the following result.

Theorem 4.3.2 *Assume that* $\mu(t)(|p(t)|+2^{\gamma-1}|Q(t)|)\le r(t)/c$ *where c is a positive constant such that* $c\ge 1$. *Suppose that* x *is a nontrivial solution of (4.3.1) and* x^Δ *does not change sign in* $(\alpha,\beta)_{\mathbb{T}}$. *If* $x(\alpha)=x^\Delta(\beta)=0$, *then*

$$2^{\gamma-1}(\gamma+1)^{\frac{\gamma}{\gamma+1}}\left(\int_\alpha^\beta \frac{|Q(t)|^{\frac{\gamma+1}{\gamma}}}{r^{\frac{1}{\gamma}}(t)}R_\alpha^\gamma(t)\Delta t\right)^{\frac{\gamma}{\gamma+1}}$$

$$+\left(\frac{\gamma}{1+\gamma}\right)^{\frac{\gamma}{\gamma+1}}\left(\int_\alpha^\beta \frac{|p(t)|^{\gamma+1}}{r^\gamma(t)}R_\alpha^\gamma(t)\Delta t\right)^{\frac{1}{\gamma+1}}\ge 1-\frac{1}{c},$$

where $Q(t) = \int_t^\beta q(s)\Delta s$. If instead $x^\Delta(\alpha) = x(\beta) = 0$, then

$$2^{\gamma-1}(\gamma+1)^{\frac{\gamma}{\gamma+1}}\left(\int_\alpha^\beta \frac{|Q(t)|^{\frac{\gamma+1}{\gamma}}}{r^{\frac{1}{\gamma}}(t)}R_\beta^\gamma(t)\Delta t\right)^{\frac{\gamma}{\gamma+1}}$$

$$+\left(\frac{\gamma}{1+\gamma}\right)^{\frac{\gamma}{\gamma+1}}\left(\int_\alpha^\beta \frac{|p(t)|^{\gamma+1}}{r^\gamma(t)}R_\beta^\gamma(t)\Delta t\right)^{\frac{1}{\gamma+1}} \geq 1 - \frac{1}{c},$$

where $Q(t) = \int_\alpha^t q(s)\Delta s$.

Remark 4.3.2 *Note that when $\mathbb{T} = \mathbb{R}$ the condition $\mu(t)(|p(t)| + 2^{\gamma-1}$ $|Q(t)|) \leq r(t)/c$ is removed since $\mu(t) = 0$.*

Next we apply Theorems 3.2.10 and 3.3.3 to determine a lower bound for the distance between consecutive generalized zeros of solutions of (4.3.1). In the following, we assume that $Q^\Delta(t) = q(t)$ and assume that there exists a unique $h \in (\alpha, \beta)_\mathbb{T}$, such that

$$R(h) := R_\alpha(h) = R_\beta(h). \tag{4.3.21}$$

Note that the best choice of h when $r(t) = 1$ is $h = (\beta + \alpha)/2$. In the following, we assume that

$$K^h(\alpha, \beta, \gamma, 1) = K_h(\alpha, \beta, \gamma, 1) < \infty, \tag{4.3.22}$$

where

$$K^h(\alpha, \beta, \gamma, 1) = \frac{2^{3\gamma-2}}{(\gamma+1)^{\frac{1}{\gamma+1}}}\left(\int_\alpha^\beta \frac{|Q(t)|^{\frac{\gamma+1}{\gamma}}}{r^{\frac{1}{\gamma}}(t)}R_\alpha^\gamma(h)\Delta t\right)^{\frac{\gamma}{\gamma+1}} + 2^{2\gamma-2}\Lambda,$$

$$K_h(\alpha, \beta, \gamma, 1) = \frac{2^{3\gamma-2}}{(\gamma+1)^{\frac{1}{\gamma+1}}}\left(\int_\alpha^\beta \frac{|Q(t)|^{\frac{\gamma+1}{\gamma}}}{r^{\frac{1}{\gamma}}(t)}R_\beta^\gamma(h)\Delta t\right)^{\frac{\gamma}{\gamma+1}} + 2^{2\gamma-2}\Lambda,$$

$$\Lambda := \sup_{\alpha \leq t \leq \beta} \mu^\gamma(t)\frac{|Q(t)|}{r(t)}, \quad \text{where } Q^\Delta(t) = q(t),$$

and

$$G^h(\alpha, \beta, 1, \gamma) = G_h(\alpha, \beta, 1, \gamma) < \infty, \tag{4.3.23}$$

where

$$G^h(\alpha, \beta, 1, \gamma) = \left(\frac{\gamma}{1+\gamma}\right)^{\frac{\gamma}{\gamma+1}}\left(\int_\alpha^\beta \frac{|p(t)|^{\gamma+1}}{r^\gamma(t)}R_\alpha^\gamma(h)\Delta t\right)^{\frac{1}{\gamma+1}},$$

$$G_h(\alpha, \beta, 1, \gamma) = \left(\frac{\gamma}{1+\gamma}\right)^{\frac{\gamma}{\gamma+1}}\left(\int_\alpha^\beta \frac{|p(t)|^{\gamma+1}}{r^\gamma(t)}R_\beta^\gamma(h)\Delta t\right)^{\frac{1}{\gamma+1}}.$$

Now, we assume that $K(\gamma, 1)$ is the solution of the equation $K(\gamma, 1) = K^h(\alpha, \beta, \gamma, 1) = K_h(\alpha, \beta, \gamma, 1)$ and given by

$$K(\gamma, 1) = \frac{2^{3\gamma - 2}}{(\gamma + 1)^{\frac{1}{\gamma+1}}} \left(\int_\alpha^\beta \frac{|Q(t)|^{\frac{\gamma+1}{\gamma}}}{r^{\frac{1}{\gamma}}(t)} R^\gamma(h) \Delta t \right)^{\frac{\gamma}{\gamma+1}} + 2^{2\gamma - 2} \Lambda, \quad (4.3.24)$$

and similarly $G(1, \gamma)$ is given by

$$G(1, \gamma) = \left(\frac{\gamma}{1 + \gamma} \right)^{\frac{\gamma}{\gamma+1}} \left(\int_\alpha^\beta \frac{|p(t)|^{\gamma+1}}{r^\gamma(t)} R^\gamma(h) \Delta t \right)^{\frac{1}{\gamma+1}}. \quad (4.3.25)$$

Theorem 4.3.3 *Assume that $Q^\Delta(t) = q(t)$ and suppose x is a nontrivial solution of (4.3.1). If $x(\alpha) = x(\beta) = 0$, then*

$$K(\gamma, 1) + G(1, \gamma) \geq 1 - \frac{1}{c}, \quad (4.3.26)$$

where $K(\alpha, \beta)$ and $K(\alpha, \beta)$ are defined as in (4.3.24) and (4.3.25).

Proof. We multiply (4.3.1) by $x^\sigma(t)$ and proceed as in Theorem 4.3.1 to obtain

$$\int_\alpha^\beta r(t) \left(x^\Delta(t) \right)^{\gamma+1} \Delta t = \int_\alpha^\beta p(t) x^\sigma(t) \left(x^\Delta(t) \right)^\gamma \Delta t + \int_\alpha^\beta Q^\Delta(t) \left(x^\sigma(t) \right)^{\gamma+1} \Delta t.$$

Integrating by parts the right-hand side, we see that

$$\begin{aligned}
\int_\alpha^\beta r(t) \left(x^\Delta(t) \right)^{\gamma+1} \Delta t &= \int_\alpha^\beta p(t) x^\sigma(t) \left(x^\Delta(t) \right)^\gamma \Delta t \\
&\quad + Q(t)(x(t))^{\gamma+1} \big|_\alpha^\beta - \int_\alpha^\beta Q(t) \left(x^{\gamma+1}(t) \right)^\Delta \Delta t. \quad (4.3.27)
\end{aligned}$$

Using $x(\alpha) = 0 = x(\beta)$ we obtain

$$\int_\alpha^\beta r(t) \left| x^\Delta(t) \right|^{\gamma+1} dt \leq \int_\alpha^\beta |p(t)| \left| x^\sigma(t) \right| \left| x^\Delta(t) \right|^\gamma \Delta t + \int_\alpha^\beta |Q(t)| \left| \left(x^{\gamma+1}(t) \right)^\Delta \right| dt.$$

We proceed as in the proof of Theorem 4.3.1 to get

$$\int_\alpha^\beta |Q(t)| \left| \left(x^{\gamma+1}(t) \right)^\Delta \right| \Delta t \leq 2^{\gamma-1} \int_\alpha^\beta |Q(t)| \, |x(t) + x^\sigma(t)|^\gamma \left| x^\Delta(t) \right| \Delta t.$$

Applying the inequality (3.3.15) with $s(t) = |Q(t)|$, $p = \gamma$ and $q = 1$, we have

$$\int_\alpha^\beta |Q(t)| \, |x^{\gamma+1}(t)|^\Delta \, dt \leq 2^{\gamma-1} K(\gamma, 1) \int_\alpha^\beta r(t) \left| x^\Delta(t) \right|^{\gamma+1} \Delta t.$$

Also, we obtain

$$\int_\alpha^\beta |p(t)|\,|x^\sigma(t)|\,\left|x^\Delta(t)\right|^\gamma \Delta t$$

$$\leq \; G(1,\gamma) \int_\alpha^\beta r(t)\left|x^\Delta(t)\right|^{\gamma+1} \Delta t + \frac{1}{c}\int_\alpha^\beta r(t)\left|x^\Delta(t)\right|^{\gamma+1} \Delta t.$$

The rest of the proof is similar to that in the proof of Theorem 4.3.1. ∎

4.4 Hamiltonian Systems

In this section we consider a linear matrix Hamiltonian dynamic system on time scales of the form

$$x^\Delta(t) = A(t)x^\sigma + B(t)u, \qquad u^\Delta(t) = -C(t)x^\sigma - A^*(t)u, \tag{4.4.1}$$

where A, B, and C are rd-continuous $n \times n$-matrix-valued functions on $\mathbb{T}$ such that $I - \mu(t)A(t)$ is invertible and $B(t)$ and $C(t)$ are positive semidefinite for all $t \in \mathbb{T}$. A corresponding quadratic functional is given by

$$\mathcal{F}(x,u) = \int_a^b \{u^*Bu - (x^\sigma)^*Cx^\sigma\}(t)\Delta t.$$

A pair (x,u) is called admissible if it satisfies the equation of motion

$$x^\Delta = A(t)x^\sigma + B(t)u.$$

Lemma 4.4.1 *If (x,u) solves (4.4.1) and if (y,v) is admissible, then*

$$\begin{aligned}
\mathcal{F}(y,v) - \mathcal{F}(x,u) &= \mathcal{F}(y-x, v-u) \\
&\quad + 2\,\mathrm{Re}\left[(y-x)^*(b)u(b) - (y-x)^*(a)u(a)\right].
\end{aligned}$$

Proof. Under the above assumption

$$\mathcal{F}(y,v) - \mathcal{F}(x,u) - \mathcal{F}(y-x, v-u)$$

$$= \int_a^b \{v^*Bv - (y^\sigma)^*Cy^\sigma - u^*Bu + (x^\sigma)^*Cx^\sigma$$

$$\quad - [(v-u)^*B(v-u) - (y^\sigma - x^\sigma)^*C(y^\sigma - x^\sigma)]\}(t)\Delta t$$

$$= \int_a^b \{-2u^*Bu + v^*Bu + u^*Bv$$

$$\quad + 2(x^\sigma)^*Cx^\sigma - (y^\sigma)^*Cx^\sigma - (x^\sigma)^*Cy^\sigma\}(t)\Delta(t)$$

$$= \int_a^b \{-2u^*Bu + 2\,\mathrm{Re}\,[u^*Bv] + 2(x^\sigma)^*Cx^\sigma - 2\,\mathrm{Re}\,[(y^\sigma)^*Cx^\sigma]\}(t)\Delta(t)$$

$$= \quad 2\operatorname{Re}\left(\int_a^b \{u^*(Bv - Bu) + [(x^\sigma)^* - (y^\sigma)^*]\, Cx^*\}\, (t)\Delta(t)\right)$$

$$= \quad 2\operatorname{Re}\left(\int_a^b \{u^*\, (y^\Delta - Ay^\sigma - x^\Delta + Ax^\sigma)\right.$$

$$\left. + [(x^\sigma)^* - (y^\sigma)^*] - u^\Delta - A^* u](t)\Delta(t)\right)$$

$$= \quad 2\operatorname{Re}\left(\int_a^b \{u^*(y^\Delta - x^\Delta) + (y^\sigma - x^\sigma)^*\, u^\Delta\right.$$

$$\left. + 2i\operatorname{Im}[u^* Ax^\sigma + (y^\sigma)^* A^* u]\, (t)\Delta(t)\right)$$

$$= \quad 2\operatorname{Re}\left(\int_a^b \{u^*(y^\Delta - x^\Delta) + (y^\sigma - x^\sigma)^* u^\Delta\}\, (t)\Delta t\right)$$

$$= \quad 2\operatorname{Re}\left(\int_a^b \{u^*(y^\Delta - x^\Delta) + (u^\Delta)^*(y^\sigma - x^\sigma)\}\, (t)\Delta t\right)$$

$$= \quad 2\operatorname{Re}\left(\int_a^b \{[u^*(y - x)]^\Delta\}\, (t)\Delta t\right)$$

$$= \quad 2\operatorname{Re}\{u^*(b)[y(b) - x(b)] - u^*(a)[y(a) - x(a)]\}.$$

$$= \quad 2\operatorname{Re}\{[y - x]^*(b)u(b) - [y - x]^*(a)u(a)\},$$

and we are finished. ∎

For the remainder of this section we denote by $W(., r)$ the unique solution of the initial value problem

$$W^\Delta = -A^*(t)W, \qquad W(r) = I, \tag{4.4.2}$$

where $r \in [a, b]$ is given. We also write

$$F(s, r) = \int_r^s W^*(t, r)B(t)W(t, r)\Delta t. \tag{4.4.3}$$

Observe that $W(t, r) \equiv I$ provided $A(t) \equiv 0$.

Lemma 4.4.2 *Let W and F be defined as in (4.4.2) and (4.4.3). If (y, v) is admissible and if $r, s \in \mathbb{T}$ with $a \le r < s \le b$ such that $F(s, r)$ is invertible, then*

$$\int_r^s (v^* Bv)(t)\Delta t \ge [W^*(s, r)y(s) - y(r)]^* F^{-1}(s, r)\, [W^*(s, r)y(s) - y(r)].$$

Proof. Let

$$x(t) = W^{*-1}(t,r)\left\{y(r) + F(t,r)F^{-1}(s,r)\left[W^*(s,r)y(s) - y(r)\right]\right\}$$

and

$$u(t) = W(t,r)F^{-1}(s,r)[W^*(s,r)y(s) - y(r)].$$

Now

$$\begin{aligned}
W(t,r)W^{-1}(\sigma(t),r) &= [W(\sigma(t),r) - \mu(t)W^\Delta(t,r)W^{-1}(\sigma(t),r) \\
&= I + \mu(t)A^*(t)W(t,r)W^{-1}(\sigma(t),r),
\end{aligned}$$

and therefore $[I - \mu(t)A^*(t)]W(t,r)W^{-1}(\sigma(t),r) = I$, so that

$$[I - \mu(t)A(t)]x^\Delta(t) = A(t)x(t) + B(t)u(t),$$

and hence

$$\begin{aligned}
x^\Delta(t) &= A(t)x(t) + \mu(t)A(t)x^2(t) + B(t)u(t) \\
&= A(t)x^\sigma(t) + B(t)u(t).
\end{aligned}$$

Thus (x,u) solves the Hamiltonian system (4.4.1) with $C = 0$ and, we may apply Lemma 4.4.1 to $\mathcal{F}_0$ defined by

$$\mathcal{F}_0(x,u) = \int_r^s (u^*Bu)(t)\Delta t,$$

to obtain

$$\begin{aligned}
\mathcal{F}_0(y,v) &= \mathcal{F}_0(x,u) + \mathcal{F}_0(y - x, v - u) \\
&\quad + 2\operatorname{Re}\left\{u^*(s)y(s) - x(s) - u^*(r)[y(r) - x(r)]\right\} \\
&= \mathcal{F}_0(x,u) + \mathcal{F}_0(y - x, v - u) \geq \mathcal{F}_0(x,u) = \int_r^s (u^*Bu)(t)\Delta t \\
&= [W^*(s,r)y(s) - y(r)]^*F^{-1}(r,s)[W^*(s,r)y(s) - y(r)].
\end{aligned}$$

which shows our claim. ∎

Remark 4.4.1 *The assumption in Lemma 4.4.2 that $F(s,r)$ is invertible if $r < s$ can be dropped if B is positive definite rather than positive semidefinite.*

We now may use Lemma 4.4.2 to derive a Lyapunov inequality for Hamiltonian systems.

Theorem 4.4.1 *Assume (4.4.1) has a solution (x,u) such that x is nontrivial and satisfies $x(a) = x(b) = 0$. With W and F introduced in (4.4.2)* and (4.4.3), suppose that $F(b, c)$ and $F(c, a)$ are invertible, where $\|x(c)\| = \max_{t \in [a,b] \cap \mathbb{T}} \|x(t)\|$. Let λ be the biggest eigenvalue of

$$F = \int_a^b W^*(t, c) B(t) W(t, c) \Delta t,$$

and let $v(t)$ be the biggest eigenvalue of $C(t)$. Then the Lyapunov inequality

$$\int_a^b v(t) \Delta t \geq \frac{4}{\lambda},$$

holds.

Proof. Suppose we are given a solution (x, u) of (4.4.1) such that $x(a) = x(b) = 0$. Lemma 4.4.1 then yields (using $y = v = 0$) that

$$\mathcal{F}(x, u) = \int_a^b \{u^* B u - (x^\sigma)^* C x^\sigma\} (t) \Delta t = 0.$$

Apply Lemma 4.4.2 twice (once with $r = a$ and $s = c$ and a second time with $r = c$ and $s = b$) to obtain

$$\int_a^b [(x^\sigma)^* C x^\sigma](t) \Delta t$$

$$= \int_a^b (u^* B u)(t) \Delta t = \int_a^c (u^* B u)(t) \Delta t + \int_a^b (u^* B u)(t) \Delta t$$

$$\geq x^*(c) W(c, a) F^{-1}(c, a) W^*(c, a) x(c) + x^*(c) F^{-1}(b, c) x(c)$$

$$= x^*(c) [F^{-1}(b, c) - F^{-1}(a, c)] x(c) \geq 4 x^*(c) F^{-1} x(c);$$

here we have used the relation $W(t, r) W(r, s) = W(t, s)$ and the inequality (see [34, Theorem 9 (i)]) and [120])

$$M^{-1} + N^{-1} \geq 4(M + N)^{-1},$$

Now, by applying the Rayleigh–Ritz Theorem (see [85, page 176]), we conclude

$$\int_a^b v(t) \Delta t \geq \int_a^b v(t) \frac{\|x^\sigma(t)\|^2}{\|x(c)\|^2} \Delta t$$

$$= \frac{1}{\|x(c)\|^2} \int_a^b v(t) (x^\sigma(t))^* x^\sigma(t) \Delta t \geq \frac{1}{\|x(c)\|^2} \int_a^b (x^\sigma(t))^* C(t) x^\sigma(t) \Delta t$$

$$\geq \frac{1}{\|x(c)\|^2} 4 x^*(c) F^{-1} x(c) \geq 4 \min_{x \neq 0} \frac{x^* F^{-1} x}{x^* x} = \frac{4}{\lambda},$$

which is the desired inequality. The proof is complete. ∎

Remark 4.4.2 *If $A \equiv 0$, then $W \equiv I$ and $F = \int_a^b B(t)\Delta t$. If, in addition*

$B \equiv 1$, then $F = b - a$. Note the Lyapunov inequality $\int_a^b v(t)\Delta t \geq (4/\lambda)$

reduces to $\int_a^b p(t)\Delta t \geq (4/b - a)$ for the scalar case.

We conclude with a result concerning the so-called right-focal boundary condition, i.e., $x(a) = u(b) = 0$.

Theorem 4.4.2 *Assume (4.4.1) has a solution (x, u) with x nontrivial and $x(a) = u(b) = 0$. With the notation as in Theorem 4.4.1, the Lyapunov inequality*

$$\int_a^b v(t)\Delta t \geq \frac{1}{\lambda},$$

holds.

Proof. Suppose (x, u) is a solution of (4.4.1) such that $x(a) = u(b) = 0$ with $a < b$. Choose the point c in $(a, b]$ where $\|x(t)\|$ is maximal. Applying Lemma 4.4.1 and we see

$$\int_a^b [(x^\sigma)^* C x^\sigma] (t)\Delta t = \int_a^b (u^* B u)(t)\Delta t \geq \int_a^b (u^* B u)(t)\Delta t.$$

Using Lemma 4.4.2 with $r = a$ and $s = c$, we get

$$
\begin{aligned}
\int_a^b (u^* B u)(t)\Delta t \;\geq\;& [W^*(c, a)x(c) - x(a)]^* F^{-1}(c, a) [W^*(c, a)x(c) - x(a)] \\
=\;& x^*(c)W(c, a)F^{-1}(c, a)W^*(c, a)x(c) \\
=\;& -x^*(c)F^{-1}(a, c)x(c) \\
=\;& x^*(c) \left(\int_a^b W^*(t, c)B(t)W(t, c)\Delta t \right)^{-1} x(c) \\
\geq\;& x^*(c) \left(\int_a^b W^*(t, c)B(t)W(t, c)\Delta t \right)^{-1} x(c) \\
=\;& x^*(c)F^{-1}x(c).
\end{aligned}
$$

Hence,

$$\int_a^b \left[(x^\sigma)^* C x^\sigma \right](t) \Delta t \geq x^*(c) F^{-1} x(c),$$

and the same arguments as in the proof of Theorem 4.4.1 completes the proof. ∎

Chapter 5

Halanay Inequalities

If people do not believe that mathematics is simple, it is only because they do not realize how complicated life is.

Von Neumann (1903–1957).

In 1966 Halanay [71] studied the stability of the delay differential equation

$$x'(t) = -px(t) + qx(t - \tau), \quad \tau > 0,$$

and proved that if

$$f'(t) \leq -\alpha f(t) + \beta \sup_{s \in [t-\tau, t]} f(s) \quad \text{for} \ t \geq t_0$$

and $\alpha > \beta > 0$, then there exist $\gamma > 0$ and $K > 0$ such that

$$f(t) \leq K e^{-\gamma(t-t_0)} \quad \text{for} \ t \geq t_0.$$

In this chapter we discuss Halanay type inequalities on time scales and we investigate the global stability of delay dynamic equations on time scales. In particular we employ the shift operators $\delta_\pm$ to construct delay dynamic inequalities on time scales and we derive Halanay type inequalities for dynamic equations on time scales.

The chapter is organized as follows. In Sect. 5.1 we give a generalized version of shift operators and in Sect. 5.2 we define the delay function by means of shift operators on time scales. In Sect. 5.3 we prove Halanay type inequalities on time scales and in Sect. 5.4 we establish some sufficient conditions guaranteeing global stability of nonlinear dynamic equations.

© Springer International Publishing Switzerland 2014
R. Agarwal et al., *Dynamic Inequalities On Time Scales*,
DOI 10.1007/978-3-319-11002-8_5

5.1   Shift Operators

In this section we give a generalized version of shift operators (see [5, 6]).

Definition 5.1.1 (Shift Operators) *Let* $\mathbb{T}^*$ *be a nonempty subset of the time scale* $\mathbb{T}$ *including a fixed number* $t_0 \in \mathbb{T}^*$ *such that there exist operators* $\delta_\pm : [t_0, \infty)_\mathbb{T} \times \mathbb{T}^* \to \mathbb{T}^*$ *satisfying the following properties:*

P.1 The functions $\delta_\pm$ are strictly increasing with respect to their second arguments, i.e., if

$$(T_0, t), (T_0, u) \in \mathcal{D}_\pm := \{(s, t) \in [t_0, \infty)_\mathbb{T} \times \mathbb{T}^* : \delta_\pm(s, t) \in \mathbb{T}^*\},$$

then

$$T_0 \le t < u \text{ implies } \delta_\pm(T_0, t) < \delta_\pm(T_0, u),$$

P.2 If $(T_1, u), (T_2, u) \in \mathcal{D}_-$ with $T_1 < T_2$, then

$$\delta_-(T_1, u) > \delta_-(T_2, u),$$

and if $(T_1, u), (T_2, u) \in \mathcal{D}_+$ with $T_1 < T_2$, then

$$\delta_+(T_1, u) < \delta_+(T_2, u),$$

P.3 If $t \in [t_0, \infty)_\mathbb{T}$, then $(t, t_0) \in \mathcal{D}_+$ and $\delta_+(t, t_0) = t$. Moreover, if $t \in \mathbb{T}^*$, then $(t_0, t) \in \mathcal{D}_+$ and $\delta_+(t_0, t) = t$ holds,

P.4 If $(s, t) \in \mathcal{D}_\pm$, then $(s, \delta_\pm(s, t)) \in \mathcal{D}_\mp$ and $\delta_\mp(s, \delta_\pm(s, t)) = t$,

P.5 If $(s, t) \in \mathcal{D}_\pm$ and $(u, \delta_\pm(s, t)) \in \mathcal{D}_\mp$, then $(s, \delta_\mp(u, t)) \in \mathcal{D}_\pm$ and

$$\delta_\mp(u, \delta_\pm(s, t)) = \delta_\pm(s, \delta_\mp(u, t)).$$

Then the operators δ_- and δ_+ associated with $t_0 \in \mathbb{T}^*$ (called the initial point) are said to be backward and forward shift operators on the set $\mathbb{T}^*$, respectively. The variable $s \in [t_0, \infty)_\mathbb{T}$ in $\delta_\pm(s, t)$ is called the shift size. The values $\delta_+(s, t)$ and $\delta_-(s, t)$ in $\mathbb{T}^*$ indicate s units translation of the term $t \in \mathbb{T}^*$ to the right and left, respectively. The sets $\mathcal{D}_\pm$ are the domains of the shift operators $\delta_\pm$, respectively.

Example 5.1.1 *Let $\mathbb{T} = \mathbb{R}$ and $t_0 = 1$. The operators*

$$\delta_-(s, t) = \begin{cases} t/s & \text{if } t \ge 0 \\ st & \text{if } t < 0 \end{cases}, \quad \text{for } s \in [1, \infty) \tag{5.1.1}$$

and

$$\delta_+(s, t) = \begin{cases} st & \text{if } t \ge 0 \\ t/s & \text{if } t < 0 \end{cases}, \quad \text{for } s \in [1, \infty) \tag{5.1.2}$$

are backward and forward shift operators (on the set $\mathbb{T}^* = \mathbb{R} - \{0\}$) associated with the initial point $t_0 = 1$. In the table below, we present different time scales with their corresponding shift operators.

$\mathbb{T}$	t_0	$\mathbb{T}^*$	$\delta_-(s,t)$	$\delta_+(s,t)$
$\mathbb{R}$	0	$\mathbb{R}$	$t-s$	$t+s$
$\mathbb{Z}$	0	$\mathbb{Z}$	$t-s$	$t+s$
$q^{\mathbb{Z}} \cup \{0\}$	1	$q^{\mathbb{Z}}$	$\frac{t}{s}$	st
$\mathbb{N}^{1/2}$	0	$\mathbb{N}^{1/2}$	$\sqrt{t^2-s^2}$	$\sqrt{t^2+s^2}$

The proof of the next lemma is a direct consequence of Definition 5.1.1.

Lemma 5.1.1 *Let δ_- and δ_+ be the shift operators associated with the initial point t_0. We have*

i. *$\delta_-(t,t) = t_0$ for all $t \in [t_0,\infty)_{\mathbb{T}}$,*

ii. *$\delta_-(t_0,t) = t$ for all $t \in \mathbb{T}^*$,*

iii. *If $(s,t) \in \mathcal{D}_+$, then $\delta_+(s,t) = u$ implies $\delta_-(s,u) = t$. Conversely, if $(s,u) \in \mathcal{D}_-$, then $\delta_-(s,u) = t$ implies $\delta_+(s,t) = u$.*

iv. *$\delta_+(t,\delta_-(s,t_0)) = \delta_-(s,t)$ for all $(s,t) \in \mathcal{D}(\delta_+)$ with $t \geq t_0$,*

v. *$\delta_+(u,t) = \delta_+(t,u)$ for all $(u,t) \in ([t_0,\infty)_{\mathbb{T}} \times [t_0,\infty)_{\mathbb{T}}) \cap \mathcal{D}_+$,*

vi. *$\delta_+(s,t) \in [t_0,\infty)_{\mathbb{T}}$ for all $(s,t) \in \mathcal{D}_+$ with $t \geq t_0$,*

vii. *$\delta_-(s,t) \in [t_0,\infty)_{\mathbb{T}}$ for all $(s,t) \in ([t_0,\infty)_{\mathbb{T}} \times [s,\infty)_{\mathbb{T}}) \cap \mathcal{D}_-$,*

viii. *If $\delta_+(s,.)$ is Δ-differentiable in its second variable, then $\delta_+^{\Delta t}(s,.) > 0$,*

ix. *$\delta_+(\delta_-(u,s),\delta_-(s,v)) = \delta_-(u,v)$ for all $(s,v) \in ([t_0,\infty)_{\mathbb{T}} \times [s,\infty)_{\mathbb{T}}) \cap \mathcal{D}_-$ and $(u,s) \in ([t_0,\infty)_{\mathbb{T}} \times [u,\infty)_{\mathbb{T}}) \cap \mathcal{D}_-$,*

x. *If $(s,t) \in \mathcal{D}_-$ and $\delta_-(s,t) = t_0$, then $s = t$.*

5.2 Delay Functions Generated by Shift Operators

In this section we define the delay function by means of shift operators on time scales. Delay functions generated by shift operators were first introduced in [5] to construct delay equations on time scales.

Definition 5.2.1 (Delay Functions) *Let $\mathbb{T}$ be a time scale that is unbounded above and $\mathbb{T}^*$ an unbounded subset of $\mathbb{T}$ including a fixed number $t_0 \in \mathbb{T}^*$ such that there exist shift operators $\delta_\pm : [t_0,\infty)_{\mathbb{T}} \times \mathbb{T}^* \to \mathbb{T}^*$ associated with t_0. Suppose that $h \in (t_0,\infty)_{\mathbb{T}}$ is a constant such that $(h,t) \in \mathcal{D}_\pm$ for all $t \in [t_0,\infty)_{\mathbb{T}}$, the function $\delta_-(h,t)$ is differentiable with an rd-continuous derivative $\delta_-^{\Delta t}(h,t)$, and $\delta_-(h,t)$ maps $[t_0,\infty)_{\mathbb{T}}$ onto $[\delta_-(h,t_0),\infty)_{\mathbb{T}}$. Then the function $\delta_-(h,t)$ is called the delay function generated by the shift δ_- on the time scale $\mathbb{T}$.*

It is obvious from P.2 in Definition 5.2.1 and (ii) of Lemma 5.1.1 that

$$\delta_-(h,t) < \delta_-(t_0,t) = t \text{ for all } t \in [t_0,\infty)_{\mathbb{T}}. \tag{5.2.1}$$

Notice that $\delta_-(h,.)$ is strictly increasing and it is invertible. Hence, by P.4 and P.5, we see that

$$\delta_-^{-1}(h,t) = \delta_+(h,t).$$

Hereafter, we shall suppose that $\mathbb{T}$ is a time scale with the delay function $\delta_-(h,.) : [t_0,\infty)_{\mathbb{T}} \to [\delta_-(h,t_0),\infty)_{\mathbb{T}}$, where $t_0 \in \mathbb{T}$ is fixed. Denote by $\mathbb{T}_1$ and $\mathbb{T}_2$ the sets

$$\mathbb{T}_1 = [t_0,\infty)_{\mathbb{T}} \text{ and } \mathbb{T}_2 = \delta_-(h,\mathbb{T}_1). \tag{5.2.2}$$

Evidently, $\mathbb{T}_1$ is closed in $\mathbb{R}$. By definition we have $\mathbb{T}_2 = [\delta_-(h,t_0),\infty)_{\mathbb{T}}$. Hence, $\mathbb{T}_1$ and $\mathbb{T}_2$ are both time scales. Let σ_1 and σ_2 denote the forward jumps on the time scales $\mathbb{T}_1$ and $\mathbb{T}_2$, respectively. By (5.2.1)–(5.2.2)

$$\mathbb{T}_1 \subset \mathbb{T}_2 \subset \mathbb{T}.$$

Thus,

$$\sigma(t) = \sigma_2(t) \text{ for all } t \in \mathbb{T}_2$$

and

$$\sigma(t) = \sigma_1(t) = \sigma_2(t) \text{ for all } t \in \mathbb{T}_1.$$

That is, σ_1 and σ_2 are the restrictions of forward jump operator $\sigma : \mathbb{T} \to \mathbb{T}$ to the time scales $\mathbb{T}_1$ and $\mathbb{T}_2$, respectively, i.e.,

$$\sigma_1 = \sigma|_{\mathbb{T}_1} \text{ and } \sigma_2 = \sigma|_{\mathbb{T}_2}.$$

Lemma 5.2.1 *The delay function $\delta_-(h,t)$ preserves the structure of the points in $\mathbb{T}_1$. That is,*

$$\sigma_1(\hat{t}) = \hat{t} \text{ implies } \sigma_2(\delta_-(h,\hat{t})) = \delta_-(h,\hat{t}).$$

$$\sigma_1(\hat{t}) > \hat{t} \text{ implies } \sigma_2(\delta_-(h,\hat{t}) > \delta_-(h,\hat{t}).$$

Using the preceding lemma and applying the fact that $\sigma_2(u) = \sigma(u)$ for all $u \in \mathbb{T}_2$ we arrive at the following result.

Corollary 5.2.1 *We have*

$$\delta_-(h,\sigma_1(t)) = \sigma_2(\delta_-(h,t)) \text{ for all } t \in \mathbb{T}_1.$$

Thus,

$$\delta_-(h,\sigma(t)) = \sigma(\delta_-(h,t)) \text{ for all } t \in \mathbb{T}_1. \tag{5.2.3}$$

By (5.2.3) we have

$$\delta_-(h, \sigma(s)) = \sigma(\delta_-(h, s)) \text{ for all } s \in [t_0, \infty)_{\mathbb{T}}.$$

Substituting $s = \delta_+(h, t)$ we obtain

$$\delta_-(h, \sigma(\delta_+(h, t))) = \sigma(\delta_-(h, \delta_+(h, t))) = \sigma(t).$$

This and (iv) of Lemma 5.1.1 imply

$$\sigma(\delta_+(h, t)) = \delta_+(h, \sigma(t)) \text{ for all } t \in [\delta_-(h, t_0), \infty)_{\mathbb{T}}.$$

Example 5.2.1 *In the following, we give some time scales with their shift operators:*

$\mathbb{T}$	h	$\delta_-(h, t)$	$\delta_+(h, t)$
$\mathbb{R}$	$\in \mathbb{R}_+$	$t - h$	$t + h$
$\mathbb{Z}$	$\in \mathbb{Z}_+$	$t - h$	$t + h$
$q^{\mathbb{Z}} \cup \{0\}$	$\in q^{\mathbb{Z}+}$	$\frac{t}{h}$	ht
$\mathbb{N}^{1/2}$	$\in \mathbb{Z}_+$	$\sqrt{t^2 - h^2}$	$\sqrt{t^2 + h^2}$

Example 5.2.2 *There is no delay function $\delta_-(h, .) : [0, \infty)_{\tilde{\mathbb{T}}} \to [\delta_-(h, 0), \infty)_{\mathbb{T}}$ on the time scale $\tilde{\mathbb{T}} = (-\infty, 0] \cup [1, \infty)$.*

Suppose on the contrary that there exists a such delay function on $\tilde{\mathbb{T}}$. Then since 0 is right scattered in $\tilde{\mathbb{T}}_1 := [0, \infty)_{\tilde{\mathbb{T}}}$ the point $\delta_-(h, 0)$ must be right scattered in $\tilde{\mathbb{T}}_2 = [\delta_-(h, 0), \infty)_{\mathbb{T}}$, i.e., $\sigma_2(\delta_-(h, 0)) > \delta_-(h, 0)$. Since $\sigma_2(t) = \sigma(t)$ for all $t \in [\delta_-(h, 0), 0)_{\mathbb{T}}$, we have

$$\sigma(\delta_-(h, 0)) = \sigma_2(\delta_-(h, 0)) > \delta_-(h, 0).$$

That is, $\delta_-(h, 0)$ must be right scattered in $\tilde{\mathbb{T}}$. However, in $\tilde{\mathbb{T}}$ we have $\delta_-(h, 0) < 0$, that is, $\delta_-(h, 0)$ is right dense. This leads to a contradiction.

5.3 Halanay Inequality on Time Scales

Let $\mathbb{T}$ be a time scale that is unbounded above and $t_0 \in \mathbb{T}^*$ an element such that there exist shift operators $\delta_\pm : [t_0, \infty) \times \mathbb{T}^* \to \mathbb{T}^*$ associated with t_0. Suppose that $h_1, h_2, \ldots, h_r \in (t_0, \infty)_{\mathbb{T}}$ are constants with

$$t_0 = h_0 < h_1 < h_2 < \ldots < h_r$$

and that there exist delay functions $\delta_-(h_i, t)$, $i = 1, 2, \ldots, r$, on $\mathbb{T}$. We define the lower Δ-derivative $\varphi^{\Delta-}(t)$ of a function $\varphi : \mathbb{T} \to \mathbb{R}$ on time scales as follows:

$$\varphi^{\Delta-}(t) = \liminf_{s \to t^-} \frac{\varphi(s) - \varphi(\sigma(t))}{s - \sigma(t)}. \tag{5.3.1}$$

Notice that

$$\varphi^{\Delta-}(t) = \varphi^{\Delta}(t)$$

provided that φ is Δ-differentiable at $t \in \mathbb{T}^{\kappa}$. Let $f(t, u, v)$ be a continuous function for all (u, v) and $t \in [t_0, \alpha)_{\mathbb{T}}$. Hereafter, we suppose that f is monotone increasing with respect to v and nondecreasing with respect to u.

Proposition 5.3.1 *Let $g(u_1, u_2, \ldots, u_r)$ be a continuous function that is monotone increasing with respect to each of its arguments. If φ and ψ are continuous functions satisfying*

$$\varphi^{\Delta-}(t) < f\left(t, \varphi(t), g\left(\varphi(\delta_-(h_1, t)), \varphi(\delta_-(h_2, t)), \ldots, \varphi(\delta_-(h_r, t))\right)\right),$$

$$\psi^{\Delta-}(t) \geq f\left(t, \psi(t), g\left(\psi(\delta_-(h_1, t)), \psi(\delta_-(h_2, t)), \ldots, \psi(\delta_-(h_r, t))\right)\right),$$

for all $t \in [t_0, \alpha)_{\mathbb{T}}$ and $\varphi(s) < \psi(s)$ for all $s \in [\delta_-(h_r, t_0), t_0]_{\mathbb{T}}$, then

$$\varphi(t) < \psi(t) \text{ for all } t \in (t_0, \alpha)_{\mathbb{T}}, \tag{5.3.2}$$

where $\alpha \in (t_0, \infty)_{\mathbb{T}}$.

Proof. Suppose that (5.3.2) does not hold for some $t \in (t_0, \alpha)_{\mathbb{T}}$. Then the set

$$M := \{t \in (t_0, \alpha)_{\mathbb{T}} : \varphi(t) \geq \psi(t)\}.$$

is nonempty. Since M is bounded below we can let $\xi := \inf M$. If ξ is left scattered (i.e., $\sigma(\rho(\xi)) = \xi$), then it follows from the definition of ξ that

$$\varphi(\rho(\xi)) < \psi(\rho(\xi)),$$

$$\varphi(\xi) \geq \psi(\xi).$$

Since $\rho(\xi)$ is right scattered, the function φ is Δ-differentiable at $\rho(\xi)$ (see [51, Theorem 1.16, (ii)]), and hence, $\varphi^{\Delta-}(\rho(\xi)) = \varphi^{\Delta}(\rho(\xi))$. Similarly we obtain $\psi^{\Delta-}(\rho(\xi)) = \psi^{\Delta}(\rho(\xi))$. Thus,

$$
\begin{aligned}
\varphi(\xi) &= \varphi(\sigma(\rho(\xi))) \\
&= \varphi(\rho(\xi)) + \mu(\rho(\xi))\varphi^{\Delta}(\rho(\xi)) \\
&= \varphi(\rho(\xi)) + \mu(\rho(\xi))\varphi^{\Delta-}(\rho(\xi)) \\
&< \varphi(\rho(\xi)) \\
&\quad + \mu(\rho(\xi))f\left(\rho(\xi), \varphi(\rho(\xi)), g\left(\varphi(\delta_-(h_1, \rho(\xi))), \varphi(\delta_-(h_2, \rho(\xi))), \ldots, \varphi(\delta_-(h_r, \rho(\xi)))\right)\right) \\
&< \psi(\rho(\xi)) \\
&\quad + \mu(\rho(\xi))f\left(\rho(\xi), \psi(\rho(\xi)), g\left(\psi(\delta_-(h_1, \rho(\xi))), \psi(\delta_-(h_2, \rho(\xi))), \ldots, \psi(\delta_-(h_r, \rho(\xi)))\right)\right) \\
&\leq \psi(\rho(\xi)) + \mu(\rho(\xi))\psi^{\Delta-}(\rho(\xi)) \\
&= \psi(\rho(\xi)) + \mu(\rho(\xi))\psi^{\Delta}(\rho(\xi)) \\
&= \psi(\sigma(\rho(\xi))) \\
&= \psi(\xi).
\end{aligned}
$$

This leads to a contradiction. If ξ is left dense, then we have $\xi > t_0$ and

$$\varphi(\xi) = \psi(\xi).$$

Since

$$\delta_-(h_r, \xi) < \xi \text{ for all } i = 1, 2, \ldots, r$$

and

$$\varphi(s) < \psi(s) \text{ for all } s \in [\delta_-(h_r, \xi), \xi)_{\mathbb{T}},$$

we obtain

$$g(\varphi(\delta_-(h_1, \xi)), \ldots, \varphi(\delta_-(h_r, \xi))) \leq g(\psi(\delta_-(h_1, \xi)), \ldots, \psi(\delta_-(h_r, \xi))),$$

and therefore,

$$
\begin{aligned}
\varphi^{\Delta^-}(\xi) &< f\left(\xi, \varphi(\xi), g\left(\varphi(\delta_-(h_1, \xi)), \varphi(\delta_-(h_2, \xi)), \ldots, \varphi(\delta_-(h_r, \xi))\right)\right) \\
&\leq f\left(\xi, \psi(\xi), g\left(\psi(\delta_-(h_1, \xi)), \psi(\delta_-(h_2, \xi)), \ldots, \psi(\delta_-(h_r, \xi))\right)\right) \\
&\leq \psi^{\Delta^-}(\xi).
\end{aligned}
$$

Also since

$$\frac{\varphi(s) - \varphi(\sigma(\xi))}{s - \sigma(\xi)} \geq \frac{\psi(s) - \psi(\sigma(\xi))}{s - \sigma(\xi)}$$

for all $s \in [\delta_-(h_r, \xi), \xi)_{\mathbb{T}}$ we get by (5.3.1) that

$$\varphi^{\Delta^-}(\xi) \geq \psi^{\Delta^-}(\xi).$$

This also leads to a contradiction and so this completes the proof. ∎

Proposition 5.3.2 *If*

$$w^{\Delta}(t) \leq f\left(t, w(t), g\left(w(\delta_-(h_1, t)), w(\delta_-(h_2, t)), \ldots, w(\delta_-(h_r, t))\right)\right)$$

for $t \in [s_0, \delta_+(\alpha, s_0))_{\mathbb{T}}$ and $y(t; s_0, w)$ is a solution of the equation

$$y^{\Delta}(t) = f\left(t, y(t), g\left(y(\delta_-(h_1, t)), y(\delta_-(h_2, t)), \ldots, y(\delta_-(h_r, t))\right)\right),$$

which coincides with w in $[\delta_-(h_r, s_0), s_0]_{\mathbb{T}}$, then, supposing that this solution is defined in $[s_0, \delta_+(\alpha, s_0))_{\mathbb{T}}$, it follows that $w(t) \leq y(t; s_0, w)$ for $t \in [s_0, \delta_+(\alpha, s_0))_{\mathbb{T}}$.

Proof. Let ε_n be a sequence of positive numbers tending monotonically to zero, and y_n be a solution of the equation

$$y^{\Delta}(t) = f\left(t, y(t), g\left(y(\delta_-(h_1, t)), y(\delta_-(h_2, t)), \ldots, y(\delta_-(h_r, t))\right)\right) + \varepsilon_n,$$

which in $[\delta_-(h_r, s_0), s_0]_{\mathbb{T}}$ coincides with $w + \varepsilon_n$. From the preceding proposition, we have

$$y_{n+1}(t) < y_n(t)$$

and

$$\lim_{n \to \infty} y_n(t) = y(t; s_0, \omega)$$

for all $t \in [s_0, \delta_+(\alpha, s_0))_{\mathbb{T}}$. From Proposition 5.3.1 we have $\omega(t) < y_n(t)$ for $t \in [s_0, \delta_+(\alpha, s_0))_{\mathbb{T}}$, and hence, $\omega(t) \leq y(t; s_0, \omega)$. The proof is complete. ∎

Hereafter, we will denote by $\tilde{\mu}$ the function defined by

$$\tilde{\mu}(t) := \sup_{s \in [\delta_-(h_r, t_0), t]_{\mathbb{T}}} \mu(s) \qquad (5.3.3)$$

for $t \in [t_0, \infty)_{\mathbb{T}}$. It is obvious that the sets $\mathbb{R}$, $\mathbb{Z}$, $\overline{q^{\mathbb{Z}}} = \{q^n : n \in \mathbb{Z} \text{ and } q > 1\} \cup \{0\}$, $h\mathbb{Z} = \{hn : n \in \mathbb{Z} \text{ and } h > 0\}$ are examples of time scales on which $\tilde{\mu} = \mu$.

Theorem 5.3.1 *Let x be a function satisfying the inequality*

$$x^{\Delta}(t) \leq -p(t)x(t) + \sum_{i=0}^{r} q_i(t)x^{\ell}(\delta_-(h_i, t)), \quad t \in [t_0, \infty)_{\mathbb{T}}, \qquad (5.3.4)$$

where $\ell \in (0, 1]$ is a constant, p and q_i, $i = 0, 1, \ldots, r$, are continuous and bounded functions satisfying $1 - \tilde{\mu}(t)p(t) \geq 0$, $q_i(t) \geq 0$, $i = 0, 1, \ldots, r-1$, $q_r(t) > 0$ for all $t \in [t_0, \infty)_{\mathbb{T}}$. Suppose that

$$p(t) - \sum_{i=0}^{r} q_i(t) > 0 \quad \text{for all } t \in [t_0, \infty)_{\mathbb{T}}. \qquad (5.3.5)$$

Then there exist a positively regressive function $\lambda : [t_0, \infty)_{\mathbb{T}} \to (-\infty, 0)$ and $K_0 > 1$ such that

$$x(t) \leq K_0 e_{\lambda}(t, t_0) \text{ for } t \in [t_0, \infty)_{\mathbb{T}} \qquad (5.3.6)$$

Proof. Consider the delay dynamic equation

$$y^{\Delta}(t) = -p(t)y(t) + \sum_{i=0}^{r} q_i(t)y^{\ell}(\delta_-(h_i, t)), \quad t \in [t_0, \infty)_{\mathbb{T}}. \qquad (5.3.7)$$

We look for a solution of Eq. (5.3.7) in the form $e_{\lambda}(t, t_0)$, where $\lambda : \mathbb{T} \to (-\infty, 0)$ is positively regressive (i.e., $1 + \mu(t)\lambda(t) > 0$) and rd-continuous. First note that

$$e_{\lambda}^{\Delta}(t, t_0) = \lambda(t)e_{\lambda}(t, t_0).$$

For a given $K > 1$, the function $Ke_{\lambda}(t, t_0)$ is a solution of (5.3.7) if and only if λ is a root of the characteristic polynomial $P(t, \lambda)$ defined by

$$P(t, \lambda) := (\lambda + p(t)) \, e_{\lambda}(t, \delta_-(h_r, t)) e_{\lambda}^{1-\ell}(\delta_-(h_r, t), t_0)$$

$$- K^{\ell-1} \sum_{i=0}^{r} q_i(t)e_{\lambda}^{\ell}(\delta_-(h_i, t), \delta_-(h_r, t)). \qquad (5.3.8)$$

For each fixed $t \in [t_0, \infty)_{\mathbb{T}}$ define the set

$$S(t) := \{k \in (-\infty, 0) : 1 + \tilde{\mu}(t)k > 0\}. \tag{5.3.9}$$

It follows from Lemma 1.1.1 that if k is a scalar in $S(t)$, then $0 < 1 + \tilde{\mu}(t)k \le 1 + \mu(u)k$ for all $u \in [\delta_-(h_r, t_0), t]_{\mathbb{T}}$ and

$$0 < e_k(\tau, s) \le \exp(k(\tau - s)), \tag{5.3.10}$$

for all $\tau \in [\delta_-(h_r, t_0), t]_{\mathbb{T}}$ with $\tau \ge s$. Now for each fixed $t \in [t_0, \infty)_{\mathbb{T}}$ the function $P(t, k)$ is continuous with respect to k in $S(t)$. Since $e_0(t, t_0) = 1$ we have

$$P(t, 0) = p(t) - K^{\ell-1} \sum_{i=0}^{r} q_i(t) > 0. \tag{5.3.11}$$

Let $t \in [t_0, \infty)_{\mathbb{T}}$ be fixed. If the interval $[\delta_-(h_r, t_0), t]_{\mathbb{T}}$ has not any right scattered points, then $\tilde{\mu}(t) = 0$ and $S(t) = (-\infty, 0)$. By (5.3.10), we get

$$\lim_{k \to -\infty} e_k(t, s) = 0,$$

and hence,

$$\lim_{k \to -\infty} P(t, k) = -K^{\ell-1} q_r(t) < 0.$$

If the interval $[\delta_-(h_r, t_0), t]_{\mathbb{T}}$ has some right scattered points (i.e., if $\tilde{\mu}(t) > 0$), then we have $S(t) = (-\frac{1}{\tilde{\mu}(t)}, 0)$. For all $k \in (-\frac{1}{\tilde{\mu}(t)}, 0)$ we have $e_k(t, s) > 0$. Since $1 - \tilde{\mu}(t)p(t) \ge 0$ for all $t \in [t_0, \infty)_{\mathbb{T}}$, we obtain

$$\lim_{k \to -\frac{1}{\tilde{\mu}(t)}^+} P(t, k) = \left(-\frac{1}{\tilde{\mu}(t)} + p(t) \right) \lim_{k \to -\frac{1}{\tilde{\mu}(t)}^+} \left[e_k(t, \delta_-(h_r, t)) e_k^{1-\ell}(\delta_-(h_r, t), t_0) \right]$$

$$- K^{\ell-1} \sum_{i=0}^{r-1} q_i(t) \lim_{k \to -\frac{1}{\tilde{\mu}(t)}^+} e_k^\ell(\delta_-(h_i, t), \delta_-(h_r, t)) - K^{\ell-1} q_r(t)$$

$$< -K^{\ell-1} q_r(t) < 0.$$

Therefore, for each fixed $t \in [t_0, \infty)_{\mathbb{T}}$, we obtain

$$0 > -K^{\ell-1} q_r(t) \ge \begin{cases} \lim\limits_{k \to -\frac{1}{\tilde{\mu}(t)}^+} P(t, k) & \text{if } \tilde{\mu}(t) > 0 \\ \lim\limits_{k \to -\infty} P(t, k) & \text{if } \tilde{\mu}(t) = 0 \end{cases}. \tag{5.3.12}$$

It follows from the continuity of P in k and (5.3.11)–(5.3.12) that for each fixed $t \in [t_0, \infty)_{\mathbb{T}}$, there exists a largest element k_0 of the set $S(t)$ such that

$$P(t, k_0) = 0.$$

Using all these largest elements we can construct a positively regressive function $\lambda : [\delta_-(h_r, t_0), \infty)_{\mathbb{T}} \to (-\infty, 0)$ by

$$\lambda(t) := \max \{k \in S(t) : P(t, k) = 0\} \qquad (5.3.13)$$

so that for a given $K > 1$, $y(t) = Ke_\lambda(t, t_0)$ is a solution to (5.3.7).

If $y(t)$ be a solution of (5.3.7), $x(t)$ satisfies (5.3.4), and $x(t) \le y(t)$ for all $t \in [\delta_-(h_r, t_0), t_0]_{\mathbb{T}}$, then by Proposition 5.3.2 the inequality $x(t) \le y(t)$ holds for all $t \in [t_0, \infty)_{\mathbb{T}}$. For a given $K > 1$, we have

$$\inf_{t \in [\delta_-(h_r, t_0), t_0]_{\mathbb{T}}} Ke_\lambda(t, t_0) = K,$$

hence, by choosing a $K_0 > 1$ with

$$K_0 > \sup_{t \in [\delta_-(h_r, t_0), t_0]_{\mathbb{T}}} x(t),$$

we get

$$x(t) < K_0 e_\lambda(t, t_0) \text{ for all } t \in [\delta_-(h_r, t_0), t_0]_{\mathbb{T}}.$$

It follows from Proposition 5.3.2 that the inequality

$$x(t) \le K_0 e_\lambda(t, t_0)$$

holds for all $t \in [t_0, \infty)_{\mathbb{T}}$. This completes the proof. ∎

Example 5.3.1 *Let* $\mathbb{T} = \mathbb{Z}$, $t_0 = 0$, $\delta_-(h_i, t) = t - h_i$, $h_i \in \mathbb{N}$, $i = 1, 2, \ldots, r - 1$, $h_r \in \mathbb{Z}^+$, *and* $0 = h_0 < h_1 < \cdots < h_r$. *Assume that* p *and* $q_i \ge 0$, $i = 0, 1, 2, \ldots, r$, *are the scalars satisfying* $q_r > 0$ *and*

$$\sum_{i=0}^{r} q_i < p \le 1.$$

Then Eq. (5.3.7) becomes

$$\Delta y(t) = -py(t) + \sum_{i=0}^{r} q_i y^\ell(t - h_i), \quad t \in \{0, 1, \ldots\}. \qquad (5.3.14)$$

The characteristic polynomial and the set $S(t)$ *given by (5.3.8) and (5.3.9) turn into*

$$P(t, \lambda) = (\lambda + p)(1 + \lambda)^{h_r}(1 + \lambda)^{(1-\ell)(t - h_r)} - K^{\ell-1} \sum_{i=0}^{r} q_i (1 + \lambda)^{\ell(h_r - h_i)}$$

and

$$S(t) = (-1, 0) \text{ for all } t \in \{0, 1, \ldots\},$$

respectively. Let $\{x(t)\}$, $t \in [-h_r, \infty)_{\mathbb{Z}}$ *be a sequence satisfying the inequality*

$$\Delta x(t) \le -px(t) + \sum_{i=0}^{r} q_i x^{\ell}(t - h_i), \quad t \in \{0, 1, \dots\}.$$

Then by Theorem 5.3.1, we conclude that there exists a constant $K_0 > 1$ *such that*

$$x(t) < K_0 \prod_{s=0}^{t-1} (1 + \lambda_0(s)), \, t \in \{0, 1, \dots\},$$

where $\lambda_0 : \mathbb{Z} \to (-1, 0)$ *is a positively regressive function defined by*

$$\lambda_0(t) = \max \Big\{ \nu \in (-1, 0) : (\nu + p)(1 + \nu)^{h_r}(1 + \nu)^{(1-\ell)(t-h_r)}$$
$$- K^{\ell-1} \sum_{i=0}^{r} q_i (1 + \nu)^{\ell(h_r - h_i)} = 0 \Big\}.$$

Theorem 5.3.2 *Let* $\tau \in [t_0, \infty)_{\mathbb{T}}$ *be a constant such that there exists a delay function* $\delta_-(\tau, t)$ *on* $\mathbb{T}$. *Let* x *be a function satisfying the inequality*

$$x^{\Delta}(t) \le -p(t)x(t) + q(t) \sup_{s \in [\delta_-(\tau,t),t]} x^{\ell}(s), \quad t \in [t_0, \infty)_{\mathbb{T}},$$

where $\ell \in (0, 1]$ *is a constant. Suppose that* p *and* q *are the continuous and bounded functions satisfying* $p(t) > q(t) > 0$ *and* $1 - \tilde{\mu}(t)p(t) \ge 0$ *for all* $t \in [t_0, \infty)_{\mathbb{T}}$. *Then there exists a constant* $M_0 > 0$ *such that*

$$x(t) \le M_0 e_{\tilde{\lambda}}(t, t_0),$$

where $\tilde{\lambda}$ *is a positively regressive function chosen as in* (5.3.16).

Proof. We proceed as in the proof of Theorem 5.3.1. Consider the dynamic equation

$$y^{\Delta}(t) = -py(t) + q \sup_{s \in [\delta_-(\tau,t),t]} y^{\ell}(s), \quad t \in [t_0, \infty)_{\mathbb{T}}. \tag{5.3.15}$$

For a given $M > 1$, $Me_{\lambda}(t, t_0)$ is a solution of (5.3.7) if and only if λ is a root of the characteristic polynomial $\tilde{P}(t, \lambda)$ defined by

$$\tilde{P}(t, \lambda) := (\lambda + p(t)) \, e_{\lambda}(t, t_0) - M^{\ell}q(t) \sup_{s \in [\delta_-(\tau,t),t]} e_{\lambda}^{\ell}(s, t_0).$$

For each fixed $t \in [t_0, \infty)_{\mathbb{T}}$ define the set

$$S(t) := \{k \in (-\infty, 0) : 1 + \tilde{\mu}(t)k > 0\}.$$

It is obvious that for each fixed $t \in [t_0, \infty)_{\mathbb{T}}$ and for all $k \in S(t)$ we have

$$\tilde{P}(t, k) = (k + p(t)) \, e_k(t, t_0) - M^{\ell}q(t)e_k^{\ell}(\delta_-(\tau, t), t_0).$$

As in the proof of Theorem 5.3.1, one may easily show that for each $t \in [t_0, \infty)_{\mathbb{T}}$, there exists a largest element of $S(t)$ such that $P(t, k) = 0$. Using these largest elements we can define a positively regressive function $\tilde{\lambda} : [\delta_-(h_r, t_0), \infty)_{\mathbb{T}} \to (-\infty, 0)$ by

$$\tilde{\lambda}(t) := \max \left\{ k \in S(t) : \tilde{P}(t, k) = 0 \right\}, \tag{5.3.16}$$

so that for a given $M > 1$, $y(t) = Me_{\tilde{\lambda}}(t, t_0)$ is a solution to (5.3.15). The rest of the proof is similar to that in Theorem 5.3.1. ∎

Finally in this section we give a result for functions satisfying the dynamic inequality

$$x^{\Delta}(t) \leq -p(t)x(t) + \prod_{i=0}^{r} \beta_i(t)x^{\alpha_i}(\delta_-(h_i, t)), \tag{5.3.17}$$

where $\alpha_i \in (0, \infty)$, $i = 0, 1, \ldots, r$, are the scalars with $\sum_{i=0}^{r} \alpha_i = 1$. Let the characteristic polynomial $Q(t, k)$ and the set $S(t)$ be defined by

$$Q(t, k) := (\lambda + p)\, e_{\lambda}(t, t_0) - \prod_{i=0}^{r} \beta_i e_{\lambda}^{\alpha_i}(\delta_-(h_i, t), t_0),$$

and (5.3.9), respectively. Using a procedure similar to that used in the proof of Theorem 5.3.1 we obtain the next result.

Theorem 5.3.3 *Let x be a Δ-differentiable function satisfying (5.3.17), where $\alpha_i \in (0, \infty)$, $i = 0, 1, \ldots, r$, are the scalars with $\sum_{i=0}^{r} \alpha_i = 1$, p and β_i, $i = 0, 1, \ldots, r$, are continuous functions with the property that $1 - \tilde{\mu}(t)p(t) \geq 0$, $\beta_i(t) > 0$, $i = 0, 1, \ldots, r$, for all $t \in [t_0, \infty)_{\mathbb{T}}$. Suppose that*

$$p(t) - \prod_{i=0}^{r} \beta_i(t) > 0$$

for all $t \in [t_0, \infty)_{\mathbb{T}}$. Then there exists a constant $L_0 > 0$ such that

$$x(t) \leq L_0 e_{\gamma}(t, t_0) \text{ for } t \in [t_0, \infty)_{\mathbb{T}},$$

where $\gamma : [t_0, \infty)_{\mathbb{T}} \to (-\infty, 0)$ is a positively regressive function given by

$$\gamma(t) := \max \left\{ k \in S(t) : Q(t, k) = 0 \right\}. \tag{5.3.18}$$

5.4   Stability of Nonlinear Dynamic Equations

In this section by means of Halanay type inequalities we propose some sufficient conditions guaranteeing global stability of nonlinear dynamic equations of the form

$$x^{\Delta}(t) = -p(t)x(t) + F(t, x(t), x(\delta_-(h_1, t)), \ldots, x(\delta_-(h_r, t))), \text{ for } t \in [t_0, \infty)_{\mathbb{T}}. \tag{5.4.1}$$

Theorem 5.4.1 *Let p and q_i, $i = 0, 1, \ldots, r$, be continuous and bounded functions satisfying $1 - \tilde{\mu}(t)p(t) > 0$, $q_i(t) \geq 0$, $i = 0, 1, \ldots, r$, $q_r(t) > 0$ and*

$$p(t) - \sum_{i=0}^{r} q_i(t) > 0$$

for all $t \in [t_0, \infty)_{\mathbb{T}}$. Let $\ell \in (0, 1]$ be a constant. Assume that there exist scalars $h_i \in [t_0, \infty)_{\mathbb{T}}$, $i = 0, 1, \ldots, r$, such that $h_0 = t_0$, $\delta_-(h_i, t)$, $i = 1, \ldots, r$, are delay functions on $\mathbb{T}$, and

$$|F(t, x(t), x(\delta_-(h_1, t)), \ldots, x(\delta_-(h_r, t)))| \leq \sum_{i=0}^{r} q_i(t)\, |x(\delta_-(h_i, t))|^{\ell} \quad (5.4.2)$$

for all $(t, x(t), x(\delta_-(h_1, t)), \ldots, x(\delta_-(h_r, t))) \in [t_0, \infty)_{\mathbb{T}} \times \mathbb{R}^{r+1}$. Then there exists a constant $M_0 > 1$ such that every solution x to (5.4.1) satisfies

$$|x(t)| \leq M_0 e_\lambda(t, t_0),$$

where λ is a positively regressive function chosen as in (5.3.13).

Proof. Let

$$\xi := \ominus(-p) = \frac{p}{1 - \mu p}.$$

Multiplying by $e_\xi(t, t_0)$ and integrating the resulting equation from t_0 to t we get that

$$x(t) = x_0 e_{\ominus\xi}(t, t_0) + \int_{t_0}^{t} F(s, x(s), x(\delta_-(h_1, s)), \ldots, x(\delta_-(h_r, s))) e_{\ominus\xi}(t, \sigma(s)) \Delta s.$$
$$(5.4.3)$$

It is straightforward to show that a solution $x(t)$ to (5.4.3) satisfies (5.4.1). This means every solution of (5.4.1) can be rewritten in the form of (5.4.3). By using (5.4.2) we obtain

$$|x(t)| \leq |x_0|\, e_{\ominus\xi}(t, t_0) + \int_{t_0}^{t} \sum_{i=0}^{r} q_i(s)\, |x(\delta_-(h_i, s))|^{\ell}\, e_{\ominus\xi}(t, \sigma(s)) \Delta s.$$

Let the function y be defined as follows:

$$y(t) = |x(t)| \text{ for } t \in [\delta_-(h_r, t_0), t_0]_{\mathbb{T}}$$

and

$$y(t) = |x_0|\, e_{\ominus\xi}(t, t_0) + \int_{t_0}^{t} \sum_{i=0}^{r} q_i(s)\, |x(\delta_-(h_i, s))|^{\ell}\, e_{\ominus\xi}(t, \sigma(s)) \Delta s \text{ for } [t_0, \infty)_{\mathbb{T}}.$$

Then we have $|x(t)| \leq y(t)$ for all $t \in [\delta_-(h_r, t_0), \infty)_{\mathbb{T}}$. By [51, Theorem 1.117] we get that

$$y^{\Delta}(t) = -p(t)\left(|x_0| \, e_{\ominus\xi}(t, t_0) + \int_{t_0}^{t}\sum_{i=0}^{r} q_i(s)\, |x(\delta_-(h_i, s))|^{\ell}\, e_{\ominus\xi}(t, \sigma(s))\Delta s\right)$$

$$+ \sum_{i=0}^{r} q_i(t)\, |x(\delta_-(h_i, t))|^{\ell}$$

$$= -p(t)y(t) + \sum_{i=0}^{r} q_i(t)\, |x(\delta_-(h_i, t))|^{\ell}$$

$$\leq -p(t)y(t) + \sum_{i=0}^{r} q_i(t)y^{\ell}(\delta_-(h_i, t))$$

for all $[t_0, \infty)_{\mathbb{T}}$. Therefore, it follows from Theorem 5.3.1 that there exists a constant $M_0 > 1$ such that

$$|x(t)| \leq M_0 e_{\lambda}(t, t_0) \text{ for } t \in [t_0, \infty)_{\mathbb{T}},$$

where $\lambda : [t_0, \infty)_{\mathbb{T}} \to (-\infty, 0)$ is a positively regressive function defined by (5.3.13). The proof is complete. ∎

Chapter 6

Wirtinger Inequalities

If a nonnegative quantity was so small that is smaller than any given one, then it certainly could not be anything but zero. To those who ask what the infinity small quantity in mathematics is, we answer that it is actually zero. Hence there are not so many mysteries hidden in this concept as they are usually believed to be. These supposed mysteries have rendered the calculus of the infinity small quite suspect to many people.

Euler (1707–1783).

The inequality of W. Wirtinger is given by

$$\int_0^1 \left(y'(t)\right)^2 dt \geq \pi^2 \int_0^1 y^2(t)\mathrm{dt} \tag{6.0.1}$$

for any $y \in C^1[0,1]$ such that $y(0) = y(1) = 0$.

In [81] Hinton and Lewis established a Wirtinger-type inequality (using the Schwarz inequality)

$$\int_a^b \frac{M^2(t)}{|M'(t)|} \left(y'(t)\right)^2 dt \geq \frac{1}{4} \int_a^b \left|M'(t)\right| y^2(t)dt \tag{6.0.2}$$

for any positive $M \in \mathrm{C}^1([a,b])$ with $M'(t) \neq 0$, $y \in \mathrm{C}^1([a,b])$ and $y(a) = y(b) = 0$.

In [119], Peňa established the discrete analogue of (6.0.2) and proved the following result. For a positive sequence $\{M_n\}_{0 \leq n \leq N+1}$ satisfying either $\Delta M > 0$ or $\Delta M < 0$ on $[0,N] \cap \mathbb{Z}$,

$$\sum_{n=0}^N \frac{M_n M_{n+1}}{|\Delta M_n|} \left(\Delta y_n\right)^2 \geq \frac{1}{\psi_J} \sum_{n=0}^N |\Delta M_n| y_{n+1}^2 \tag{6.0.3}$$

© Springer International Publishing Switzerland 2014

R. Agarwal et al., *Dynamic Inequalities On Time Scales*,

DOI 10.1007/978-3-319-11002-8_6

holds for any sequence $\{y_n\}_{0 \le n \le N+1}$ with $y_0 = y_{N+1} = 0$, where

$$\psi_J = \left(\sup_{0 \le n \le N} \frac{M_n}{M_{n+1}} \right) \left[1 + \left(\sup_{0 \le n \le N} \frac{|\Delta M_n|}{|\Delta M_{n+1}|} \right)^{1/2} \right]^2 . \tag{6.0.4}$$

In this chapter we present Wirtinger-type inequalities on time scales. In Sects. 6.1–6.3 a variety of Wirtinger-type inequalities will be established. Section 6.4 will present an application.

6.1 Wirtinger-Type Inequality I

In this section we prove a Wirtinger-type inequality on time scales. The results are adapted from [82].

Theorem 6.1.1 *For a positive function* $M \in C^1_{rd}(\mathcal{I})$ *satisfying either* $M^\Delta > 0$ *or* $M^\Delta < 0$ *on* $\mathcal{I}^\kappa$, *we have*

$$\int_a^b \frac{M(t)M(\sigma(t))}{|M^\Delta(t)|} \left(y^\Delta(t) \right)^2 \Delta t \ge \frac{1}{\psi^2} \int_a^b |M^\Delta(t)| \, (y^\sigma(t))^2 \Delta t \tag{6.1.1}$$

for any $y \in C^1_{rd}(\mathcal{I})$ *with* $y(a) = y(b) = 0$, $\mathcal{I} = [a,b]_\mathbb{T} \subset \mathbb{T}$, *and*

$$\psi = \left(\sup_{t \in \mathcal{I}^\kappa} \frac{M(t)}{M(\sigma(t))} \right)^{1/2} + \left[\left(\sup_{t \in \mathcal{I}^\kappa} \frac{\mu(t) |M^\Delta(t)|}{M(\sigma(t))} \right) + \left(\sup_{t \in \mathcal{I}^\kappa} \frac{M(t)}{M(\sigma(t))} \right) \right]^{1/2} . \tag{6.1.2}$$

Proof. Let M and y be as in the theorem. We will omit the subscript t in the computations. Let

$$A \; := \; \int_a^b |M^\Delta| \, (y^\sigma)^2 \, \Delta t, \quad B := \int_a^b \frac{MM^\sigma}{|M^\Delta|} (y^\Delta)^2 \Delta t,$$

$$\alpha \; := \; \left(\sup_{t \in \mathcal{I}^k} \frac{M_t}{M_t^\sigma} \right)^{1/2}, \quad \beta := \left(\sup_{t \in \mathcal{I}^k} \frac{\mu_t |M_t^\Delta|}{M_t^\sigma} \right).$$

Suppose that $M^\Delta > 0$ (the other case is treated similarly). Then we have

$$\begin{aligned} A \; &= \; \int_a^b M^\Delta (y^\sigma)^2 \Delta t = -\int_a^b My^\Delta (y + y^\sigma) \Delta t = -\int_a^b My^\Delta (2y^\sigma - \mu y^\Delta) \Delta t \\ &\le \; 2 \int_a^b M |y^\sigma| |y^\Delta| \, \Delta t + \int_a^b \mu M \left(y^\Delta \right)^2 \Delta t \\ &= \; 2 \int_a^b \sqrt{\frac{MM^\sigma}{|M^\Delta|}} |y^\Delta| \sqrt{\frac{M}{M^\sigma}} |M^\Delta| \Delta t + \int_a^b \frac{MM^\sigma}{|M^\Delta|} \frac{\mu |M^\Delta|}{M^\sigma} (y^\Delta)^2 \Delta t. \end{aligned}$$

As a result we have

$$
\begin{aligned}
A &= \int_a^b M^\Delta (y^\sigma)^2 \Delta t \le 2 \left(\int_a^b \frac{MM^\sigma}{|M^\Delta|} \left| y^\Delta \right|^2 \Delta t \right)^{1/2} \left(\int_a^b \frac{M}{M^\sigma} \left| M^\Delta \right| (y^\sigma) \Delta t \right)^{1/2} \\
&\quad + \left(\sup_{t \in \mathcal{I}^k} \frac{\mu_1 \left| M_t^\Delta \right|}{M_t^\sigma} \right) \int_a^b \frac{MM^\sigma}{|M^\Delta|} \left| y^\Delta \right|^2 \Delta t \\
&\le 2\sqrt{B} \left(\sup_{t \in \mathcal{I}^k} \frac{M_t}{M_t^\sigma} \right)^{1/2} \left(\int_a^b \left| M^\Delta \right| (y^\sigma)^2 \Delta t \right)^{1/2} + \beta B = 2\alpha \sqrt{AB} + \beta B.
\end{aligned}
$$

Dividing both sides of the above inequality by $\sqrt{AB}$ we obtain

$$
\frac{\sqrt{A}}{\sqrt{B}} \le 2\alpha + \beta \frac{\sqrt{B}}{\sqrt{A}}.
$$

By relabeling with $C := \sqrt{A/B}$ we get a quadratic inequality $C^2 \le 2\alpha C + \beta$, i.e., $(C - \alpha)^2 \le \alpha^2 + \beta$. Hence,

$$
\alpha - \sqrt{\beta + \alpha^2} \le C \le \alpha + \sqrt{\beta + \alpha^2},
$$

and since $C \ge 0$, we have $C^2 \le (\alpha + \sqrt{\beta + \alpha^2})^2$. The proof is complete. ∎

6.2 Wirtinger-Type Inequality II

In this section we derive some Wirtinger-type inequalities on time scales where we will use $y^{\gamma+1}$ instead of the term $(y^\sigma)^{\gamma+1}$. The results are adapted from [13]. Throughout this section and in the next section we assume that the function $M \in C_{rd}^1(\mathcal{I})$ is positive, where $\mathcal{I} = [a, b]_{\mathbb{T}}$, and let

$$
\alpha := \sup_{t \in \mathcal{I}^\kappa} \left(\frac{M(\sigma(t))}{M(t)} \right)^{\frac{\gamma}{\gamma+1}} \quad \text{and} \quad \beta := \sup_{t \in \mathcal{I}^\kappa} \left(\frac{\mu(t) \left| M^\Delta(t) \right|}{M(t)} \right)^\gamma. \tag{6.2.1}
$$

Theorem 6.2.1 *Suppose $\gamma \ge 1$ is an odd integer. For a positive $M \in C_{rd}^1(\mathcal{I})$ satisfying either $M^\Delta > 0$ or $M^\Delta < 0$ on $\mathcal{I}^\kappa$, we have*

$$
\int_a^b \frac{M^\gamma(t)M(\sigma(t))}{|M^\Delta(t)|^\gamma} \left(y^\Delta(t) \right)^{\gamma+1} \Delta t \ge \frac{1}{\Psi^{\gamma+1}(\alpha, \beta, \gamma)} \int_a^b \left| M^\Delta(t) \right| (y(t))^{\gamma+1} \Delta t \tag{6.2.2}
$$

for any $y \in C_{rd}^1(\mathcal{I})$ with $y(a) = y(b) = 0$, where $\Psi(\alpha, \beta, \gamma)$ is the largest root of the equation

$$
x^{\gamma+1} - 2^{\gamma-1}(\gamma + 1)\alpha x^\gamma - 2^{\gamma-1}\beta = 0. \tag{6.2.3}
$$

Proof. Let y and M be defined as above and let

$$
A := \int_a^b M^\Delta(t)(y(t))^{\gamma+1} \Delta t \quad \text{and} \quad B := \int_a^b \frac{M^\gamma(t)M^\sigma(t)}{|M^\Delta(t)|^\gamma} \left(y^\Delta(t) \right)^{\gamma+1} \Delta t.
$$

Using the integration by parts formula and the fact that $y(a) = y(b) = 0$, we have

$$A = \int_a^b M^\Delta(t)(y(t))^{\gamma+1}\Delta t$$

$$= \left\{ \mathrm{sgn}\left(M^\Delta(a)\right)\right\} \int_a^b M^\Delta(t) y^{\gamma+1}(t)\Delta t$$

$$= \left\{ \mathrm{sgn}\left(M^\Delta(a)\right)\right\} \left\{ \left[M(t) y^{\gamma+1}(t)\right]_a^b - \int_a^b M(\sigma(t))\left(y^{\gamma+1}\right)^\Delta (t)\Delta t\right\}$$

$$= -\left\{ \mathrm{sgn}\left(M^\Delta(a)\right)\right\} \int_a^b M(\sigma(t))\left(y^{\gamma+1}\right)^\Delta (t)\Delta t$$

$$\leq \int_a^b M(\sigma(t))\left|\left(y^{\gamma+1}\right)^\Delta (t)\right|\Delta t.$$

$$(6.2.4)$$

By the Pötzsche chain rule we obtain

$$\left|\left(y^{\gamma+1}\right)^\Delta (t)\right| = (\gamma+1)\left|\int_0^1 \left[y(t) + \mu(t) h y^\Delta(t)\right]^\gamma dh\right| \left|y^\Delta(t)\right|$$

$$\leq (\gamma+1)\left|y^\Delta(t)\right| \int_0^1 \left|y(t) + \mu(t) h y^\Delta(t)\right|^\gamma dh. \qquad (6.2.5)$$

Applying the inequality (see [110, page 500])

$$|u + v|^\gamma \leq 2^{\gamma-1}\left(|u|^\gamma + |v|^\gamma\right), \quad \text{where} \quad u, v \in \mathbb{R}$$

with $u = y(t)$ and $v = \mu(t) h y^\Delta(t)$, we have from (6.2.5) that

$$\left|\left(y^{\gamma+1}\right)^\Delta (t)\right| \leq (\gamma+1)\left|y^\Delta(t)\right| \int_0^1 \left|y(t) + \mu(t) h y^\Delta(t)\right|^\gamma dh$$

$$\leq 2^{\gamma-1}(\gamma+1)\left|y^\Delta(t)\right| \left\{\int_0^1 |y(t)|^\gamma dh + \int_0^1 \left|\mu(t) h y^\Delta(t)\right|^\gamma dh\right\}$$

$$= 2^{\gamma-1}(\gamma+1)\left|y^\Delta(t)\right| |y(t)|^\gamma + 2^{\gamma-1}\left|y^\Delta(t)\right| \left|\mu(t) y^\Delta(t)\right|^\gamma.$$

$$(6.2.6)$$

Substituting (6.2.6) into (6.2.4), we have

$$A \leq 2^{\gamma-1}(\gamma+1)\int_a^b M^\sigma(t)\left|y^\Delta(t)\right| |y(t)|^\gamma \Delta t$$

$$+2^{\gamma-1}\int_a^b M^\sigma(t)\mu^\gamma(t)\left|y^\Delta(t)\right|^{\gamma+1}\Delta t$$

$$
\begin{aligned}
= \quad & 2^{\gamma-1}(\gamma+1) \int_a^b \left(\frac{M^\sigma(t) M^\gamma(t)}{(M^\Delta(t))^\gamma} \right)^{\frac{1}{\gamma+1}} |y^\Delta(t)| \\
& \times \left(\frac{M^\sigma M^\Delta(t)}{M(t)} \right)^{\frac{\gamma}{\gamma+1}} |y(t)|^\gamma \, \Delta t \\
& + 2^{\gamma-1} \int_a^b \left(\frac{\mu(t) M^\Delta(t)}{M(t)} \right)^\gamma \frac{M^\sigma(t) M^\gamma(t)}{(M^\Delta(t))^\gamma} |y^\Delta(t)|^{\gamma+1} \, \Delta t \\
= \quad & 2^{\gamma-1}(\gamma+1) \int_a^b \left(\frac{M^\sigma(t) M^\gamma(t)}{(M^\Delta(t))^\gamma} |y^\Delta(t)|^{\gamma+1} \right)^{\frac{1}{\gamma+1}} \\
& \times \left(\frac{M^\sigma M^\Delta(t)}{M(t)} |y(t)|^{\gamma+1} \right)^{\frac{\gamma}{\gamma+1}} \, \Delta t \\
& + 2^{\gamma-1} \int_a^b \left(\frac{\mu(t) M^\Delta(t)}{M(t)} \right)^\gamma \left(\frac{M^\sigma(t) M^\gamma(t)}{(M^\Delta(t))^\gamma} |y^\Delta(t)|^{\gamma+1} \right) \, \Delta t.
\end{aligned}
$$

(6.2.7)

Applying the Hölder inequality with

$$
f = \left(\frac{M^\sigma M^\gamma}{|M^\Delta|^\gamma} |y^\Delta|^{\gamma+1} \right)^{\frac{1}{\gamma+1}}, \quad g = \left(\frac{M^\sigma |M^\Delta|}{M} |y|^{\gamma+1} \right)^{\frac{\gamma}{\gamma+1}}, \quad p = \gamma+1, \ q = \frac{\gamma+1}{\gamma},
$$

we obtain from (6.2.7) that

$$
\begin{aligned}
A \leq \ & 2^{\gamma-1}(\gamma+1) \left\{ \int_a^b \left(\frac{M^\sigma M^\gamma}{|M^\Delta|^\gamma} |y^\Delta|^{\gamma+1} \right) (t)\Delta t \right\}^{\frac{1}{\gamma+1}} \\
& \times \left\{ \int_a^b \left(\frac{M^\sigma |M^\Delta|}{M} |y|^{\gamma+1} \right) (t)\Delta t \right\}^{\frac{\gamma}{\gamma+1}} \\
& + 2^{\gamma-1} \int_a^b \left\{ \left(\frac{\mu |M^\Delta|}{M} \right)^\gamma \left(\frac{M^\sigma M^\gamma}{|M^\Delta|^\gamma} |y^\Delta|^{\gamma+1} \right) \right\} (t)\Delta t \\
\leq \ & 2^{\gamma-1}(\gamma+1)\alpha \left\{ \int_a^b \left(\frac{M^\sigma M^\gamma}{|M^\Delta|^\gamma} |y^\Delta|^{\gamma+1} \right) (t)\Delta t \right\}^{\frac{1}{\gamma+1}} \\
& \times \left\{ \int_a^b \left(|M^\Delta| |y|^{\gamma+1} \right) (t)\Delta t \right\}^{\frac{\gamma}{\gamma+1}} \\
& + 2^{\gamma-1}\beta \int_a^b \left(\frac{M^\sigma M^\gamma}{|M^\Delta|^\gamma} |y^\Delta|^{\gamma+1} \right) (t)\Delta t
\end{aligned}
$$

holds, i.e.,

$$
A \leq 2^{\gamma-1} \left((\gamma+1)\alpha B^{\frac{1}{\gamma+1}} A^{\frac{\gamma}{\gamma+1}} + \beta B \right).
$$

(6.2.8)

Dividing both sides of (6.2.8) by $B^{\frac{1}{\gamma+1}}A^{\frac{\gamma}{\gamma+1}}$, we obtain

$$\frac{A^{\frac{1}{\gamma+1}}}{B^{\frac{1}{\gamma+1}}} \leq 2^{\gamma-1}\left((\gamma+1)\alpha + \beta\frac{B^{\frac{\gamma}{\gamma+1}}}{A^{\frac{\gamma}{\gamma+1}}}\right),$$

and putting $C := A^{\frac{1}{\gamma+1}}/B^{\frac{1}{\gamma+1}} > 0$, we get

$$C \leq 2^{\gamma-1}\left((\gamma+1)\alpha + \frac{\beta}{C^\gamma}\right),$$

which gives

$$C^{\gamma+1} - 2^{\gamma-1}(\gamma+1)\alpha C^\gamma - 2^{\gamma-1}\beta \leq 0 \quad \text{with} \quad C = \frac{A^{\frac{1}{\gamma+1}}}{B^{\frac{1}{\gamma+1}}}. \qquad (6.2.9)$$

By (6.2.9), we have

$$C \leq \Psi(\alpha, \beta, \gamma),$$

where $\Psi(\alpha, \beta, \gamma)$ is the largest root of Eq. (6.2.3), and thus we have, replacing C in terms of A and B as in (6.2.9),

$$A^{\frac{1}{\gamma+1}} \leq \Psi(\alpha, \beta, \gamma)B^{\frac{1}{\gamma+1}},$$

i.e.,

$$A \leq \Psi^{\gamma+1}(\alpha, \beta, \gamma)B,$$

which is the desired inequality (6.2.2). The proof is complete. ∎

Using Theorem 6.2.1 with $\gamma = 1$, we obtain the following result.

Corollary 6.2.1 *For a positive function* $M \in C^1_{rd}(\mathcal{I})$ *satisfying either* $M^\Delta > 0$ *or* $M^\Delta < 0$ *on* $\mathcal{I}^\kappa$, *we have*

$$\int_a^b \frac{M(t)M(\sigma(t))}{|M^\Delta(t)|}\left(y^\Delta(t)\right)^2 \Delta t \geq \frac{1}{\varphi^2}\int_a^b |M^\Delta(t)|\,(y(t))^2\Delta t \qquad (6.2.10)$$

for any $y \in C^1_{rd}(\mathcal{I})$ *with* $y(a) = y(b) = 0$, *where*

$$\varphi := \sqrt{\sup_{t\in\mathcal{I}^\kappa}\frac{M(\sigma(t))}{M(t)}} + \sqrt{\sup_{t\in\mathcal{I}^\kappa}\frac{M(\sigma(t))}{M(t)} + \sup_{t\in\mathcal{I}^\kappa}\frac{\mu(t)\,|M^\Delta(t)|}{M(t)}}.$$

Proof. When $\gamma = 1$, we see that the largest root $\Psi(\alpha, \beta, 1)$ of the quadratic equation $x^2 - 2x\alpha - \beta = 0$ is given by

$$\Psi(\alpha, \beta, 1) = \alpha + \sqrt{\alpha^2 + \beta} = \varphi,$$

and then (6.2.10) follows from Theorem 6.2.1. ∎

6.3   Further Wirtinger Inequalities

In this section we apply different algebraic inequalities to establish some Wirtinger-type inequalities. The results in this section are adapted from [13]. To prove the main results, we will use the inequality (see [110, page 518])

$$(u + v)^{\gamma+1} \leq u^{\gamma+1} + (\gamma + 1)u^{\gamma}v + L_{\gamma}vu^{\gamma}, \quad \text{where} \quad u > v > 0. \quad (6.3.1)$$

Note that $L_{\gamma} > 1$ in (6.3.1) is a constant which does not depend on u or v.

Theorem 6.3.1 *Suppose $\gamma \geq 1$ is an odd integer. For a positive $M \in C^1_{rd}(\mathcal{I})$ satisfying either $M^{\Delta} > 0$ or $M^{\Delta} < 0$ on $\mathcal{I}^{\kappa}$, we have*

$$\int_a^b \frac{M^{\gamma}(t)M(\sigma(t))}{|M^{\Delta}(t)|^{\gamma}} \left(y^{\Delta}(t)\right)^{\gamma+1} \Delta t \geq \frac{1}{\Phi^{\gamma+1}(\alpha, \beta, \gamma)} \int_a^b \left|M^{\Delta}(t)\right| (y(t))^{\gamma+1} \Delta t$$

$$(6.3.2)$$

for any $y \in C^1_{rd}(\mathcal{I})$ with $y(a) = y(b) = 0$, where $\Phi(\alpha, \beta, \gamma)$ is the largest root of the equation

$$(\gamma+1)x + \frac{(\gamma+1)\gamma}{2}x^2 - (\gamma+1)x^{\gamma} - 2^{\gamma-1}(\gamma+1)(1+\alpha)x^{\gamma} - 2^{\gamma-1}\beta = 0. \quad (6.3.3)$$

Proof. We proceed as in the proof of Theorem 6.2.1 to get (6.2.9), i.e.,

$$C^{\gamma+1} \leq 2^{\gamma-1}(\gamma + 1)\alpha C^{\gamma} + 2^{\gamma-1}\beta, \quad \text{with} \quad C = \frac{A^{\frac{1}{\gamma+1}}}{B^{\frac{1}{\gamma+1}}}. \quad (6.3.4)$$

Applying the inequality (6.3.1) with $u = C$, $v = 1$, and $L_{\gamma} = 2^{\gamma-1}(\gamma+1) > 1$, we have

$$C^{\gamma+1} \geq (C + 1)^{\gamma+1} - (\gamma + 1)C^{\gamma} - 2^{\gamma-1}(\gamma + 1)C^{\gamma},$$

which together with (6.3.4) implies that

$$(C + 1)^{\gamma+1} - (\gamma + 1)C^{\gamma} - 2^{\gamma-1}(\gamma + 1)(1 + \alpha)C^{\gamma} - 2^{\gamma-1}\beta \leq 0. \quad (6.3.5)$$

Using the inequality

$$(1 + u)^{\gamma+1} \geq (\gamma + 1)u + \frac{(\gamma + 1)\gamma}{2}u^2, \quad \text{where} \quad u > 0 \quad (6.3.6)$$

for $u = C$, we have from (6.3.5) that

$$(\gamma + 1)C + \frac{(\gamma + 1)\gamma}{2}C^2 - (\gamma + 1)C^{\gamma} - 2^{\gamma-1}(\gamma + 1)(1 + \alpha)C^{\gamma} - 2^{\gamma-1}\beta \leq 0.$$

Therefore

$$C \leq \Phi(\alpha, \beta, \gamma),$$

where $\Phi(\alpha, \beta, \gamma)$ is the largest root of the equation (6.3.3), and thus we have, replacing C in terms of A and B as in (6.3.4),

$$A^{\frac{1}{\gamma+1}} \leq \Phi(\alpha, \beta, \gamma) B^{\frac{1}{\gamma+1}},$$

i.e.,

$$B \geq \frac{1}{\Phi^{\gamma+1}(\alpha, \beta, \gamma)} A,$$

which is the desired inequality (6.3.2). The proof is complete. ∎

Using the inequality

$$(1+u)^{\gamma+1} \geq \frac{(\gamma+1)\gamma}{2} u^2, \quad \text{where} \quad u > 0$$

instead of the inequality (6.3.6) that was used in the proof of Theorem 6.3.1, we have the following result.

Theorem 6.3.2 *Suppose $\gamma \geq 1$ is an odd integer. For a positive $M \in C_{rd}^1(\mathcal{I})$ satisfying either $M^\Delta > 0$ or $M^\Delta < 0$ on $\mathcal{I}^\kappa$, we have*

$$\int_a^b \frac{M^\gamma(t) M(\sigma(t))}{|M^\Delta(t)|^\gamma} \left(y^\Delta(t) \right)^{\gamma+1} \Delta t \geq \frac{1}{\Omega^{\gamma+1}(\alpha, \beta, \gamma)} \int_a^b |M^\Delta(t)| \, (y(t))^{\gamma+1} \Delta t$$
(6.3.7)

for any $y \in C_{rd}^1(\mathcal{I})$ with $y(a) = y(b) = 0$, where $\Omega(\alpha, \beta, \gamma)$ is the largest root of the equation

$$\frac{(\gamma+1)\gamma}{2} x^2 - (\gamma+1)x^\gamma - 2^{\gamma-1}(\gamma+1)(1+\alpha)x^\gamma - 2^{\gamma-1}\beta = 0. \quad (6.3.8)$$

In the following, we apply the inequality

$$u^{\gamma+1} + \gamma v^{\gamma+1} - (\gamma+1)uv^\gamma \geq 0, \quad \text{where} \quad u, \, v \geq 0 \qquad (6.3.9)$$

and derive the following result.

Theorem 6.3.3 *Suppose $\gamma \geq 1$ is an odd integer. For a positive $M \in C_{rd}^1(\mathcal{I})$ satisfying either $M^\Delta > 0$ or $M^\Delta < 0$ on $\mathcal{I}^\kappa$, we have*

$$\int_a^b \frac{M^\gamma(t) M(\sigma(t))}{|M^\Delta(t)|^\gamma} \left(y^\Delta(t) \right)^{\gamma+1} \Delta t \geq \frac{1}{\Lambda^{\gamma+1}(\alpha, \beta, \gamma)} \int_a^b |M^\Delta(t)| \, (y(t))^{\gamma+1} \Delta t$$
(6.3.10)

for any $y \in C_{rd}^1(\mathcal{I})$ with $y(a) = y(b) = 0$, where $\Lambda(\alpha, \beta, \gamma)$ is the largest root of the equation

$$(\gamma+1)(x+1) - \gamma - (\gamma+1)x^\gamma - 2^{\gamma-1}(\gamma+1)(1+\alpha)x^\gamma - 2^{\gamma-1}\beta = 0. \quad (6.3.11)$$

Proof. We proceed as in the proof of Theorem 6.3.1 to get (6.3.5), i.e.,

$$(C+1)^{\gamma+1} - (\gamma+1)C^{\gamma} - 2^{\gamma-1}(\gamma+1)C^{\gamma} \le 2^{\gamma-1}(\gamma+1)\alpha C^{\gamma} + 2^{\gamma-1}\beta. \quad (6.3.12)$$

Applying the inequality (6.3.9) with $u = C + 1$ and $v = 1$, we have from (6.3.12) that

$$(\gamma+1)(C+1) - \gamma - (\gamma+1)C^{\gamma} - 2^{\gamma-1}(\gamma+1)(1+\alpha)C^{\gamma} - 2^{\gamma-1}\beta \le 0.$$

Therefore $C \le \Lambda(\alpha, \beta, \gamma)$, where $\Lambda(\alpha, \beta, \gamma)$ is the largest root of Eq. (6.3.11). This implies from the definition of C that

$$A^{\frac{1}{\gamma+1}} \le \Lambda(\alpha, \beta, \gamma)B^{\frac{1}{\gamma+1}}.$$

Thus $A \le \Lambda^{\gamma+1}(\alpha, \beta, \gamma)B$, which is the desired inequality (6.3.10). The proof is complete. ■

Note that in the case $\mathbb{T} = \mathbb{R}$, we have $\sigma(t) = t$ and $\mu(t) = 0$ so that

$$\alpha = 1 \quad \text{and} \quad \beta = 0. \quad (6.3.13)$$

Then we have from Theorems 6.2.1, 6.3.1–6.3.3 the following results.

Theorem 6.3.4 *Let $\gamma \ge 1$ be an odd integer. For a positive function $M \in C^1([a, b])$ satisfying either $M' > 0$ or $M' < 0$ on $[a, b]$, we have*

$$\int_a^b \frac{M^{\gamma+1}(t)}{|M'(t)|^{\gamma}} (y'(t))^{\gamma+1} \, dt \ge \frac{1}{2^{\gamma^2-1}(\gamma+1)^{\gamma+1}} \int_a^b \left| M'(t) \right| (y(t))^{\gamma+1} dt,$$

for any $y \in C^1([a, b])$ with $y(a) = y(b) = 0$.

Proof. Using (6.3.13), Eq. (6.2.3) becomes

$$x^{\gamma+1} - 2^{\gamma-1}(\gamma+1)x^{\gamma} = 0.$$

Hence $\Psi(1, 0, \gamma)$ from Theorem 6.2.1 is given by

$$\Psi(1, 0, \gamma) = 2^{\gamma-1}(\gamma+1),$$

and the proof is completed by applying Theorem 6.2.1. ■

Theorem 6.3.5 *Let $\gamma \ge 1$ be an odd integer. For a positive $M \in C^1([a, b])$ satisfying either $M' > 0$ or $M' < 0$ on $[a, b]$, we have*

$$\int_a^b \frac{M^{\gamma+1}(t)}{|M'(t)|^{\gamma}} (y'(t))^{\gamma+1} \, dt \ge \frac{1}{\Phi^{\gamma+1}(1, 0, \gamma)} \int_a^b \left| M'(t) \right| (y(t))^{\gamma+1} dt,$$

for any $y \in C^1([a, b])$ with $y(a) = y(b) = 0$, where $\Phi(1, 0, \gamma)$ is the largest root of the equation

$$(\gamma+1)x + \frac{(\gamma+1)\gamma}{2}x^2 - (\gamma+1)x^{\gamma} - 2^{\gamma}(\gamma+1)x^{\gamma} = 0.$$

Theorem 6.3.6 *Let* $\gamma \geq 1$ *be an odd integer. For a positive* $M \in \mathrm{C}^1([a,b])$ *satisfying either* $M' > 0$ *or* $M' < 0$ *on* $[a, b]$, *we have*

$$\int_a^b \frac{M^{\gamma+1}(t)}{|M'(t)|^\gamma} \left(y'(t)\right)^{\gamma+1} dt \geq \frac{1}{\Omega^{\gamma+1}(1,0,\gamma)} \int_a^b \left|M'(t)\right| \left(y(t)\right)^{\gamma+1} dt,$$

for any $y \in \mathrm{C}^1([a,b])$ *with* $y(a) = y(b) = 0$, *where* $\Omega(1,0,\gamma)$ *is the largest root of the equation*

$$\frac{(\gamma+1)\gamma}{2}x^2 - (\gamma+1)x^\gamma - 2^\gamma(\gamma+1)x^\gamma = 0.$$

Theorem 6.3.7 *Let* $\gamma \geq 1$ *be an odd integer. For a positive* $M \in \mathrm{C}^1([a,b])$ *satisfying either* $M' > 0$ *or* $M' < 0$ *on* $[a, b]$, *we have*

$$\int_a^b \frac{M^{\gamma+1}(t)}{|M'(t)|^\gamma} \left(y'(t)\right)^{\gamma+1} dt \geq \frac{1}{\Lambda^{\gamma+1}(1,0,\gamma)} \int_a^b \left|M'(t)\right| \left(y(t)\right)^{\gamma+1} dt,$$

for any $y \in \mathrm{C}^1([a,b])$ *with* $y(a) = y(b) = 0$, *where* $\Lambda(1,0,\gamma)$ *is the largest root of the equation*

$$(\gamma+1)(x+1) - \gamma - (\gamma+1)x^\gamma - 2^\gamma(\gamma+1)x^\gamma = 0.$$

Note that in the case $\mathbb{T} = \mathbb{Z}$, we have $\sigma(t) = t+1$ and $\mu(t) = 1$ so that

$$\alpha = \sup_{0 \leq n \leq N} \left(\frac{M_{n+1}}{M_n}\right)^{\frac{\gamma}{\gamma+1}} \quad \text{and} \quad \beta = \sup_{0 \leq n \leq N} \left(\frac{|\Delta M_n|}{M_n}\right)^\gamma. \tag{6.3.14}$$

Then we have from Theorems 6.2.1, 6.3.1–6.3.3 the following results.

Theorem 6.3.8 *Let* $\gamma \geq 1$ *be an odd integer. For a positive sequence* $\{M_n\}_{0 \leq n \leq N+1}$ *satisfying either* $\Delta M > 0$ *or* $\Delta M < 0$ *on* $[0,N] \cap \mathbb{Z}$, *we have*

$$\sum_{n=0}^N \frac{M_n^\gamma M_{n+1}}{|\Delta M_n|^\gamma} \left(\Delta y_n\right)^{\gamma+1} \geq \frac{1}{\Psi^{\gamma+1}(\alpha,\beta,\gamma)} \sum_{n=0}^N |\Delta M(n)| \left(y(n)\right)^{\gamma+1},$$

for any sequence $\{y_n\}_{0 \leq n \leq N+1}$ *with* $y_0 = y_{N+1} = 0$, *where* $\Psi(\alpha,\beta,\gamma)$ *is the largest root of the equation*

$$x^{\gamma+1} - 2^{\gamma-1}(\gamma+1)\alpha x^\gamma - 2^{\gamma-1}\beta = 0.$$

Theorem 6.3.9 *Let* $\gamma \geq 1$ *be an odd integer. For a positive sequence* $\{M_n\}_{0 \leq n \leq N+1}$ *satisfying either* $\Delta M > 0$ *or* $\Delta M < 0$ *on* $[0,N] \cap \mathbb{Z}$, *we have*

$$\sum_{n=0}^N \frac{M_n^\gamma M_{n+1}}{|\Delta M_n|^\gamma} \left(\Delta y_n\right)^{\gamma+1} \geq \frac{1}{\Phi^{\gamma+1}(\alpha,\beta,\gamma)} \sum_{n=0}^N |\Delta M(n)| \left(y(n)\right)^{\gamma+1},$$

for any sequence $\{y_n\}_{0 \leq n \leq N+1}$ with $y_0 = y_{N+1} = 0$, where $\Phi(\alpha, \beta, \gamma)$ is the largest root of the equation

$$(\gamma + 1)x + \frac{(\gamma + 1)\gamma}{2}x^2 - (\gamma + 1)x^\gamma - 2^{\gamma-1}(\gamma + 1)(1 + \alpha)x^\gamma - 2^{\gamma-1}\beta = 0.$$

Theorem 6.3.10 *Let $\gamma \geq 1$ be an odd integer. For a positive sequence $\{M_n\}_{0 \leq n \leq N+1}$ satisfying either $\Delta M > 0$ or $\Delta M < 0$ on $[0, N] \cap \mathbb{Z}$, we have*

$$\sum_{n=0}^{N} \frac{M_n^\gamma M_{n+1}}{|\Delta M_n|^\gamma} (\Delta y_n)^{\gamma+1} \geq \frac{1}{\Omega^{\gamma+1}(\alpha, \beta, \gamma)} \sum_{n=0}^{N} |\Delta M(n)| (y(n))^{\gamma+1},$$

for any sequence $\{y_n\}_{0 \leq n \leq N+1}$ with $y_0 = y_{N+1} = 0$, where $\Omega(\alpha, \beta, \gamma)$ is the largest root of the equation

$$\frac{(\gamma + 1)\gamma}{2}x^2 - (\gamma + 1)x^\gamma - 2^{\gamma-1}(\gamma + 1)(1 + \alpha)x^\gamma - 2^{\gamma-1}\beta = 0.$$

Theorem 6.3.11 *Let $\gamma \geq 1$ be an odd integer. For a positive sequence $\{M_n\}_{0 \leq n \leq N+1}$ satisfying either $\Delta M > 0$ or $\Delta M < 0$ on $[0, N] \cap \mathbb{Z}$, we have*

$$\sum_{n=0}^{N} \frac{M_n^\gamma M_{n+1}}{|\Delta M_n|^\gamma} (\Delta y_n)^{\gamma+1} \geq \frac{1}{\Lambda^{\gamma+1}(\alpha, \beta, \gamma)} \sum_{n=0}^{N} |\Delta M(n)| (y(n))^{\gamma+1}$$

for any sequence $\{y_n\}_{0 \leq n \leq N+1}$ with $y_0 = y_{N+1} = 0$, where $\Lambda(\alpha, \beta, \gamma)$ is the largest root of the equation

$$(\gamma + 1)(x + 1) - \gamma - (\gamma + 1)x^\gamma - 2^{\gamma-1}(\gamma + 1)(1 + \alpha)x^\gamma - 2^{\gamma-1}\beta = 0.$$

6.4   An Application of Wirtinger Inequalities on Time Scales

In this section we show how Opial and Wirtinger inequalities may be used to find a lower bound for the smallest eigenvalue of a Sturm–Liouville eigenvalue problem. The results are adapted from [13, 123].

Consider the Sturm–Liouville eigenvalue problem

$$- y^{\Delta\Delta}(t) + q(t)y^\sigma(t) = \lambda y^\sigma(t), \quad y(0) = y(\beta) = 0, \qquad (6.4.1)$$

and assume that λ_0 is the smallest eigenvalue of (6.4.1). To find a lower bound, we will apply an Opial type inequality and the Wirtinger inequality

$$\int_\alpha^\beta \frac{M(t)M^\sigma(t)}{|M^\Delta(t)|} (y^\Delta(t))^2 \Delta t \geq \frac{1}{\psi^2} \int_\alpha^\beta |M^\Delta(t)| (y^\sigma(t))^2 \Delta t, \qquad (6.4.2)$$

for a positive function $M \in C^1_{rd}(\mathbb{I})$ with either $M^\Delta(t) > 0$ or $M^\Delta(t) < 0$ on $\mathbb{I}$, $y \in C^1_{rd}(\mathbb{I})$ with $y(\alpha) = 0 = y(\beta)$, for $\mathbb{I} = [\alpha, \beta]_\mathbb{T} \subset \mathbb{T}$ and

$$\psi = \left(\sup_{t \in \mathbb{I}^k} \frac{M(t)}{M^\sigma(t)} \right)^{1/2} + \left[\left(\sup_{t \in \mathbb{I}^k} \frac{\mu(t) \, |M^\Delta(t)|}{M^\sigma(t)} \right) + \left(\sup_{t \in \mathbb{I}^k} \frac{M(t)}{M^\sigma(t)} \right) \right]^{1/2}.$$

We let

$$A(Q) = \sqrt{(\beta - \alpha)} \left(\int_\alpha^\beta Q^2(t)\Delta t \right)^{\frac{1}{2}} + \sup_{\alpha \leq t \leq \beta} \mu(t) \, |Q(t)|.$$

Theorem 6.4.1 *Assume that λ_0 is the smallest eigenvalue of (6.4.1) and assume that $q(t) = Q^\Delta(t) + \gamma$, where $\gamma < \lambda_0$. Then*

$$|\lambda_0 - \gamma| \geq \frac{1 - A(Q)}{(1 + \sqrt{2})^2 \sigma^2(\beta)}. \tag{6.4.3}$$

Proof. Let $y(t)$ be the eigenfunction of (6.4.1) corresponding to λ_0. Multiplying (6.4.1) by $y^\sigma(t)$ and proceeding as in the proof of Theorem 4.1.6, we have

$$- \int_0^\beta y^{\Delta\Delta} y^\sigma(t)\Delta t + \int_0^\beta q(t) \, (y^\sigma(t))^2 \, \Delta t = \lambda_0 \int_0^\beta (y^\sigma(t))^2 \, \Delta t.$$

This implies, after integrating by parts and using the fact that $y(0) = y(\beta) = 0$, that

$$(\lambda_0 - \gamma) \int_0^\beta (y^\sigma(t))^2 \, \Delta t$$

$$= \int_0^\beta (y^\Delta(t))^2 \, \Delta t + \int_0^\beta Q^\Delta(t) \, (y^\sigma(t))^2 \, \Delta t$$

$$= \int_0^\beta (y^\Delta(t))^2 \, \Delta t - \int_0^\beta Q(t) \, [y(t) + y^\sigma(t)] \, y^\Delta(t)\Delta t$$

$$\geq \int_0^\beta (y^\Delta(t))^2 \, \Delta t - \int_0^\beta |Q(t)| \, |y(t) + y^\sigma(t)| \, |y^\Delta(t)| \, \Delta t.$$

Proceeding as in the proof of Theorem 4.1.6 by applying an Opial type inequality on the term

$$\int_0^\beta |Q(t)| \, |y(t) + y^\sigma(t)| \, |y^\Delta(t)| \, \Delta t,$$

we obtain

$$|\lambda_0 - \gamma| \int_0^\beta (y^\sigma(t))^2 \, \Delta t \geq \int_0^\beta (y^\Delta(t))^2 \, \Delta t - A(Q) \int_0^\beta |y^\Delta(t)|^2 \, \Delta t.$$

Now, applying Wirtinger's inequality (6.4.2) by putting $M(t) = t$, we have that

$$|\lambda_0 - \gamma| \, \psi_1^2 \sigma^2(\beta) \int_0^\beta (y^\Delta(t))^2 \Delta t \geq \int_0^\beta \left(y^\Delta(t)\right)^2 \Delta t - A(Q) \int_0^\beta \left(y^\Delta(t)\right)^2 \Delta t,$$

where

$$\psi_1 = \left(\sup_{t \in \mathbb{I}^k} \frac{t}{\sigma(t)}\right)^{1/2} + \left[\left(\sup_{t \in \mathbb{I}^k} \frac{\mu(t)}{\sigma(t)}\right) + \left(\sup_{t \in \mathbb{I}^k} \frac{t}{\sigma(t)}\right)\right]^{1/2} \leq 1 + \sqrt{2}.$$

This implies that

$$|\lambda_0 - \gamma| \, (1 + \sqrt{2})^2 \sigma^2(\beta) \geq 1 - A(Q).$$

From this we obtain a lower bound of λ_0 as given in (6.4.3). The proof is complete. ∎

Bibliography

[1] S. Abramovich, G. Jameson, and G. Sinnamon, Inequalities for averages of convex and superquadratic functions, J. Inequal. Pure Appl. Math. 5 (2004), no. 4, Article 91, 14 pages.

[2] S. Abramovich, G. Jameson, and G. Sinnamon, Refining Jensen's inequality, Bull. Math. Soc. Sci. Math. Roumanie (N.S.) 47(95) (2004), no. 1–2, 3–14.

[3] M. Abramowitz and I. A. Stegun, *Handbook of Mathematical Functions with Formulas, Graphs, and Mathematical Tables*, 9th printing. New York: Dover, 1972.

[4] M. Adıvar and Y. N. Raffoul, Stability and Periodicity in dynamic delay equations, Comput. Math. Appl. 58 (2009), 264–272.

[5] M. Adıvar and Y.N. Raffoul, Shift operators and stability in delayed dynamic equations, Rendiconti del Seminario Matematico Università e Politecnico di Torino, 68 (2010), 369–397.

[6] M. Adıvar and Y. N. Raffoul, Existence of resolvent for Volterra integral equations on time scales, Bull. of Aust. Math. Soc., 82 (2010), 139–155.

[7] M. Adıvar, Function bounds for solutions of Volterra integro dynamic equations on time scales, E. J. Qualitative Theory of Diff. Equ., 7 (2010), 1–22.

[8] R. P. Agarwal, *Difference Equations and Inequalities: Theory, Methods and applications*, Marcel Dekker Inc., New York, 1992.

[9] R. P. Agarwal and P. Y. H. Pang, *Opial Inequalities with Applications in Differential and Difference Equations*, Kluwer, Dordrechet (1995).

[10] R. P. Agarwal and P. Y. H. Pang, Remarks on the generalization of Opial's inequality, J. Math. Anal. Appl. 190 (1995), 559–557.

© Springer International Publishing Switzerland 2014
R. Agarwal et al., *Dynamic Inequalities On Time Scales*,
DOI 10.1007/978-3-319-11002-8

[11] R. Agarwal, M. Bohner, and A. Peterson, Inequalities on time scales: A survey, Mathl. Inequal. Appl. 7 (2001), 535–557.

[12] R. P. Agarwal, M. Bohner and P. J. Y. Wong, Sturm-Liouville eigenvalue problems on time scales, Appl. Math. Comp. 99 (1999), 153–166.

[13] R. P. Agarwal, M. Bohner, D. O'Regan, and S. H. Saker, Some Wirtinger-type inequalities on time scales and their applications, Pacific J. Math. 252 (2011), 1–18.

[14] R. P. Agarwal, M. Bohner, D. O'Regan and A. Peterson, Dynamic equations on time scales: A survey, *J. Comp. Appl. Math., Special Issue on Dynamic Equations on Time Scales*, edited by R. P. Agarwal, M. Bohner, and D. O'Regan, (Preprint in Ulmer Seminare 5) **141**(1–2) (2002) 1–26.

[15] R. P. Agarwal, M. Bohner, Basic calculus on time scales and some of its applications, Results Math. 35 (1999), 3–22.

[16] R. P. Agarwal, Y. H. Kim, and S. K. Sen, New discrete Halanay inequalities: Stability of difference equations, Communications in Applied Analysis, 12 (2008), 83–90.

[17] R. P. Agarwal, Y. H. Kim, and S. K. Sen, Advanced discrete Halanay type inequalities: Stability of difference equations, Journal of Inequalities and Applications, 2009, Article ID 535849.

[18] R. P. Agarwal, V. Otero-Espinar, K. Perera, and D. R. Vivero, Wirtinger's inequalities on time scales, Canad. Math. Bull. 51 (2008), 161–171

[19] E. Akın, Y. N. Raffoul, and C. Tisdell, Exponential stability in functional dynamic equations on time scales, Communications in Mathematical Analysis, 9 (2010), 93–108.

[20] E. Akın, Y. N. Raffoul, Boundedness in functional dynamic equations on time scales, *Advances in Difference Equations*, Vol. 2006, Article ID 79689, Pages 1–18 DOI 10.1155/ADE/2006/79689.

[21] E. Akın, L. Erbe, B. Kaymakçalan, and A. Peterson, Oscillation results for a dynamic equation on a time scale, On the occasion of the 60th birthday of Calvin Ahlbrandt. J. Differ. Equations Appl. (6), 793–810, 2001.

[22] E. Akın Bohner, M. Bohner, and Faysal Akın, Pachpatte inequalites on time scales, Journal of Inequalities in Pure and Applied Mathematics, 6 (1), Article 6, 2005.

[23] M. R. S. Ammi, R. A. Ferreira and D. F. M. Torres, Diamond-α Jensen's Inequality on Time Scales, Journal of Inequalities and Applications, Volume 2008, Article ID 576876, 13 pages.

[24] M. R. S Ammi and D. F. M. Torres, Hölder's and Hardy's two dimensional diamond-alpha inequalities on time scales, Ann. Univ. Craiova Math. Comp. Sci. Series 37 (2010), 1–11.

[25] D. R. Anderson, Young's integral inequality on time scales revisted, J. Inequal. Pure and Appl. Math. 8 (2007), article 64, 1–5.

[26] D. R. Anderson, Time scale integral inequalities, J. Ineq. in Pure Appl. Math. 6 (3) (2006), Art. 66, 1–15.

[27] D. R. Anderson, J. Bullock, L. Erbe, A. Peterson, and H. Tran, Nabla dynamic equations on time scales, Panamer. Math. J., 13:1, 1–47, (2003).

[28] D. R. Anderson, R. J. Krueger and A. Peterson, Delay dynamic equations with stability, Advances in Difference Equations vol. 2006 (2006) 19 p. doi:10.1155/ADE/2006/94051 Article ID 94051.

[29] D. R. Anderson, S. Noren and B. Perreault, Young's integral inequality with upper and lower bounds, Elect. J. Diff. Eqns. 2011 (2011), no. 74, 1–10.

[30] M. Anwar, R. Bibi, M. Bohner and J. Pečarić, Integral inequalities on time scales via the theory of isotonic linear functionals, Abst. Appl. Anal. 2011 (2011), Article Id. 48359, 16 pages.

[31] N. Atasever, On Diamond-Alpha Dynamic Equations and Inequalities, (2011). *Georgia Southern University*, USA, *Electronic Theses & Dissertations*. Paper 667

[32] N. Atasever, B. Kaymakçalan, G. Lešaja and K. Tas, Generalized diamond-α dynamic Opial inequalities, Adven. Diff. Equns. 2012, (2012), 109, 1–9.

[33] B. Aulbach, Analysis auf Zeitmengen, Lecture Notes, Universitat Augsburg, (1990).

[34] B. Aulbach and S. Hilger, Linear dynamic processes with inhomogeneous time scales, Nonlinear Dynamics and Quantum Dynamical Systems, Akademie-Verlag (Berlin) (1990).

[35] B. Aulbach and S. Hilger, A Unified Approach to Continuous and Discrete Dynamics, In Differential Equations: Qualitative Theory, 37–56, Colloq. Math. Soc. Janos Bolyai, North-Holland, (1988).

[36] C. T. H. Baker, Development and application of Halanay-type theory: Evolutionary differential and difference equations with time lag, J. Comp. Appl. Math., 234 (2010), 2663–2682.

[37] C. T. H. Baker and A. Tang, 'Generalized Halanay inequalities for Volterra functional differential equations and discretised versions', Numerical Analysis Report 229, (Manchester Center for Computational Mathematics, University of Manchester, England, 1996).

[38] S. Banić and S. Varošanec, Functional inequalities for superquadratic functions, Int. J. Pure Appl. Math. 43 (2008), no. 4, 537–549.

[39] J. Barić, R. Bibi, M. Bohner and J. Pečarić, Time scale integral inequalities for superquadratic functions, J. Korean. Math. Soc. 50 (2013), 465–477.

[40] P. R. Beesack, Hardy's inequality and its extensions, Pacific J. Math. 11 (1961), 39–61.

[41] P. R. Bessack, On an integral inequality of Z.Opial, Trans. Amer. Math. Soc. 104 (1962), 470–475.

[42] P. R. Beesack and K. M. Das, Extensions of Opial's inequality, Pacific J. Math 26 (1968), 215–232.

[43] M. Bohner, Some oscillation criteria for first order delay dynamic equations, Far East J. Appl. Math., 18 (2005), 289–304.

[44] L. Bi, M. Bohner, and M. Fan. Periodic solutions of functional dynamic equations with infinite delay. Nonlinear Anal., 68(2008), 1226–1245.

[45] R. Bibi, M. Bohner, J. Pečarić and S. Varošanec, Minkowski and Beckenbach-Dresher inequalities and functionals on time scales, J. Math. Ineq. 7 (2013), 299–312.

[46] W. Blaschke, Kreis und Kugel, Leipzig, 1916.

[47] M. Bohner, O. Duman, Opial-type inequalities for diamond-alpha derivatives and integrals on time scales, Diff. Eqns.& Dyn. Syst. 18 (2010), 229–237.

[48] M. Bohner, S. Clark and J. Ridenhour, Lyapunov inequalities for time scales, J. Ineq. Appl. 7 (2002), 61–77.

[49] M. Bohner and B. Kaymakçalan, Opial Inequalities on time scales, Ann. Polon. Math.77 (1) (2001), 11–20.

[50] M. Bohner, R. R. Mahmoud and S. H. Saker, Discrete, continuous, delta, nabla and diamond-alpha Opial inequalities, Math. Ineq. Appl. (submitted).

[51] M. Bohner and A. Peterson. *Dynamic equations on time scales.* Birkhäuser Boston Inc., Boston, MA, 2001.

[52] M. Bohner and A. Peterson. *Advances in dynamic equations on time scales.* Birkhäauser, Boston, 2003.

[53] M. Bohner and G. Sh. Guseinov, Multiple integration on time scales, Dyn. Syst. and Appl. 14, (2005), 579–606.

[54] D. Boyd and J. S. W. Wong, An extension of Opial's inequality, J. Math. Anal. Appl. 19 (1967), 100–102.

[55] M. Bohner and A. Zafer, Lyapunov-type inequalities for planer linear dynamic Hamiltonian systems, Appl. Anal. Discr. Math.

[56] I. Brnetić and J. Pečarić, Some new Opial-type inequalities, Math. Ineq. Appl. 3 (1998), 385–390.

[57] R. C. Brown and D. B. Hinton, Opial's inequality and oscillation of 2nd order equations, Proc. Amer. Math. Soc. 125 (1997), 1123–1129.

[58] S. Chen, Lyapunov inequalities for differential and difference equations, Fasc. Math. 41 (1991), 23–25.

[59] Erbe, T. S. Hassan, A. Peterson and S. H. Saker, Oscillation criteria for half-linear delay dynamic equations on time scales, Nonlinear Dyn.. Sys. Th. 9 (2009). 51–68.

[60] L. Erbe, A. Peterson, and S. H. Saker, Oscillation criteria for second-order nonlinear dynamic equations on time scales. J. London Math. Soc. 76 (2003) 701–714.

[61] L. Erbe, A. Peterson and S. H. Saker, Hille-Kneser type criteria for second-order dynamic equations on time scales, Advances in Difference Eqns. 2006 (2006), 1–18.

[62] J. B. Diaz and F. T. Metcalf; An analytic proof of Young's inequality, American Math. Monthly, 77 (1970) 603–609.

[63] C. Dinu, Hermite-Hadamard inequality on time scales, J. Ieq. Appl. 2008, (2008), Article ID 287947, 24 pages.

[64] C. Dinu, A weighted Hermite Hadamard inequality for Steffensen-Popoviciu and Hermite-Hadamard weights on time scales, An. Şt. Univ. Ovidius Constanţa 17 (2009), 77–90.

[65] K. Fan, O. Taussky and J. Todd, Discrete analogs of inequalities of Wirtinger, Montash. Math. 59 (1955), 73–90.

[66] A. M. Fink and D. F. St. Mary, On an inequality of Nehari, Proc. Amer. Math. Soc. 21 (1969), 640–641.

[67] G. Foland, *Real Analysis: Modern Techniques and Their Applications*, John Wiley & Sons, Inc., New York, Second Edition 1999.

[68] K. Gopalsamy, *Stability and oscillations in delay differential equations of population dynamics*, Kluwer Academic Publishers, Dordrecht, The Netherlands, 1992.

[69] C. Ha, Eigenvalues of a Sturm-Liouville problem and inequalities of Lyapunov type, Proc. Amer. Math. Soc. 126 (1998), 1123–1129.

[70] J. Hadamard, Étude sur les propriétés des fonctions entiéres et en particulier d'une fonction considérée par Riemann, J. Math. Pures Appl., 58 (1893), 171–215.

[71] A. Halanay, Differential Equations: Stability, Oscillations, Time lags, Academic Press, New York, NY USA, 1966.

[72] G. H. Hardy, J.E. Littlewood and G. Pólya, *Inequalities*, 2nd ed. Cambridge, England: Cambridge University Press, 1988.

[73] B. J. Harris and Q. Kong, On the oscillation of differential equations with an oscillatory coefficient, Tran. Amer. Math. Soc. 347 (1995), 1831–1839.

[74] P. Hartman and A. Wintner, On an oscillation criterion of de la valee Poussion, Quart. Appl. Math. 13 (1955), 330–332.

[75] R. J. Higgins and A. Peterson, Cauchy functions and Taylor's formula for time scale $\mathbb{T}$, Proceeding of six Inter. Conf. Diff. Eqns., Roca Raton, FL, CRC. (2004), 299–308.

[76] S. Hilger, Ein Ma kettenkalkul mit Anwendung auf Zentrumsmannigfaltigkeiten, PhD thesis, Universitat Wurzburg, (1988).

[77] S. Hilger, Analysis on measure chains—a unified approach to continuous and discrete calculus, *Results Math.* 18 (1990) 18–56.

[78] S. Hilger, Differential and Difference Calculus, Nonlinear Anal. Proceedings of Second World Congress of Nonlinear Analysts, 30(5), 2683–2694, (1997).

[79] R. Hilscher, Linear Hamiltonian systems on time scales: Positivity of quadratic funetionals. Math. Comput. Modelling 32 (2000), 507–527.

[80] R. Hilscher, Positivity of quadratic funetionals on time scales: Necessity. Math. Naehr. (2000).

[81] D. B. Hinton and R.T. Lewis, Discrete spectra criteria for singular differential operators with middle terms, Math. Proc. Cambridge Philos. Soc. 77 (1975) 337–347.

[82] R. Hilscher, A time scale version of a Wirtinger-type inequality and applications, J. Comp. Appl. Math. 141 (2002), 219–226.

[83] J. Hoffacker and C. Tisdell, Stability and instability for dynamic equations on time scales, *Comput. Math. Appl.* 49 (2005), pp. 1327–1334.

[84] O. Hölder, Uber einen Mittelwerthssatz, Nachr. Ges. Wiss. Gottingen (1889) 38–47.

[85] R. A. Horn and C. R. Johnson, *Matrix Analysis*, Cambridge University Press, Cambridge 1991).

[86] L. K. Hua, On an inequality of Opial, Sci. Sinica. 14 (1965), 789–790.

[87] B. Jessen, Bemærkninger om konvekse Funktioner og Uligheder imellem Middelværdier. I, Matematisk Tidsskrift B. 2 (1931). 17–28.

[88] A. Ivanov, E. Liz, and S. Trofimchuk, Halanay inequality, Yorke 3/2 stability criterion, and differential equations with maxima, Tokohu Math., 54 (2002), 277–295.

[89] V. Kac and P. Cheung, *Quantum Calculus*, Springer, New York, 2001.

[90] B. Karpuz, B. Kaymakçalan and Ö. Öclan, A generalization of Opial's inequality and applications to second order dynamic equations, Diff. Eqns. Dyn. Sys. 18 (2010), 11–18.

[91] B. Karpuz and U. M. Özkan, Some generalizations for Opial inequality involving several functions and their derivatives of arbitrary order on arbitrary time scales, Math. Ineq. Appl. (preprint).

[92] E. Kaufmann and Y.N. Raffoul, Periodicity and stability in neutral nonlinear dynamic equations with functional delay on a time scale, Electron. J. Differential Equations, 27 (2007), pp. 1–12.

[93] E. Kaufmann and Y.N. Raffoul, Periodic solutions for a neutral nonlinear dynamical equations on time scale, J. Math. Anal. Appl. 319 (1) (2006), pp. 315–325.

[94] B. Kaymakçalan, Existence and Comparison Results for Dynamic Systems on Time Scale, J. Math. Anal. Appl., 172, 243–255, (1993).

[95] B. Kaymakçalan and V. Lakshmikantham, Monotone Flows and Fixed Points for Dynamic Systems on Time Scales, Comput. Math. Model., 28), 185–189, (1994).

[96] B. Kaymakalan and S. Leela, A Survey of Dynamic Systems on Time Scales, Nonlinear Times Digest, 1, 37–60, (1994).

[97] B. Kaymakçalan, S. A. Ozgun and A. Zafer, Gronwall-Bihari type Inequalities on Time Scales, Advances in Difference Equations, Proceeding of the Second International Conference on Difference Equations and Applications, Veszprem-Hungary, 481–490, (1997).

[98] B. Kaymakçalan, S.A. Ozgun and A. Zafer, Asymptotic behavior of higher order equations on time scale, Comput. Math. Model., Special Issue on Difference Equations, 36(10–12), 299–306, (1998).

[99] B. Kaymakçalan, V. Lakshmikantham, and S. Sivasundaram, *Dynamic Systems on Measure Chains*, Mathematics and its Applications 370, Kluwer Academic Publishers (Dordrecht) (1996).

[100] W.G. Kelley and A. C. Peterson, *Difference Equations; An Introduction with Applications*, Academic Press, New York 1991.

[101] A. A. Lasota, A discrete boundary value problem, Ann Polon. Math. 20 (1968), 183–190.

[102] C-F. Lee, C. C. Yeh, C. H. Hong and R. P. Agarwal, Lyapunov and Wirtinger inequalities, Appl. Math. Lett. 17 (2004), 847–853.

[103] N. Levinson, On an inequality of Opial and Beesack, Proc. Amer. Math. Soc. 15 (1964), 565–566.

[104] E. Liz and J. B. Ferreiro, A note on the global stability of generalized difference equations, Appl. Math. Lett. 15 (2002), 655–659.

[105] A. M. Lyapunov, Problème Général de la Stabilitié du mouvement, Ann. Fac. Sci Toulouse Math. 9 (1907), 203–274.

[106] C. L. Mallows, An even simpler proof of Opial inequality,Proc. Amer. Math. Soc. 16 (1965), 173.

[107] P. Maroni, Sur l' inegalité d'Opial-Beesack, C. R. Acad. Sci. Paris Ser A-B. 264 (1967), A62-A64.

[108] S. Mohamad and K. Gopalsamy, Continuous and discrete Halanay-type inequalities, Bull. Aust. Math. Soc. 61 (2000), 371–385.

[109] D. Mozyrska and D. F. M. Torres, A study of diamond-alpha dynamic equations on regular time scales, Afr. Diaspora J. Math., 8, No. 1, 35–47, (2009).

[110] D. S. Mitinović, J. E. Pečarić and A. M. Fink, *Classical and New Inequalties in Analysis, Klwer Academic Publisher*, 1993.

[111] D. S. Mitrinovic, *Analytic Inequalities*, Springer-Verlag, New York, 1970.

[112] C. Olech, A simple proof of a certain result of Z.Opial, Ann. Polon. Math. 8 (1960), 61–63.

[113] Z. Opial, Sur uné inegalité, Ann. Polon. Math. 8 (1960), 92–32.

[114] U. M. Özkan and H. Yildirim, Steffensen's integral inequality on time scales, J. Ineq. Appl. 2007 (2007), Article ID 46524, 10 pages.

[115] U. M. Özkan, M. Z. Sarikaya and H Yildirim, Extensions of certain integral inequalities on time scales, Applied Mathematics Letters, 21, (2008) 993–1000.

[116] B. G. Pachpatte, Lyapunov type integral inequalities for certain differential equations, Georgian Math. J. 4 (1997), 139–148.

[117] J. E. Pečarić, F. Proschan, and Y. L. Tong, *Convex Functions, Partial Orderings, and Statistical Applications*, vol. 187 of Mathematics in Science and Engineering, Academic Press, Boston, Mass, USA, 1992.

[118] R. N. Pederson, On an inequality of Opial, Beesack and Levinson, Proc. Amer. Math. Soc. 16 (1965), 174.

[119] S. Peňa, Discrete spectra criteria for singular difference operators, Math. Bohem. 124 (1) (1999) 35–44.

[120] A. Peterson and J. Ridenhour, A disconjugacy criterion of W. T. Reid for difference equations. Proe. Amer. Math. Soc. 114 (1992), 459–468.

[121] J. W. Rogers Jr., Q. Sheng, Notes on the diamond-alpha dynamic derivative on time scales, J. Math. Anal. Appl. 326, 228–241, (2007).

[122] S. H. Saker, *Oscillation Theory of Dynamic Equations on Time Scales: Second and Third Orders*, Lambert Academic Publishing, Germany (2010).

[123] S. H. Saker, Opial's type inequalities on time scales and some applications, Annales Polonici Mathematici 104 (2012), 243–260.

[124] S. H. Saker, Some new inequalities of Opial's type on time scales, Abstract and Applied Analysis 2012, art. no. 683136.

[125] S. H. Saker, New inequalities of Opial's type on time scales and some of their applications, Discrete Dynamics in Nature and Society 2012, art. no. 362526.

[126] S. H. Saker, Dynamic inequalities on time scales: A Survey, Journal of Fractional Calculus and Applications, Vol. 3(S). July, 11, 2012 (Proc. of the 4th. Symb. of Fractional Calculus and Applications), No. 2, pp. 1–36.

[127] S. H. Saker, Some Opial dynamic inequalities involving higher order derivatives on time scales, Discrete Dynamics in Nature and Society (2012), art. no. 157301.

[128] S. H. Saker, Applications of Opial inequalities on time scales on dynamic equations with damping terms, Mathl. Comp. Modelling 58 (2013) (11–12), 1777–1790

[129] S. H. Saker, Some nonlinear dynamic inequalities on time scales and applications, J. Math. Ineq. 4 (2010), 561–579.

[130] S. H. Saker, Lyapunov inequalities for half-linear dynamic equations on time scales and disconjugacy, Dyn. Contin. Discr. Impuls. Syst. Series B: Applications & Algorithms 18 (2011), 149–161.

[131] S. H. Saker, Some new inequalities of Opial's type on time scales, Abstract and Applied Analysis 2012, art. no. 683136.

[132] S. H. Saker, Some Opial-type inequalities on time scales, Abstr. Appl. Anal. 2011 (2011), Art. no. 265316, 19 pages.

[133] S. H. Saker, New inequalities of Opial's type on time scales and some of their applications, Discrete Dynamics in Nature and Society 2012, art. no. 362526.

[134] S. H. Saker, Lyapunov type inequalities for a second order differential equations with a damping term, Annal. Polon. Math. 103 (2012), pp. 37–57.

[135] S. H. Saker, Oscillation criteria of second-order half-linear dynamic equations on time scales, J. Comp. Appl. Math. 177 (2005), 375–387.

[136] S. H. Saker, R. P. Agarwal, D O'Regan, Higher order dynamic inequalities on time scales, Math. Ineq. Appl. 17 (2014), 461–472.

[137] I. J. Schoenberg, The finite Fourier series and elementary geometry, Amer. Math. Monthly 57 (1950), 390–404.

[138] H. M. Sirvastava, K. L. Tseng, S. J. Tseng and J. C. Lo, Some Weighted Opial type inequalities on time scales, Taw. J. Math. 14 (2010), 107–122.

[139] H. M. Sirvastava, K. L. Tseng, S. J. Tseng and J. C. Lo, Some generalizations of Maroni's inequality on time scales, Math. Ineq. Appl. 14 (2012), 469–480.

[140] Q. Sheng, M. Fadag, J. Henderson and J. M. Davis, An exploration of combined dynamic derivatives on time scales and their applications, Nonlinear Anal. Real World Appl. 7, 395–413, (2006).

[141] V. Spedding, Taming Nature's Numbers, New Scientist, July 19, 2003, 28–31.

[142] J. F. Steffensen, On certain inequalities between mean values, and their application to actuarial problems, Skandinavisk Aktuarietidskrift 1, (1918), 82–97.

[143] A. Tiryaki, M. Ünal and D. Çakmak, Lyapunov-type inequalities for nonlinear systems, J. Math. Anal. Appl. 332 (2007), 497–511.

[144] E. Tolsted; An elementary derivation of the Cauchy, Hölder, and Minkowski inequalities from Young's inequality, Math. Mag. 37 (1964), 2–12.

[145] A. Tuna and S. Kutukeu, Some integral inequalities on time scales, Appl. Math. Math. Eng. Ed. 29 (2008), 23–29.

[146] S. Udpin and P. Niamsup, New discrete type inequalities and global stability of nonlinear difference equations, Appl. Math, Lett. 22 (2009), 856–859.

[147] M. Ünel and D. Çakmak, Lyapunov-type inequalities for certain nonlinear systems on time scales, Turk. J. Math. 32 (2008), 255–275.

[148] W. Wang, A generalized Halanay inequality for stability of nonlinear neutral functional differential equations, Journal of Inequalities and Applications, 2010, Article ID 475019.

[149] F. H. Wong, W. C. Lin, S. L. Yu and C. C. Yeh, Some generalization of Opial's inequalities on time scales, Taw. J. Math. 12 (2008), 463–471.

[150] Fu-H. Wong, C. C. Yeah and W. C. Lian, An extension of Jensens' inequality on time scales, Advn. Dynm. Sys. Appl., 1 (2006), 113–120.

[151] Fu-H. Wong, C. Yeh, Shiueh-L. Yuc, Chen-H. Hong, Young's inequality and related results on time scales, Applied Mathematics Letters 18 (2005) 983–988.

[152] G. S. Yang, On a certain result of Z. Opial, Proc. Japan Acad. 42 (1966), 78–83.

[153] G. S. Yang, A note on some integrodifferential inequalities, Soochow J. Math. 9 (1983), 231–236.

[154] Y. Yang and J. Cao, Stability and periodicity in delayed cellular neural networks with impulsive effects, Nonlinear Analysis: Real World Applications, 8 (2007), 362–374.

[155] B. I. Yaşar and A. Tuna, Some results related to integral inequalties on time scales, Inter. Math. Forum 2 (2007), 1157–1161.

[156] C. Yeh, Fu-H. Wong and H. Li, Čebyšev's inequality on time scales, J. Ineq. Pure and Appl. Math. 6 (2005), Article 7, 10 pages.

[157] W. H. Young, On class of summable functions and there Fourier series, Proc. Roy. Soc. London A 87 (1912) 225–229.

[158] Z. Zhao, Y. Xu and Y. Li, Dynamic inequalities on time scalse, Inter. J. Pure Appl. Math. 22 (2005), 49–56.

Index

© Springer International Publishing Switzerland 2014
R. Agarwal et al., *Dynamic Inequalities On Time Scales*,
DOI 10.1007/978-3-319-11002-8

Printed in the United States
By Bookmasters